KYOTO AREA STUDIES ON ASIA
Center for Southeast Asian Studies, Kyoto University
Volume 18

Farming with Fire and Water

KYOTO AREA STUDIES ON ASIA

CENTER FOR SOUTHEAST ASIAN STUDIES, KYOTO UNIVERSITY

The Nation and Economic Growth: Korea and Thailand
YOSHIHARA Kunio

One Malay Village: A Thirty-Year Community Study
TSUBOUCHI Yoshihiro

Commodifying Marxism:
The Formation of Modern Thai Radical Culture, 1927–1958
Kasian TEJAPIRA

Gender and Modernity: Perspectives from Asia and the Pacific
HAYAMI Yoko, TANABE Akio, TOKITA-TANABE Yumiko

Practical Buddhism among the Thai-Lao:
Religion in the Making of a Region
HAYASHI Yukio

The Political Ecology of Tropical Forests in Southeast Asia:
Historical Perspectives
LYE Tuck-Po, Wil DE JONG, ABE Ken-ichi

Between Hills and Plains:
Power and Practice in Socio-Religious Dynamics among Karen
HAYAMI Yoko

Ecological Destruction, Health and Development:
Advancing Asian Paradigms
FURUKAWA Hisao, NISHIBUCHI Mitsuaki, KONO Yasuyuki, KAIDA Yoshihiro

Searching for Vietnam:
Selected Writings on Vietnamese Culture and Society
A. Terry RAMBO

Laying the Tracks: The Thai Economy and its Railways 1885–1935
KAKIZAKI Ichiro

After the Crisis:
Hegemony, Technocracy and Governance in Southeast Asia
SHIRAISHI Takashi, Patricio N. ABINALES

Dislocating Nation-States: Globalization in Asia and Africa
Patricio N. ABINALES, ISHIKAWA Noboru and TANABE Akio

People on the Move: Rural–Urban Interactions in Sarawak
SODA Ryoji

Living on the Periphery: Development and Islamization among the Orang Asli
NOBUTA Toshihiro

Myths and Realities: The Democratization of Thai Politics
TAMADA Yoshifumi

East Asian Economies and New Regionalism
ABE Shigeyuki and Bhanupong NIDHIPRABA

The Rise of Middle Classes in Southeast Asia
SHIRAISHI Takashi and Pasuk PHONGPAICHIT

Farming with Fire and Water: The human ecology of a composite swiddening community in Vietnam's northern mountains
Trần Đức VIÊN, A. Terry RAMBO and Nguyễn Thanh LÂM

KYOTO AREA STUDIES ON ASIA
CENTER FOR SOUTHEAST ASIAN STUDIES, KYOTO UNIVERSITY
VOLUME 18

Farming with Fire and Water

The Human Ecology of a Composite Swiddening Community in Vietnam's Northern Mountains

Edited by
Trần Đức VIÊN
A. Terry RAMBO
Nguyễn Thanh LÂM

With the assistance of
Neil L. Jamieson and Stephen J. Leisz

Kyoto University Press

TRANS PACIFIC PRESS

The publication of this volume was partially funded by a grant to CARES from the Ford Foundation. The volume also received financial support from the JSPS Global COE Program (E-04): In Search of Sustainable Humanosphere in Asia and Africa.

First published in 2009 jointly by:

Kyoto University Press
Kyodai Kaikan
15-9 Yoshida Kawara-cho
Sakyo-ku, Kyoto 606-8305, Japan
Telephone: +81-75-761-6182
Fax: +81-75-761-6190
Email: sales@kyoto-up.or.jp
Web: http://www.kyoto-up.or.jp

Trans Pacific Press
PO Box 164, Balwyn North, Melbourne
Victoria 3104, Australia
Telephone: +61 3 9859 1112
Fax: +61 3 9859 4110
Email: tpp.mail@gmail.com
Web: http://www.transpacificpress.com

Figures set by Kyoto University Press.

Printed in Melbourne by BPA Print Group

Distributors

Australia and New Zealand
DA Information Services/Central Book Services
648 Whitehorse Road
Mitcham, Victoria 3132
Australia
Telephone: +61(0)3-9210-7777
Fax: + 61(0)3-9210-7788
Email: books@dadirect.com
Web: www.dadirect.com

USA and Canada
International Specialized Book Services (ISBS)
920 NE 58th Avenue, Suite 300
Portland, Oregon 97213-3786
USA
Telephone: (800) 944-6190
Fax: (503) 280-8832
Email: orders@isbs.com
Web: http://www.isbs.com

Asia and the Pacific
Kinokuniya Company Ltd.

Head office:
38-1 Sakuragaoka 5-chome
Setagaya-ku, Tokyo 156-8691
Japan
Telephone: +81(0)3-3439-0161
Fax: +81(0)3-3439-0839
Email: bkimp@kinokuniya.co.jp
Web: www.kinokuniya.co.jp

Asia-Pacific office:
Kinokuniya Book Stores of Singapore Pte., Ltd.
391B Orchard Road #13-06/07/08
Ngee Ann City Tower B
Singapore 238874
Telephone: +65 6276 5558
Fax: +65 6276 5570
Email: SSO@kinokuniya.co.jp

ISSN 1445–9663 (Kyoto Area Studies on Asia)
ISBN 978–1–920901–29–5

Contents

List of Figures vii
List of Plates x
List of Tables xii
Preface xvii
Acknowledgments xxi
List of Contributors xxv
List of Major Government Policies and Programs Affecting Farming Communities in the Northern Mountain Region xxxi

PART I: BACKGROUND OF RESEARCH IN TAT HAMLET

Introduction to Part I 3
1 Research on Composite Swidden Agriculture in Tat Hamlet *Trần Đức Viên and A. Terry Rambo* 5
2 Hamlet of the Waterfalls:An Overview of the Environmental and Social Characteristicsof Tat Hamlet *A. Terry Rambo, Trần Đức Viên, and Nguyễn Thanh Lâm* 17

PART II: STRUCTURE AND FUNCTIONING OF THE COMPOSITE SWIDDEN AGROECOSYSTEM

Introduction to Part II 45
3 The Composite Swidden Agroecosystem of Tat Hamlet *A. Terry Rambo and Nguyễn Thanh Lâm* 47
4 Paddy Fields *Patma Vityakon* 66
5 Fallow Swidden Fields: Floral Composition, Successional Dynamics, and Farmer Management *Mai Văn Thành and Trần Đức Viên* 76
6 Livestock *Nguyễn Thanh Lâm and Trần Đức Viên* 90
7 Fishponds *Nguyễn Thanh Lâm and Trần Đức Viên* 104
8 Homegardens and Tree Gardens *Deanna Donovan, Đào Trọng Hưng, and Nguyễn Thanh Lâm* 112
9 Secondary Forests *Deanna Donovan, Đào Trọng Hưng, Trịnh Thị Tiến,Trần Chí Trung, and Stephen J. Leisz* 119
10 Recent Changes in the Composite Swiddening Farming System in Tat Hamlet *Nguyễn Thanh Lâm, Aran Patanothai, and A. Terry Rambo* 134

PART III: THE SOCIAL SYSTEM OF TAT HAMLET

Introduction to Part III 157

11 Social Organization and the Management of Natural Resources in Tat Hamlet *A. Terry Rambo and Trần Đức Viên* 159

12 Diversification of Household Livelihood Strategies in Tat Hamlet *Nguyễn Thị Hải Ninh, Đặng Việt Quang, Nguyễn Đình Tiến, and Ingrid Foik* 184

13 The Evolving Role of the Market *Phạm Thị Mai Hương, Nguyễn Thanh Lâm,Vũ Thị Thao, and Trần Đức Viên* 202

14 Gender Roles in the Household Economy in Tat Hamlet *Nguyễn Thế Phương, Phạm Thị Dung, Nguyễn Thanh Lâm, and Nguyễn Thị Minh Hiền* 226

PART IV: THE ECOLOGICAL SUSTAINABILITY OF COMPOSITE SWIDDEN AGRICULTURE

Introduction to Part IV 239

15 A Nutrient Balance Analysis of the Composite Swiddening Agroecosystem *Nguyễn Văn Dung, Trần Đức Viên, Georg Cadisch, Nguyễn Thanh Lâm, Aran Patanothai, A. Terry Rambo, and Trần Mạnh Tường* 243

16 Does Composite Swiddening Cause Deforestation? Evidence from Analysis of Land Cover Change in Tat Hamlet from 1952 to 2003 *Stephen J. Leisz, Nguyễn Thị Thu Hà, Võ Hữu Công, Trần Nguyên Bằng, and Nông Hữu Dương* 284

PART V: THE COMPOSITE SWIDDEN AGROECOSYSTEM OF TAT HAMLET IN WIDER PERSPECTIVE

Introduction to Part V 307

17 Distribution of Composite Swiddening Agroecosystems in Vietnam's Northern Mountain Region *Stephen J. Leisz, Nguyễn Thị Thu Hà, and Nguyễn Thị Bích Yên* 309

18 Tat Hamlet in Comparative Perspective *A. Terry Rambo, Stephen J. Leisz, Trần Đức Viên, Đào Minh Trường, Đào Trọng Hưng, Lê Quang Trung,and Nghiêm Phương Tuyến* 317

19 Forest Interrupted: Tat Hamlet and the Political Ecology of Swidden Agriculture *Jefferson Fox* 342

Appendix 353
Bibliography 438
Index 454

List of Figures

1.1 Location of Tat hamlet in northern Vietnam 6
2.1 Map of Tat hamlet 18
2.2 Monthly mean temperature and rainfall in Tat hamlet 21
2.3 Land use map of Tat hamlet 26
4.1 Sketch map of Tat hamlet showing locations of paddy fields in relation to physical structures 68
5.1 Diagram comparing the sizes of the A, B, C, and D standard plots 78
6.1 Changes in number of species of livestock owned by households in Tat hamlet between 1998 and 2004 99
8.1 Map and profile of a typical homegarden in Tat hamlet 113
8.2 Map and profile of a typical tree garden in Tat hamlet 116
9.1 Map and profile of a secondary forest community. 122
9.2 Map and profile of the high secondary forest at Pù Cọ Hiên. 123
9.3 Map and profile of a shrub and grass community in a fallowed swidden field. 125
10.1 Changes in type of individual households between 1998-99 and 2003 147
13.1 Goods flows via the marketing channels in Tat hamlet during the colonial era 205
13.2 Marketing channels in Tat hamlet from 1960 to 1969 206
13.3 Goods flows via the marketing channels in Tat hamlet from 1969 to 1987 207
13.4 Material flows in the marketing channels in Tat hamlet after 1987 209
13.5 Changes in the numbers of collectors over the past 50 years in Tat hamlet 212
13.6 Marketing channels for fresh cassava 213
13.7 Marketing channels for plantation forest products in Tat hamlet 218
13.8 Marketing channels for timber products in Tat hamlet 219
13.9 Marketing channels for non-timber forest products in Tat hamlet 221
13.10 Marketing channels for agricultural crops in Tat hamlet 223
13.11 Marketing channels for large livestock in Tat hamlet 223

13.12 Marketing channels for pigs, poultry, and fish in Tat hamlet 224
15.1 Comprehensive model of nutrient inputs and outputs for a cropping field 245
15.2 Map of the experimental site 253
15.3 Monthly rainfall distribution from 2000-2004 261
15.4 Number of days with daily rainfall > 30 mm/day 262
15.5 Cumulative yearly amounts of rainfall lost in runoff from the different erosion plots 263
15.6 Effect of fallow time on runoff 264
15.7 Daily erosion events during 2003 in the different erosion plots 264
15.8 Yearly erosion from different land uses 265
15.9 Soil organic carbon (SOC) losses by erosion under different land uses 267
15.10 Nutrient ratio lost in different upland land uses by erosion, runoff and leaching in 2000 268
15.11 Changes in soil texture over time in the upland field experiments 269
15.12 Changes in soil pH from 1999 to 2004 in the upland experimental plots 270
15.13 Effect of number of cropping cycles on aggregated N and P nutrient system balances 272
15.14 Effect of number of cropping cycles on aggregated K nutrient system balances 272
15.15 Regeneration patterns of systems N balance in different swidden systems 273
15.16 Regeneration of soil organic matter with ongoing fallow time 274
15.17 Yearly total fine sediment flow from uplands to lowlands measured at the outlet of the watershed 275
15.18 Yearly amount of sediment retained in the 3 upper paddy fields 276
15.19 Changes in soil texture over time in paddy fields 277
15.20 Effect of sedimentation on distribution of soil texture at different distances from the base of upstream discharge inlet into the paddy fields 277
15.21 Yearly nutrient flows from upland fields to the paddy field 278
15.22 Yearly nutrient outflows from paddy fields 278

15.23	Yearly net balance of inflow-outflow of nutrients in the paddy field	280
15.24	Average yearly total nutrient flows from upland fields (via run-off and erosion) into and out from the paddy	281
16.1 (a)	Land cover map of Tat hamlet in 1995	288
16.1 (b)	Land cover map of Tat hamlet in 1998	288
16.1 (c)	Land cover map of Tat hamlet in 2000	289
16.1 (d)	Land cover map of Tat hamlet in 2003	289
16.2	Land cover dynamics 1995-2003	295
16.3	Land cover transect 1995 to 2003	296
16.4	Transect map of land cover 1998–2003	297
16.5	Changes in population and land cover from 1995–2003	300
17.1	Districts included in the study done by Leisz et al. (2005)	310
17.2	Communes with composite swidden agriculture in 2000	312
18.1	Location of study communities in the Northern Mountain region	318
18.2	Development status diagrams of the five communities	326
18.3	Profiles of emergent property rankings of the five NMR communities	339
19.1	Map of Montane Mainland Southeast Asia	343

List of Plates

1.1	The composite swidden agroecosystem in Tat hamlet	6
1.2	Tat hamlet as it appeared during our first visit in January 1992	10
2.1	Tat hamlet as seen from the surrounding mountains	19
2.2	Swidden landscape in Tat hamlet	24
2.3	Woman wearing traditional style dress	29
2.4	Woman in traditional dress cooking in her kitchen	29
2.5	Women in Tat hamlet wearing a mixture of traditional and modern style clothing	30
2.6	House built on ground in Kinh style	31
2.7	Burial offerings at new grave	32
2.8	Ritual symbols at a man's grave	32
2.9	Woman threshing rice with her feet	34
2.10	Old woman husking rice using mortar and pestle	35
2.11	Woman carrying firewood	36
2.12	Television arrives in Tat hamlet	39
3.1	Paddy fields in Tat hamlet	49
3.2	Buffalo harrowing paddy fields	50
3.3	Women cultivating steep swidden field	52
3.4	Rice swidden in Tat hamlet	53
3.5	Truck carrying logs to the lowlands	60
3.6	Hunter with the trap he used to catch a bird in the forest	61
4.1	Bamboo pipeline carrying irrigation water to the paddy fields	70
4.2	Woman broadcasting chemical fertilizer	74
5.1	Broom grass being dried in the sun after collection from fallowed swiddens	85
5.2	Melia planted as successional crop in rice swidden	87
6.1	Buffalo being led back to the settlement after grazing in the mountains.	93
6.2	Livestock are often kept under the houses at night	93
6.3	Farmer feeding his flock of chickens and ducks	97
6.4	Buffalo grazing on stubble after the rice is harvested in paddy field	101
7.1	Farmer feeding his fish with taro leaves	108
7.2	A couple catching fish in their fishpond	109
8.1	Homegarden protected from livestock by a split bamboo fence	113
8.2	Woman watering household vegetable plot	114

9.1 Mature secondary forest on hill top 121
9.2 Women boiling bamboo shoots collected from the forest 128
9.3 Couple returning to house after collecting firewood in the forest 130
11.1 New post office in the village center 177
11.2 New primary school in village center 177
11.3 Neighbors assist family in re-roofing house 181
11.4 An exchange labor group planting a swidden field 181
13.1 Bamboo stems stacked by the roadside waiting to be picked up by a dealer 215
13.2 Woman selling bamboo shoots to buyer along the road 215
13.3 Women carrying bamboo shoots in pack baskets 221
14.1 Man setting fire to slash in swidden field 229
14.2 Women weeding their rice swidden field 229
14.3 Woman cutting firewood in her swidden field 230
14.4 Women sewing clothes 235
14.5 A community meeting with men as the dominant participants 235
15.1 Burning swidden fields, Tat hamlet 247
15.2 Newly burned swidden field with ash on top of soil 247
15.3 Woman carrying rice seedlings 249
15.4 Nursery for rice seedlings with fence to protect it from livestock 256
16.1 Photo comparisons of points on transect in 1997 and 2003 298

All photographs were taken by A. Terry Rambo unless otherwise noted.

List of Tables

1.1 List of household surveys conducted in Tat hamlet 14
2.1 Monthly mean temperatures in Tat hamlet 20
2.2 Monthly rainfall in Tat hamlet 20
2.3 Physical and chemical properties of upland and paddy soil profiles in Tat hamlet 23
2.4 Land use in Tat hamlet 25
2.5 Changes in the population of Tat hamlet from 1985 to 2005 27
3.1 Changes in the abundance of wild game species in Tat hamlet 61
3.2 Resource flows and other interactions among components of the composite swidden agroecosystem 64
4.1 pH of different layers of the soil profile of *Na Kor Kok* (Kok-tree paddy field) 69
4.2 Comparative properties of topsoils of paddy fields of Tat hamlet 72
5.1 Distribution of flora life forms in fallows in Tat hamlet 79
5.2 Useful fallow plant species in Tat hamlet 80
5.3 Changes in number of species in floral life forms in different fallow stages 81
5.4 Biomass of different life forms in different fallow stages (kg/ha) 81
5.5 The number of useful plant species in different fallow stages in Tat hamlet 82
5.6 Rating of relative importance of plants collected from fallows during the cooperative period 83
5.7 Rating of relative importance of plants collected from fallows in the present time 84
5.8 Farmer perceptions of the value of different fallow management systems 89
6.1 Ownership of buffalo and cattle by households in Tat hamlet 91
6.2 Division of labor in caring for large livestock among household members 92
6.3 Farmer perceptions of problems in raising large livestock in Tat hamlet 94
6.4 Vaccinations and disease among large livestock in Tat hamlet 94
6.5 Household ownership of poultry in Tat hamlet 97
6.6 Number of different species of livestock kept by households in Tat hamlet 99
6.7 Main sources of household cash income in 1998 and 2004 in Tat hamlet 100

7.1 Ownership of fishponds by households 105
7.2 Stocking density of fishponds in Tat hamlet 107
9.1 Collection and utilization of non-timber forest products by households 127
9.2 The most commonly traded non-timber forest products in Tat hamlet 131
9.3 Contribution of NTFPs to household income in Tat hamlet 132
10.1 Major characteristics of Tat hamlet and its farming systems 136
10.2 Characteristics of households of Types 1, 2 and 3 in 1998 145
10.3 Changes in crop components of farming systems of 40 households in Tat hamlet from 1998–99 to 2003 148
10.4 Changes in livestock ownership by households in Tat hamlet between 1998–99 and 2003 149
12.1 Agricultural landholdings of different groups of households in 2005 188
12.2 Average area of crops grown by different groups of households 188
12.3 Labor force characteristics of groups of households 189
12.4 Average number of domestic animals per household 189
12.5 Income of households from different sources by groups in Tat hamlet 191
12.6 Income diversification index (IDI) of households in different groups 193
12.7 Percentage of farm households participating in off-farm income generating activities in Tat hamlet in 1999 and 2004 195
12.8 Productivity of crops by household group 195
12.9 Gross margin to land of crop production by household group 197
12.10 Gross margin to labor 197
12.11 Average gross margin for pigs, poultry and fish by household group 198
12.12 Farm households employing different strategies of income diversification 199
13.1 Major events in the historical development of marketing channels in Tat hamlet 204
13.2 Percentage of households selling their products at different places 214
13.3 Wholesale and retail selling prices of agricultural and forest products in Tat hamlet 216
14.1 Gender division of labor in crop production 228
14.2 Division of labor among household members in caring for livestock 231

14.3 Gender division of labor in forest activities 232
14.4 Villager assessments of the comparative workloads of men and women 233
14.5 Responsibilities of men and women in household decision-making about farming activities 233
14.6 Gender roles in decision-making about forest management 234
14.7 Responsibilities of spouses in decision-making about household management 234
14.8 Decision-making about children 234
IV.1 Substitution effect of paddy field area for rice swidden field area in Tat hamlet 241
15.1 Physical and chemical properties of the surface soil in the experimental site in Tat hamlet 251
15.2 Vegetation communities in the upland study site 252
15.3 Cropping sequences of the upland experimental plots 254
15.4 Nutrient losses by erosion, runoff and leaching in the upland fields 267
15.5 Vegetation biomass and loss of nutrients from burning in the swidden field 269
15.6 Changes in the chemical characteristics of topsoil in the upland field after five years 271
15.7 N, P and K nutrient balances of secondary forest 273
15.8 Average yearly contribution of available nutrients from the upland field to the paddy field 279
16.1 Changes in area of different land cover types (ha) in Tat Hamlet 290
16.2 Detected change in land cover from 1995 to 1998 in Tat hamlet 291
16.3 Detected change in land cover in Tat hamlet from 1998 to 2000 293
16.4 Detected change in land cover in Tat hamlet from 2000 to 2003 294
16.5 Comparison of general land cover in Tat hamlet from 1952 to 2003 301
17.1 Distribution of farming system type in the five districts 311
17.2 Indicators of composite swidden agriculture in a commune 315
18.1 Key characteristics of the five NMR communities 319
18.2 Development status of the five NMR communities 324
18.3 Productivity rankings for the five NMR communities 331
18.4 Diversity rankings of the five NMR communities 331
18.5 Sustainability rankings for the five NMR communities 333
18.6 Equitability rankings for the five NMR communities 335
18.7 Integration rankings for the five NMR communities 337
18.8 Solidarity rankings for the five NMR communities 337

18.9 Combined rankings for the emergent properties of the five NMR communities 339

Appended Tables

5.A1 List of plants identified in fallow fields in Tat hamlet with information on their utilization by the villagers 354

5.A2 Genus and species of plants in Tat hamlet fallow fields according to family names 382

5.A3 Plant species found in fields in different fallow stages 386

8.A1 Useful plants species grown in home and tree gardens in Tat hamlet in 1998 and 2004 388

9.A1 Checklist of plant species in secondary-evergreen broad leaf forest of medium to tall trees 391

9.A2 Checklist of plant species in bamboo forest mixed with woody trees 397

15.A1 Nutrient balance of the swidden fields 402

15.A2 Nutrient balances in paddy fields adjacent to upland stream outlet 432

17.A1 Rules used for classifying general farming system type by commune 435

17.A2 Land cover, landscape, topographic and population density information for each of the general farming system types identified in the NMR 436

Preface

This volume presents a detailed description of the composite swidden system employed by the Đà Bắc Tày ethnic minority people of Tat hamlet in Hòa Bình Province in Vietnam's Northern Mountain Region (NMR). The farming system in Tat hamlet combines slash-and-burn agriculture in swidden fields on the hill slopes (farming with fire) with the cultivation of irrigated paddy fields in the valleys (farming with water). In this book, we look at this complex agroecosystem from the perspective of human ecology, which is the scientific study of human relations with the natural environment. The overall goal of human ecology research is to understand both how the natural environment influences human society and how human activities affect the natural environment (Rambo 1983). Because of the breadth of its concerns, human ecology research is necessarily an interdisciplinary endeavor that involves both natural and social scientists (Rambo and Sajise 1984).

As a consequence of its interdisciplinary character, human ecology occupies an uncertain place in most area studies centers. It is represented, if present at all, by geographers and anthropologists, but rarely by ecologists or agricultural scientists. Only two major area studies centers concerned with Southeast Asia, the East-West Center in Honolulu and the Center for Southeast Asian Studies of Kyoto University, have consistently had natural scientists as well as social scientists as members of their regular faculties.

Although some of the pioneering studies that established human ecology as a respectable field of study were carried out in Vietnam during the French colonial era (Rambo 1997c), the institutional structure for scientific research during the socialist era was not conducive to such an interdisciplinary approach. Instead, following the Soviet model, Vietnamese natural scientists and social scientists were based in wholly separate institutions with minimal cross-disciplinary communication. Thus, it was truly an innovative step when, in 1989, scientists from the East-West Center and the Southeast Asian Universities Agroecosystem Network (SUAN) worked together with Vietnamese scientists from the Center for Natural Resources and Environmental Studies (CRES) to make a human ecology research study in what was then Vĩnh Phú Province in the Midlands of northern Vietnam (Lê Trọng Cúc *et al.* 1990).

This was followed by a series of additional studies in other critical ecosystems in Vietnam (e.g., Lê Trọng Cúc and Rambo 1993; Lê Trọng Cúc *et al.* 1996; Donovan *et al.* 1997), the writing of the first Vietnamese language textbook on human ecology (Lê Trọng Cúc and Rambo 1995), and, ultimately, to the setting-up of the long-term research project on the human ecology of composite swidden agriculture in Tat hamlet on which this book is based.

Our research in Tat hamlet began in 1992 as part of a project on the human ecology of shifting cultivation organized by the East-West Center in collaboration with CRES. In its early stages, several individual scientists from the Hanoi University of Agriculture's Center for Agricultural Research and Ecological Studies (CARES) and the Faculty of Agriculture of Khon Kaen University (KKU) also took part in this project. In later stages of the research, many academic staff and students of the Hanoi University of Agriculture (HUA) carried out extensive fieldwork on many different topics in Tat hamlet. These studies have formed the basis for the majority of the chapters in this book. Scientists from other institutions also continued to do research in Tat hamlet, with those from the EWC and KKU having an especially deep involvement, as is shown by their contributions to this volume.

This project is one of a handful of long-term interdisciplinary human ecology research projects carried out in mainland Southeast Asia in the past 30 years. Perhaps the only comparable projects in terms of scope, duration, and the complex mix of researchers from different disciplines and institutions involved, are the Center for Southeast Asian Studies of Kyoto University project in Don Daeng village in Northeast Thailand in the 1980s (Fukui 1993) and the project organized by Professor Sakurai Yumio of Tokyo University to study Bạch Cốc village in Vietnam's Red River Delta, in which some 176 Japanese scholars have participated since it was started in 1993 (Sakurai 2003: 11).

Although this volume is intended to be a contribution of human ecology research to area studies, the research itself was undertaken to meet very practical needs associated with the development of ethnic minority communities in the NMR. Within the NMR, the pressure of growing populations, degrading soils, loss of forest cover, and the reduction of biodiversity continue to generate poverty, further environmental degradation, and contribute to a host of social, economic, and political problems (Jamieson *et al.* 1998). The work discussed in the following pages was undertaken in full awareness of this huge national problem and in the hope of making some small contribution to efforts to resolve it. The Government of Vietnam is in the midst of a long and on-going effort to develop the northern mountains economically, to integrate their ethnic minority populations into the nation, and to eliminate hunger, reduce poverty, and protect the environment. Many government laws, policies, programs, and efforts to introduce new technologies and to build the physical and institutional infrastructure have, taken together, had a powerful impact upon the NMR, and they continue to do so. While the achievements of this massive development effort have been considerable, there have also been some failures and disappointments.

Of particular concern is the fact that, despite the vast sums of money being spent on their behalf, ethnic minority farmers in the uplands are still economically and socially marginalized.

There are many unanswered questions about the social, economic, and environmental changes that are taking place as upland farmers struggle to earn a living and to manage the complex agroecosystems upon which their livelihoods depend in the face of rapid integration into the market system. Is their way of life sustainable? How are they responding to the many environmental and social stresses impinging upon their lives? What strategies are they devising to cope with the pressing problems that confront them and to take advantage of the new opportunities that emerge as new roads and better communications bring images of the modern world into their lives, along with new ideas, new problems, and new desires?

This volume represents our attempt to answer some of these questions based upon our research on composite swidden agriculture over the past 15 years at Tat hamlet. It is primarily an effort to describe Tat hamlet and to analyze the changes that are taking place there. We describe changes in the farming systems as they relate to a changing biophysical and social environment on the one hand; and, on the other hand, to the livelihood strategies of individual households, each with its own characteristics and circumstances that influence the choices made in production and consumption activities.

The study site was carefully selected as representing what may be one of the most sustainable upland agricultural systems in the NMR. The Đà Bắc Tày people of Tat hamlet have practiced their own distinctive composite swidden agriculture for a century or more. The agroecosystem they have evolved shows considerable promise for serving as a model that can enhance sustainability and improve living standards in many other upland ethnic minority villages in Vietnam and beyond. The project was designed to generate detailed empirical data on the structure, functioning, and dynamics of the human ecosystem of Tat hamlet, especially the composite swidden agroecosystem on which the hamlet's inhabitants rely for their survival. Research was done on a great variety of topics including soil erosion, nutrient balances, land use, forest management, community social organization, household economics, marketing, and other relevant environmental and social aspects of the community. This volume brings together a series of individual research reports generated by many of the different researchers who have participated in the human ecological study of Tat hamlet in the years since 1992.

The book is divided into five parts:

Part I includes a general introduction to our research in Tat hamlet (Chapter 1) and a description of the study site (Chapter 2).

Part II describes the structure and functioning of the composite swidden agroecosystem. Chapter 3 provides an overview of the system as a whole. The following chapters focus on different system components, including paddy fields (Chapter 4), fallow swidden fields and their vegetation (Chapter 5), livestock (Chapter 6), fishponds (Chapter 7), homegardens and tree gardens (Chapter 8) and secondary forest (Chapter 9). The final chapter in this part reviews changes in the composite swidden farming system in recent years (Chapter 10).

Part III focuses on the social system of Tat hamlet. Chapter 11 offers an account of the social organization of the hamlet as that affects management of natural resources. Chapter 12 looks at the diversification of household livelihood strategies and the economic performance of composite swiddening, and Chapter 13 examines the evolving role of the market for agricultural and forest products. Finally, chapter 14 discusses gender roles in the household economy.

Part IV offers an assessment of the ecological sustainability of composite swidden agriculture. It presents a nutrient balance analysis of swidden and paddy fields (Chapter 15) and an analysis of changes in land use and vegetation cover in the hamlet since 1952 (Chapter 16).

Part V looks at composite swiddening in wider perspective. Chapter 17 describes the occurrence of composite swidden agroecosystems in other parts of the NMR and Chapter 18 compares the environmental and social situation of Tat hamlet with four other upland farming communities in that region. Finally, Chapter 19 relates the research on composite swiddening in Tat hamlet to broader issues of the status of swidden agriculture in montane mainland Southeast Asia as a whole.

It is our hope that this book will help to make the composite swidden agroecosystem of Tat hamlet more widely known. This wonderfully complex and adaptable system, which has evolved over many generations of trial and error experimentation by the Đà Bắc Tày farmers, should be of special interest to academics engaged in Southeast Asian area studies as well as to applied researchers and development planners who are directly involved in efforts to facilitate rural development in the uplands of Vietnam and elsewhere in the mountains of Southeast Asia.

Trần Đức Viên, A. Terry Rambo, and Nguyễn Thanh Lâm

Acknowledgments

This book is the product of the work of scientists from several different institutions and countries who have done research in Tat hamlet since we began our studies there in 1992. Their research in Tat hamlet has been carried out as part of a series of collaborative projects organized by the Center for Natural Resources and Environmental Studies (CRES) of the Vietnam National University, Hanoi, the Hanoi University of Agriculture (HUA), and the East-West Center (EWC), Honolulu, USA.

Funding for this research was provided by grants to CRES from the Ford Foundation and the Swedish International Development Cooperation Agency (SIDA); to the EWC from the Ford Foundation, the Keidanren Nature Conservation Fund, the John D. and Catherine T. MacArthur Foundation, the US National Science Foundation (DEB 96-13613), the Rockefeller Brothers Fund, and SIDA; and to HUA from the Ford Foundation and the Toyota Foundation. We want to thank John Ambler, formerly the Ford Foundation Representative to Vietnam and Thailand, for having taken such a keen interest in developing this project. His successor, Charles Bailey, continued and expanded the Foundation's support to HUA for this research.

Our research has greatly benefited from the continued interest of Lê Trọng Cúc, formerly the Director of the Center for Natural Resources and Environmental Studies (CRES) of the Vietnam National University, Hanoi. Dr. Cúc, together with Terry Rambo and Trần Đức Viên, first identified Tat hamlet as the suitable site for this project, and he played a key role in organizing several subsequent studies there.

Valuable assistance in designing and conducting the nutrient balance research was provided by scientists from the Faculty of Agriculture of Khon Kaen University (KKU), Thailand, including Aran Patanothai, Patma Vityakon, Vichai Sriboonlue, Viriya Limpinuntana, and Vidhaya Tre-loges. Jefferson Fox and Deanna Donovan from the EWC made important contributions to several studies on forests and land use change.

Nguyễn Viết Tùng and Đặng Vũ Bình, both former Rectors of HUA, provided very strong administrative support for the project within the university. Nguyễn Ngọc Hồng, formerly Director of the Department of Science and Technology of Hòa Bình province, has also been generous with his assistance in overcoming administrative problems associated with this effort.

We would like to thank the many staff members of the HUA Center for Agricultural Research and Ecological Studies (CARES) who have contributed to this research and provided help in the preparation of this book. Over the years many CARES scientists have participated in this research project and much of the data processing and preparation of the tables, maps, diagrams used in the compilation of this book was carried out by staff members of CARES, including Phạm Tiến Dũng, Nguyễn Văn Dung, Phạm Thị Hương, Phạm Văn Phê, Nguyễn Xuân Thành, Đặng Việt Quang, Nguyễn Thị Thu Hà, Nguyễn Thị Bích Yên, Đinh Thị Hải Vân, Nguyễn Thị Phương Mai, Phạm Thị Mai Hương, Trần Mai Hương, Vũ Thị Thao, Đỗ Thị Hường, Trần Nam Anh, Trần Nguyên Bằng, Võ Hữu Công, Nông Hữu Dương, Trần Trung Kiên, Nguyễn Thế Phương, Trần Mạnh Tường, Trần Hữu Cường, Nguyễn Khắc Tích, Phạm Tân Tiến, Mai Văn Thành, Lăng Văn Trung, and Phạm Thị Thu Thủy. A number of foreign scientific collaborators of CARES, including George Cadisch (the University of Hohenheim), Ingrid Foik, Neil Jamieson, Stephen Leisz (formerly of Copenhagen University, currently at Colorado State University), as well as several volunteers, including Tyler McKinley, Amanda Allbritton (Volunteers in Asia), Don Miller (Fauna and Flora International), and Claire Rambo, have also made valuable contributions to this project.

We would also like to thank the generations of HUA undergraduate and graduate students from class 35 (1994) to class 45 (2004), including Đỗ Minh Phan, Doãn Minh Phương, Nguyễn Danh Đàn, Trần Bình Đà, Trần Văn Kỳ, Nguyễn Mạnh Giao, Mạc Đăng Mạnh, and Nguyễn Thị Thu Hà who have helped in this research. Some of these students lived in Tat hamlet and worked with farmers there for months at a time.

Nguyễn Thanh Lâm served as project manager, with administrative assistance from Đỗ Thị Hường and Nguyễn Thị Thu Hà. Terry Rambo was responsible for final editing of all chapters with assistance from Nguyễn Thanh Lâm, Neil Jamieson, and Stephen Leisz. Nguyễn Thanh Lâm oversaw the final assembly of the manuscript with the assistance of Nguyễn Thị Hải Ninh. Hayami Yoko, Chair of the Publications Committee of the Center for Southeast Asian Studies (CSEAS) of Kyoto University, and Yonezawa Mariko, CSEAS General Editor, provided continuing invaluable advice during the extended editorial process. Suzuki Tetsuya, Executive Director and Managing Editor of the Kyoto University Press, oversaw preparation of the manuscript for publication. We would especially like to acknowledge the very valuable contribution of the three anonymous referees who acted as peer reviewers of the initial draft of this volume for the CSEAS Publications Committee. It took us nearly a year of hard effort to make the hundreds of revisions they suggested, but we know that this is a far better book than it would have been without their involvement.

Above all others, we would like to give our most special thanks to the farmers in Tat hamlet who have been so hospitable to us over the years. We would also like to thank the many local officials of Tat hamlet, Tân Minh village, Đà Bắc District, and Hòa Bình Province who have given us so much help and support while we carried out our studies.

We would like to emphasize that we have always conceived of our research in Tat hamlet as being aimed at generating better understanding of the human ecology of the composite swidden agricultural system of Tat hamlet. Although this work has obvious relevance to the overall development situation in Vietnam's mountain regions, it was not undertaken to evaluate the success, or lack thereof, of any development programs by the Vietnamese government or any other development agencies. And, of course, the views expressed in this book are those of the authors and are not necessarily shared by any of the agencies that sponsored this research.

List of Abbreviations

ANOVA	Analysis of variance
CARES	Center for Agricultural Research and Ecological Studies, Hanoi University of Agriculture, Vietnam
CRES	Center for Natural Resources and Environmental Studies, Vietnam National University, Hanoi
CSA	Composite swidden agroecosystem
CSEAS	Center for Southeast Asian Studies, Kyoto University, Japan
CV	Coefficient of variance
DAES	Department of Agroecology and Environmental Sciences, Hanoi University of Agriculture, Vietnam
EWC	East-West Center, Honolulu, USA
FAO	Food and Agriculture Organization of the United Nations
GIS	Geographical information system
GM	Gross margin
GO	Gross output
GPS	Global positioning system
GOV	Government of Vietnam
KKU	Khon Kaen University, Thailand
IDI	Income Diversification Index by Simpson
IFPRI	International Food Policy Research Institute
JICA	Japan International Cooperation Agency
HUA	Hanoi University of Agriculture, Vietnam
NGO	Non-governmental organization
NMR	Northern Mountain Region
NTFP	Non-timber forest product
PAM	World Food Program of the United Nations
RS	Remote sensing
SIDA	Swedish International Development Cooperation Agency
SOC	Soil organic carbon
SUAN	Southeast Asian Universities Agroecosystem Network
USEPAM	University Support to Environmental Planning and Management
VAC	Vietnamese acronym for integrated household agricultural system based on home gardens, fishponds, and stables for livestock
VND	Vietnamese Dong (US$1 = 16,097 VND)

List of Contributors

Editors

Trần Đức Viên is an Associate Professor of the Hanoi University of Agriculture, where he has served as the Rector since late 2006. He teaches courses on agroecology, farming systems research and development, and environmental management. He played a key role in the creation of the HUA Center for Agricultural Research and Environmental Studies (CARES) and serves as its director. CARES is a multi-disciplinary center with a mandate to do research designed to help farmers enhance their knowledge, increase agricultural productivity, and encourage environmental protection and sustainable development of agriculture and rural areas. He has conducted research on upland development in Vietnam for many years, often working in collaboration with researchers from different disciplines and institutions, both national and international. While trained as a crop and land management scientist (Ph.D. from HUA, 1998), his research has focused on the relationship between the different cultures, farming systems, and the environment of Vietnam's mountains. He is particularly concerned with the relationship between the diverse indigenous cultures found in the different parts of the uplands and the farming systems on which they rely for their subsistence. He has published numerous articles, reports, and books of which the most recent are: (1) Trần Đức Viên, A.Terry Rambo, Nguyễn Thanh Lâm, eds, 2008, *Canh tác nương rẫy: Một góc nhìn* (Swidden Agriculture: A Point of View). Hanoi: Agricultural Publishing House; (2) Trần Đức Viên and Nguyễn Thanh Lâm, 2006. *Sinh thái học đồng ruộng: Giáo trình dùng cho sinh viên đại học và sau đại học* (Crop Ecology: Textbook for undergraduate and graduate levels), Hanoi: Agricultural Publishing House; and (3) Trần Đức Viên, Nguyễn Vinh Quang, Mai Văn Thành, 2005, *The Impacts of Decentralization of Forest Management on Livelihoods of Ethnic Minority Groups in the Vietnam's Uplands*, Hanoi: Agricultural Publishing House.

Email: tdvien@hua.edu.vn

A. Terry Rambo is Special Professor in the Program on System Approaches in Agriculture, Faculty of Agriculture, Khon Kaen University, Thailand. He is also an Adjunct Senior Fellow of the East-West Center and a Visiting Professor at the Hanoi University of Agriculture. From 2000 to 2004 he was a Professor of the Center for Southeast Asian Studies at Kyoto University. Prior to joining

CSEAS he was a senior fellow of the East-West Center in Honolulu, where he served as Director of the Program on Environment from 1992 to 1997 and as the EWC Representative in Vietnam from 1997 to 2000. He was a lecturer in the Department of Anthropology and Sociology of the University of Malaya (1975-1980) and Visiting Professor of Social Science at the Dalat University Graduate School of Politics and Economics in Saigon (1973-1975). He has a Ph.D. in anthropology from the University of Hawaii (1972) and has conducted field research in Belize, Vietnam, and Malaysia. He has published extensively on Vietnamese culture and society, the Orang Asli of Malaysia, problems of upland development, and human ecology in Southeast Asia. His most recent book is *Searching for Vietnam*, 2004, Kyoto University Press and Trans Pacific Press.
Email: rambo@kku.ac.th.

Nguyễn Thanh Lâm is the Executive Director of the Center for Agricultural Research and Ecological Studies (CARES), Hanoi University of Agriculture, and the Chair of the Department of Environmental Technology. He holds a Ph.D. in Agronomy from Khon Kaen University (2005). His areas of specialization are agroecology, soil science, biodiversity conservation, and environmental protection. He is a country coordinator for number of regional projects on upland farming systems, neglected and underutilized plant species, plant genetic resources, cross-boundary watershed management between Vietnam and Laos, and environmental planning and management for Cambodia, Laos, and Vietnam. He has published papers in several international and Vietnamese journals, including *Agricultural Sustainability*, *Agriculture, Ecosystems and Environment*, *Agricultural Systems*, *Southeast Asian Studies*, and *Biodiversity Conservation*. His most recent books include: (1) Trần Đức Viên, A.Terry Rambo, Nguyễn Thanh Lâm, eds, 2008, *Canh tác nương rẫy: Một góc nhìn* (Swidden Agriculture: A Point of View). Hanoi: Agricultural Publishing House; (2) Trần Đức Viên and Nguyễn Thanh Lâm, 2006, *Sinh thái học đồng ruộng: Giáo trình dùng cho sinh viên đại học và sau đại học* (Crop Ecology: Textbook for undergraduate and graduate levels), Hanoi: Agricultural Publishing House.
Email: lamnt@cares.org.vn

Associate Editors

Neil L. Jamieson has spent eighteen years working in Asia, thirteen of them in Vietnam. Much of his work has been related to the intersection of culture, environment, and development in Southeast Asia. His interests and writings extend into continuity and change in Vietnamese culture, art and literature, the sociology of development, and ethnic minority issues. He served for over two years as a Provincial Representative for the U.S. Aid mission in South Vietnam (1963-1965), including 18 months in upland areas; participated in the study of the effects of herbicide use in Vietnam for the National Academy of Sciences; worked on environmental and development issues at the East West Center (1982-1988); headed the Winrock International office in Hanoi (1993-1998); and has carried out numerous consultancies for the World Bank, FAO, the Asia Development Bank, the Swedish International Development Agency, and other organizations. He has taught at the University of Hawaii and Chaminade University in Hawaii and conducted many teaching and training seminars on the human dimensions of upland development in Vietnam in conjunction with Vietnam National University, Hanoi, and the Hanoi University of Agriculture. He co-authored, with Lê Trọng Cúc and A. Terry Rambo, *The Development Crisis in Vietnam's Mountains*, published in both Vietnamese and English. His other writings include *Understanding Vietnam* (University of California Press, 1993) and numerous articles for scholarly journals and encyclopedias, as well as working papers and consultant reports concerned with issues of upland development in Vietnam.

Email: njamieson@verizon.net

Stephen J. Leisz is Assistant Professor of Geography at Colorado State University in Fort Collins, Colorado. His area of specialization is the monitoring of land use / land cover change in montane mainland Southeast Asia and recent farming systems changes in the region. He has a Ph.D. in Geography from the University of Copenhagen (2007) and an M.Sc. in Environmental Monitoring, with an emphasis on remote sensing and GIS, from the University of Wisconsin-Madison (1996). He has done extended field research in Madagascar and Vietnam, where between 1997 and 2007 he carried out investigations on land use and land cover change in the Northern Mountain Region. From 1997 to 1999 he was an East-West Center Researcher working in conjunction with CRES, of the Vietnam National University, Hanoi. From 2000 to 2007 he was employed by the Institute of Geography, University of Copenhagen, to work in collaboration with the Center for Agricultural Research and Ecological Studies of Hanoi University of Agriculture in north-central and northern Vietnam. In 2007 he worked for the Bishop Museum in Honolulu, Hawaii, and continues to collaborate with museum personnel on several projects related to climate change in Melanesia. In 2008 he was a visiting fellow at Kyoto University's Center for Southeast Asian Studies and he co-organized the Demise of Swidden in Southeast Asia workshop and conference in Hanoi, Vietnam. As part of that conference he also co-organized a session assessing the recent trends in the extent of swidden in the region. He has published several papers on farming system and land use/land cover changes in Vietnam and Yunnan, China.

Email: steve.leisz@colostate.edu

Authors

Georg Cadisch, Professor of Plant Production in the Tropics and Subtropics, Institute of Plant Production and Agroecology, and Director of the Tropenzentrum, University Hohenheim, Germany

Đặng Việt Quang, Fellow for Agricultural Economics, Center for Agricultural Research and Ecological Studies, Hanoi University of Agriculture, Vietnam

Đào Minh Trường, Researcher, Center for Natural Resources and Environmental Studies, Vietnam National University, Hanoi

Đào Trọng Hưng, Senior Researcher, The Institute of Ecology and Biological Resources, Hanoi, Vietnam

Deanna Donovan, Environmental Economics Adviser, Joint Nature Conservation Committee (JNCC), Peterborough, UK

Ingrid Foik, independent scholar, Chestnut Ridge, N. Y., USA, formerly visiting lecturer in agricultural economics, Hanoi University of Agriculture, Vietnam

Jefferson Fox, Coordinator of Environmental Studies and Senior Fellow, the East-West Center, Honolulu, Hawaii 96848-1601, USA

Lê Quang Trung, Researcher, Center for Natural Resources and Environmental Studies, Vietnam National University, Hanoi, and Research Assistant, Vietnam Ministry of Agriculture and Rural Development - Swedish International Development Cooperation Agency Cooperation Programme

Mai Văn Thành, Lecturer, Faculty of Natural Resources and Environment, Hanoi University of Agriculture, Vietnam

Nghiêm Phương Tuyến, Researcher, the Center for Natural Resources and Environmental Studies, Vietnam National University, Hanoi

Nguyễn Đình Tiến, Researcher for Agricultural Economics, Center for Agricultural Research and Ecological Studies, Hanoi University of Agriculture, Vietnam

Nguyễn Thế Phương, Researcher for Forestry, Center for Agricultural Research and Ecological Studies, Hanoi University of Agriculture, Vietnam

Nguyễn Thị Bích Yên, Lecturer, Faculty of Natural Resources and Environment, Hanoi University of Agriculture, Vietnam, and Ph.D. student, The University of Tokyo, Japan

Nguyễn Thị Hải Ninh, Researcher for Economic and Policy Analysis, Center for Agricultural Research and Ecological Studies, Hanoi University of Agriculture, Vietnam

Nguyễn Thị Minh Hiền, Research Fellow, Agricultural Economics and Rural Development, Hanoi University of Agriculture, Vietnam

Nguyễn Thị Thu Hà, Research Fellow, Center for Agricultural Research and Ecological Studies, Hanoi University of Agriculture, Vietnam, and Ph.D. Fellow, International Institute for Geo-Information Science and Earth Observation (ITC), The Netherlands

Nguyễn Văn Dung, Associate Professor for Hydrology and Irrigation, Center for Agricultural Research and Ecological Studies, Hanoi University of Agriculture, Vietnam

Nông Hữu Dương, Researcher for Natural Research Management, Center for Agricultural Research and Ecological Studies, Hanoi University of Agriculture, Vietnam

Aran Patanothai, Professor Emeritus, Department of Plant Science and Agricultural Resources, Faculty of Agriculture, Khon Kaen University, Khon Kaen 40002, Thailand

Phạm Thị Dung, Researcher for Crop Science, Center for Agricultural Research and Ecological Studies, Hanoi University of Agriculture, Vietnam

Phạm Thị Mai Hương, Researcher for Agricultural Economics, Center for Agricultural Research and Ecological Studies, Hanoi University of Agriculture, Vietnam

Trần Chí Trung, Research Fellow, the Center for Natural Resources and Environmental Studies, Vietnam National University, Hanoi

Trần Mạnh Tường, Researcher for Plant Sciences, Center for Agricultural Research and Ecological Studies, Hanoi University of Agriculture, Vietnam

Trần Nguyên Bằng, Researcher for Land Resources, Center for Agricultural Research and Ecological Studies, Hanoi University of Agriculture, Vietnam

Trịnh Thị Tiện, Researcher, International Relations Office, Nông Lâm University, Ho Chi Minh City, Vietnam

Patma Vityakon, Associate Professor of Soil Science, Land Resources and Environment Section, Department of Plant Science and Agricultural Resources, Faculty of Agriculture, Khon Kaen University, Khon Kaen 40002, Thailand

Võ Hữu Công, Researcher for Natural Research Management, Center for Agricultural Research and Ecological Studies, Hanoi University of Agriculture, Vietnam

Vũ Thị Thao, Ph.D. Fellow, Institute of Geography & Geology, University of Copenhagen, Denmark

List of Major Government Policies and Programs Affecting Farming Communities in the Northern Mountain Region

Year	Government policies/program	Objectives
1954	Land reform	Land expropriated from feudal landlords was redistributed to individual peasant households
1959	Formation of mutual aid teams (*Tổ đổi công)*	Teams of villagers were organized to help households cultivate their paddy fields
1959	Formation of agricultural cooperatives (*Hợp tác xã*)	Paddy land, buffalo, and tools owned by households were collectivized and all production activities managed by the cooperative
1968	Fixed Cultivation and Sedentarisation Program (*Định canh định cư*)	National program designed to end slash-and-burn farming in the uplands while reducing poverty, promoting access to education, and promoting national security. By 1990, this program had reached 1.9 million ethnic minority people
1981	Directive 100	Farmers make contracts with cooperatives to take over the planting, tending, and harvesting of paddy fields in exchange for delivering a specified quantity of grain to the cooperative
1986	*Đổi mới* economic reform policies	Shift from central planning to free market economic management
1988	Resolution 10	Allocation of agricultural land directly to household land users
1991	Forest law	Law on forest protection and management
1992	327 program	Provision of subsidies to farmers to reforest "barren hills" (1993–1998)
1993	Land law	Granting of the rights to households to transfer, exchange, lease, inherit and mortgage agricultural plots
1994	Resolution No. 02/CP	Forestland allocation for forestry purposes
1995	Resolution No. 01/CP	Allocation of contract land for agriculture, forestry, and aquaculture
1998	Circular No. 346/1998/TT-TCDC	Revisions to the Land Law

YEAR	GOVERNMENT POLICIES/PROGRAM	OBJECTIVES
1998	Five Million Hectare Reforestation Program (661 Program)	Follow-on to 327 Program for reforestation
1998	135 Program	Provides development assistance to improve the living standards of remote, isolated and disadvantaged communities, especially those of ethnic minorities, in the mountains and border zones

PART I

BACKGROUND OF RESEARCH IN TAT HAMLET

Introduction

This part consists of two chapters that are intended to provide the reader with an introduction to composite swiddening as an agricultural system and to our study site of Tat hamlet. Chapter 1 describes composite swiddening as a distinctive type of agroecosystem. It points out that, while agricultural systems containing many of the elements of composite swiddening are actually quite widespread in montane mainland Southeast Asia, this type of agricultural system had not previously been recognized as a distinctive type of swidden farming.

The chapter also sets out the objectives of the research project, discusses the reasons for the selection of Tat hamlet as the research site, reviews the history of our research there, and describes the various data sets employed in the analyses presented in the various chapters in this volume. Finally, it examines the significance of our research in Tat hamlet for Southeast Asian area studies, agricultural systems research, and rural development policy making for upland areas in Vietnam and montane mainland Southeast Asia as a whole.

Chapter 2 presents a brief overview of Tat hamlet, the site of our research for the past 15 years. Called 'the hamlet of the waterfalls' by its Đà Bắc Tày inhabitants, Tat hamlet is a small rural settlement located deep in the mountain and valley realm of Đà Bắc district in Hòa Bình province. The hamlet, which is located in a narrow river valley cut through steep mountain slopes, is quite similar in topography, climate, and other physical characteristics to hundreds of other isolated settlements of ethnic minority people in Vietnam's Northwestern Mountain range. What makes it distinctive is the extent to which its hill slopes are still covered by forest vegetation and this despite the fact that its human occupants have been practicing swidden farming for more than a century.

The first section of the chapter describes the hamlet's physical and biological characteristics, including its location, topography, climate, soils, vegetation cover, and patterns of land use; the second section presents a description of the social and cultural characteristics of the community, including settlement pattern, population, ethnic composition, cultural patterns, social organization, socioeconomic differentiation, administrative organization and public services, transportation and communications, and the economy; and the third section reviews some recent changes in Tat hamlet.

1
Research on Composite Swidden Agriculture in Tat Hamlet

Trần Đức Viên and A. Terry Rambo

Introduction

This volume presents the findings of a long-term interdisciplinary research project on the human ecology of the composite swiddening agriculture of the Đà Bắc Tày ethnic minority people of Tat hamlet. Tat hamlet is located in Tân Minh Village in Đà Bắc district of Hòa Bình province, the gateway to the vast Northwestern Mountains (Figure 1.1).

In this introduction, we first briefly describe composite swiddening as a distinctive type of agroecosystem. We next set out the objectives of the research project, discuss the reasons for the selection of Tat hamlet as the research site, review the history of our research there, and describe the various data sets employed in the analyses presented in the various chapters in this volume. Finally, we look at the significance of our research for Southeast Asian area studies, agroecology and agricultural systems research, and rural development policy making for upland areas in Vietnam and montane mainland Southeast Asia as a whole.

What is Composite Swidden Agriculture?

Composite swiddening is defined as an agricultural system that integrates permanent wet rice fields and rotating swidden plots into a single household resource system (Rambo 1998) (Plate 1.1). It might be more precisely labeled as the "composite paddy-swidden agroecosystem" (Rambo 2007) but, for the sake of brevity, we will usually refer to it simply as composite swiddening.

Although composite swiddening was first identified as a distinctive type of agroecosystem on the basis of research in Tat hamlet, it is actually quite a common and widespread agroecosystem in upland areas in northern Vietnam. For example, Cuisinier (1948) described a similar agricultural system that was employed by the Mường ethnic minority group in Hòa Bình province during

Figure 1.1 Location of Tat hamlet in northern Vietnam

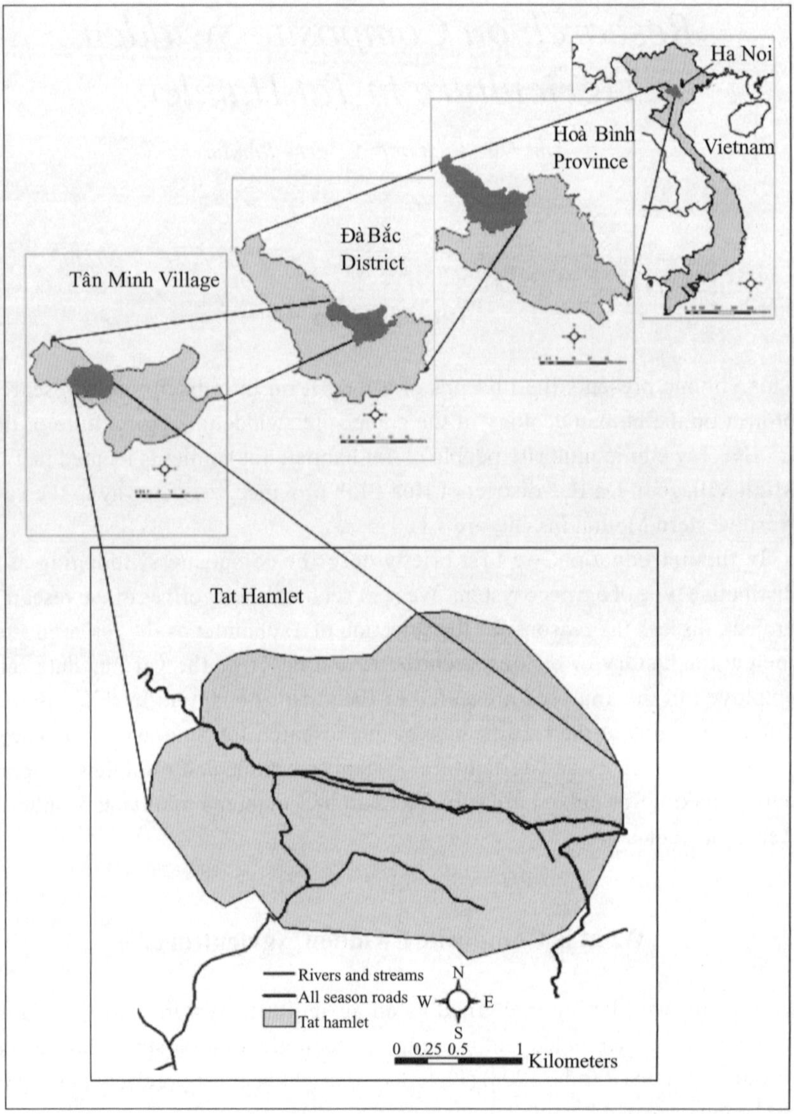

the French colonial era. A recent study by CARES researchers (see Chapter 17) found that composite swiddening type farming systems are currently found in 50 percent of 124 communes (villages) in five randomly selected districts in the NMR. Seemingly similar kinds of composite agroecosystems involving both swiddens and paddy fields have also been reported as being used by

Plate 1.1: The composite swidden agroecosystem in Tat hamlet with paddy fields in the valley and swidden fields on the hill slopes (Nguyễn Thanh Lâm)

mountain minority groups in other parts of montane mainland Southeast Asia, including among the Shan of Burma and northern Thailand (Durrenberger 1981), the Lua' and Karen of northern Thailand (Kunstadter 1978a), and the Hani of Xishuangbanna Prefecture in Southwestern China (Pei Shengji: personal communication, January 1992). The farming system of the Ifugao of the Cordillera in the Philippines also incorporates a combination of swiddens and paddy fields (Dove 1983a).

It is the integration of paddy fields and swidden fields together to form a single complex household farming system that differentiates composite swidden agriculture from other more widely practiced and better-known swidden systems in Southeast Asia. In the case of Tat hamlet, the complexity of the system is further increased by incorporation of many additional components, including homegardens, tree gardens, livestock, and fish ponds. Households also make heavy use of resources from the fallow swidden fields and the secondary forest. It is the simultaneous cultivation of both paddy fields and swiddens by each individual household, however, that distinguishes this farming system from other forms of swidden agriculture. Moreover, it is this combining of paddy fields and swiddens into a single system that we hypothesized might make composite swidden agriculture a more sustainable agroecosystem than pure swiddening systems. This research project was initiated in order to find out if composite swidden agriculture really was a relatively sustainable system of land use for the uplands, to better understand how this system functioned to meet the needs of the farmers, and to see if it was possible to further improve it.

Project Objectives

Our research in Tat hamlet had both scientific and applied objectives. The four major scientific objectives were:

1) Description of the structure and functioning of the agroecosystem
The first objective of our research was to describe the structure and functioning of the key individual components (paddy fields, swiddens, fallowed swiddens, secondary forests, homegardens, tree gardens, fishponds, livestock) that in combination make up the composite swiddening agroecosystem of Tat hamlet. This involved making detailed studies of each of the key components, including mapping, botanical species inventories, and recording information on productivity and farmer management practices.

2) Description of social and economic aspects of composite swiddening
The second research objective was to describe some of the key social and economic aspects of Tat hamlet as these affect the management of the composite swidden agricultural system. Studies were carried out on various aspects of the social system of the hamlet, including social organization, the role of women in natural resource management, household economics, labor utilization, and marketing channels for crops and forest products.

3) Assessment of agroecosystem sustainability
Our third objective was to assess the sustainability of this system. The main approach employed was to measure nutrient flows within and among the swidden and paddy field components of the composite swidden agroecosystem, as these powerfully affect the productivity and sustainability of the entire system. In our research on nutrient balance we have tried to understand the nutrient flows between hillside fields and paddy fields in the valley throughout the complete swidden cycle. GIS analysis of aerial photographs and satellite images was also employed to measure changes in vegetation cover within the hamlet in order to provide an ecological assessment of sustainability.

4) Examining composite swiddening in a wider regional perspective
The fourth objective of this research was to look at the wider relevance of composite swiddening in Vietnam's northern mountains. This included

investigating the geographical distribution of this type of agroecosystem in the NMR and developing criteria for determining where composite swiddening is likely to be an appropriate agroecosystem and where it is not. We also looked at Tat hamlet in comparison to several other communities in the NMR in order to assess the relative performance of its agroecosystem in terms of the emergent properties of productivity, diversity, sustainability, autonomy, equitability, and solidarity.

In addition to these scientific goals, researchers from HUA devoted considerable time and effort to working together with the farmers on applied experiments intended to improve the productivity and sustainability of various components of the agroecosystem. This applied research, although producing numerous valuable results, many of which have been adopted by the farmers for continued use, lies outside the scope of this volume.

Why Was Tat Hamlet Selected as the Research Site?

We selected Tat hamlet as the site for this research because it was a long-inhabited settlement where swidden agriculture had been practiced for many years but where good quality forest still covered much of the landscape. Our goal was to identify the factors underlying the evident sustainability of swidden agriculture in Tat hamlet and to understand why swiddening there had not resulted in the serious environmental degradation that could be observed in many other farming settlements in the NMR. Thus, we were under no illusion that Tat hamlet was somehow typical of the thousands of farming settlements in the NMR; quite to the contrary, it was selected precisely because it was so visibly atypical.

It was in January of 1992 that we made our first brief visit to Tat hamlet in the course of a reconnaissance survey of swiddening communities in the mountainous zone of Hòa Bình province. We almost failed to find our future research site. On the day of our first visit, it was raining hard and the narrow dirt road we were following into the mountains of Đà Bắc district was wet and slippery and becoming increasingly dangerous as the mountain slopes became steeper. We had just decided that we would drive only a few more kilometers before giving up our search for the day when our Land Rover rounded a sharp bend on the winding mountain road and we had our first view of Tat hamlet (Plate 1.2). As soon as we saw it we knew that we had found the type of site we were looking for. As was noted in the unpublished report of our reconnaissance (Rambo 1992: 10):

> Of the sites that we visited, [Tat hamlet in] Tân Minh village appears the most suitable for our purposes. It offers a diverse landscape with maximum biological

Plate 1.2: Tat hamlet as it appeared during our first visit in January 1992

> and cultural diversity. The economy is based on swidden agriculture but there is still a large area of intact primary forest. The Tay villagers are friendly and cooperative. It does not appear that they are engaged in significant illegal use of resources that would make full reporting of research findings impossible. Improved strategies for resource management and conservation of biological diversity developed and tested at Tân Minh would have a wide area of application in the Northwestern highlands. It is thus an ideal site for a long-term applied eco-development project....

Of course, not all of our initial assumptions about Tat hamlet were confirmed by subsequent research but, overall, it has proved to be a highly suitable site for our studies. In particular, our impression of the villagers as being friendly and cooperative has been repeatedly validated in our continuing close interactions with them.

A Brief History of Research in Tat Hamlet

Following our initial visit to Tat hamlet, teams of researchers from CRES, HUA, the EWC, and KKU made a total of six field trips to Tat hamlet between 1992 and 1995. In addition to making observations of different components of the agroecosystem and social system on all of these visits, a questionnaire survey of a small sample of households was completed in August 1992 to determine the productivity and labor requirements of the swiddens and paddy fields. Information collected on these visits was used to prepare the first description of composite swidden agriculture as a distinctive type of upland agroecosystem that was submitted in a report to the MacArthur Foundation in 1995 (subsequently revised and published as Rambo [1998] and Rambo and Lê Trọng Cúc [1997]). In the rainy season of 1995, HUA initiated research using the pin method on soil erosion and nutrient balance in swidden fields in Tat hamlet (Trần Đức Viên 1998). This study was continued until 1999.

In 1995-1998, the EWC, with support from the US National Science Foundation, organized a joint project with CRES and HUA on forest fragmentation and hydrology in Tat hamlet. This project revealed that the share of the hamlet area covered by woody vegetation had not declined over the 50-year period for which data were available, despite rapid population growth and very marked shortening of the fallow cycle (Fox *et al.* 2000; Fox *et al.* 2003). Researchers involved in this project also produced several publications on the hydrological effects of forest fragmentation resulting from swidden agriculture (Giambelluca 2002; Giambelluca *et al.* 2003; Ziegler *et al.* 2004, 2006, 2007a) and on the impact of weeding of steep swidden fields on soil translocation (Ziegler *et al.* 2007b).

In late 1998 and early 1999, a team of CARES researchers carried-out a separate study of social and economic conditions in Tat hamlet under a contract with CRES as part of a larger project to monitor changes in Vietnam's Northern Mountain Region funded by the Swedish International Development Cooperation Agency (SIDA). A random sample of 42 households was interviewed using a standardized questionnaire designed by Terry Rambo, who was the Technical Consultant to the SIDA study. In addition to providing the basis for several publications (Lê Trọng Cúc and Rambo 1999; 2001; Rambo and Trần Đức Viên 2001a and 2001b), the data collected in this survey have provided a baseline against which to compare more recent studies on social and economic aspects of Tat hamlet.

In 1998, the Ford Foundation made a grant to HUA to support CARES research in Tat hamlet with a particular focus on assessment of the sustainability of the composite swidden system by means of measurement of long-term nutrient balances in its components. Because the study of nutrient balances was a new and largely untested approach, CARES organized a series of workshops bringing together HUA scientists with scientists from Khon Kaen University (KKU) and the East-West Center to design a detailed research plan and study methodology. The first workshop was held at HUA and Tat hamlet in May 1999 and was followed by a second workshop held at KKU in July 1999. The output of these workshops was a detailed data collection plan for studying nutrient balance in paddy fields and swidden plots in Tat hamlet. The experimental plots and an automatic weather station were then set up at Tat hamlet with valuable labor contributions from local people. In the rainy season of 2000, a third HUA-KKU workshop was held in Tat hamlet to give the researchers the opportunity to see what was going on in the experimental plots and make adjustments in the study design and methods of data collection to better meet the research goals.

Beginning in 2000, with additional support from the Ford Foundation, CARES greatly broadened the scope of its work in Tat hamlet. The research on soil erosion and nutrient balances in the experimental swidden and paddy fields was continued, while a new series of studies on economics, marketing, forest management, and changes in land cover and land use was initiated. A number of publications grew out of this research (e.g., Nguyễn Thanh Lâm *et al.* 2004; Trần Đức Viên *et al.* 2006; Nguyễn Văn Dung *et al.* 2008). Various aspects of this research continued until 2006, after which a final technical report on project findings was compiled by CARES (Trần Đức Viên *et al.* 2006). Most of the chapters in this volume are based on sections of the CARES technical report. Several additional chapters, written by researchers involved in earlier studies by CRES and the EWC, have also been included. A few chapters have been previously published as journal articles. These are included in this volume with permission from the original publishers.

Data Bases Employed in Research on Tat Hamlet

The nineteen chapters in this volume were written by many different authors, who worked in Tat hamlet at different times between 1992 and 2006, and who enjoyed often quite different individual experiences of fieldwork there. Not surprisingly, these authors are not always in full agreement about every aspect of what is a very diverse, complex and dynamic ecological and social situation. Although the editors have attempted to resolve any glaring contradictions, we have not sought to impose a unitary view on contributors.

Because different chapters are based on multiple different data sets that were collected from different samples of households at different times and for different purposes, they may sometimes present somewhat different statistics for what are seemingly the same phenomena. Such differences are not errors that the editors failed to detect; instead they are simply the artifacts of the use of different data sets. To aid the reader in keeping track, all of the households surveys made in Tat hamlet between 1992 and 2006 used in writing this volume are listed in Table 1.1 along with the chapters in which these data sets are cited.

Other types of data employed in chapters describing ecological surveys and species inventories, nutrient balances, and land use and vegetation cover are described in detail in the methodology sections of these chapters.

Significance of Research on Composite Swiddening in Tat Hamlet

The research on the human ecology of composite swiddening described in this volume should be useful from a number of perspectives, including those of area studies, studies of agroecology and agricultural systems, and upland development policymaking:

Southeast Asian area studies

This volume is an addition to the growing pool of human ecological studies on rural communities in Southeast Asia. Community studies of this type are an important resource for area studies specialists. The chapters in this book document and analyze in great detail the human ecology of one small hamlet in a remote corner of Hòa Bình province. Very detailed information is provided about the people of Tat hamlet and the adjustments that they have made in their farming system in order to adapt to changing environmental and economic circumstances. In contrast to some other Southeast Asian countries, notably Indonesia, Malaysia, the Philippines, and Thailand, it is rare to have so much information about any rural community in Vietnam, let alone a relatively remote ethnic minority settlement in the uplands. What makes this study even more unusual in the Vietnamese context is that it describes changes in Tat hamlet over a fifteen year period, offering a truly diachronic view of the rapidly evolving situation in this upland community since Vietnam abandoned its command economy system of management and embarked on widespread institutional reforms based on acceptance of market mechanisms.

Studies of agroecology and agricultural systems

The research described in this volume should be of interest to scholars concerned with the study of agricultural systems because it presents the first detailed description of the previously little known system of agriculture that we have labeled 'composite swiddening.' Although systems of a similar type occur in many different parts of Southeast Asia, no detailed descriptions of the structure, functioning, and dynamics of this type of system have previously been available in the literature of agricultural ecology. This research thus adds another model to the existing set employed by researchers to classify and understand upland agricultural systems. The analysis of nutrient balances in swidden fields and

Table 1.1 List of household surveys conducted in Tat hamlet between 1992–2006

Name of survey	Date collected	Sample size	Sample type	Method of collection	Responsible institution (individual)	Chapter(s) where used
Agroecosystem productivity and labor allocation	8/1992 (corrected 2/93)	9 households	Random	Questionnaire	CRES-EWC (Nghiêm Phương Tuyến, Terry Rambo, Keith Fahrney)	10
Household socio-economic status (SIDA NMR baseline study)	12/1998–2/1999	42 households	Random	Questionnaire	CARES-CRES-EWC (Trần Đức Viên)	10, 12, 14, 18
Collection and utilization of NTFPs	3–4/2002	60 households	Stratified random sample of poor, middle and better-off households	Questionnaire	CRES (Trịnh Thị Tiện)	
Household farming systems	12/2003	41 households	Restudy of 41 households from 1998–1999 sample of 42 households	Questionnaire	CARES (Nguyễn Thanh Lâm)	10, 12
Production and marketing of agricultural and forest products	1/2004	38 households	Random	Questionnaire, group discussion	CARES (Phạm Tiến Dũng, Bùi Thị Gia, Đặng Việt Quang)	14
Homegardens	11/2004	30 households	Random	Questionnaires, observation	CARES (Nông Hữu Dương)	8
Forest management (including role of women)	11/2004	30 households	Random	Questionnaire	CARES (Nguyễn Thế Phương)	13

Farm household budgets (including agricultural inputs)	2/2005	26 households	Purposive	Questionnaires, group discussions	CARES (Đặng Việt Quang)	4
Household economics	3/2005	39 households	Restudy of 39 households from1998–99 sample of 42 households	Questionnaire, group discussions	CARES (Đặng Việt Quang, Phạm Thị Mai Hương, Vũ Thị Thao, Trần Nam Anh)	12, 13
Household economics	1/2006	48 households (49 households were interviewed, but 1 was excluded from analysis because of incomplete information)	Stratified random sample of households with 3 types of farming system. Swidden only, composite swidden, and modified composite swidden	Questionnaires, semi-structured interviews, group discussions	CARES (Nguyễn Đình Tiến, Nguyễn Thị Hải Ninh, Phạm Thanh Lan)	12
Gender roles	1/2006	42 households	Same sample as 1/2006 household economics survey but 6 households headed by widows were excluded from analysis.	Questionnaire	CARES (Nguyễn Đình Tiến, Nguyễn Thị Hải Ninh, Phạm Thanh Lan)	14
Livestock and fish	3–4/2006	30 households	Random	Questionnaire, group discussion	CARES (Phạm Thị Dung)	6, 7

paddy fields (Chapter 15) also offers new insights into how upland-lowland interactions influence agroecosystem sustainability in mountainous areas.

Upland development policymaking

The findings of our research in Tat hamlet may also be of value to those concerned with formulation and implementation of development policies for the NMR, and even for other parts of montane mainland Southeast Asia. Although Tat hamlet was not selected for study because we thought it was a "typical" community in the NMR, indeed, as was discussed earlier, it was selected precisely because its Đà Bắc Tày inhabitants practiced an atypically sustainable type of swidden agriculture, recent changes in the environmental conditions and economic situation of Tat hamlet are in many ways indicative of broader changes in the Northern mountains (Jamieson *et al.* 1998, Rambo and Jamieson 2003, Rambo 2004) and in the wider expanses of montane mainland Southeast Asia (Kono and Rambo 2004). Thus, our research in Tat hamlet sheds considerable light upon the nature of both the problems and the opportunities that millions of upland people, especially disadvantaged ethnic minority people, are encountering as they strive to maintain their livelihoods and to improve their standard of living in the face of rapid population growth and degradation of natural resources.

In detailing what has happened in Tat hamlet, what the various actors have done—especially farm households, but also policymakers, local officials, researchers, and extension workers—and the results of those actions, this volume demonstrates both the potential positive contributions of development assistance and the limits of current approaches to upland development. It also presents a general model of the composite swidden agroecosystem that we believe can be successfully adapted as an improvement on existing models in many other sites in the NMR. We believe that this model holds great promise for helping other upland farming communities to make the difficult transition to more sustainable livelihood strategies.

2
Hamlet of the Waterfalls: An Overview of the Environmental and Social Characteristics of Tat Hamlet

A. Terry Rambo, Trần Đức Viên, and Nguyễn Thanh Lâm

Introduction

This chapter presents a brief description of Tat hamlet (*Ban Tạt* in the language of its inhabitants, *Bản Tát* in Vietnamese), which is the main site where our research on composite swiddening has been conducted. The first section describes the site's physical and biological characteristics, including its location, topography, climate, soils, vegetation cover, and patterns of land use; the second section presents a description of the social and cultural characteristics of the community, including settlement pattern, population, ethnic composition, cultural patterns, social organization, socioeconomic differentiation, administrative organization and public services, transportation and communications, and the economy; and the third section reviews some recent changes in the hamlet.

Physical and Environmental Characteristics of Tat Hamlet

In the T'ai language of its Đà Bắc Tày inhabitants (Tày Đà Bắc in Vietnamese), *Bǎn Tạt* means 'the hamlet of the waterfalls,' which is an apt choice given the many small streams that cascade down over the steep rocky mountain slopes that fringe the narrow valley where the hamlet is centered.

Location

The central point of Tat hamlet is located at N 20.95 and E 105.12. It is one of ten hamlets that make up Tân Minh village, which is one of the 21 villages in Đà Bắc district, in Hòa Bình province. Figure 2.1 is a map showing the hamlet boundaries, major rivers and roads, and location of houses and other cultural features.

Figure 2.1 Map of Tat hamlet

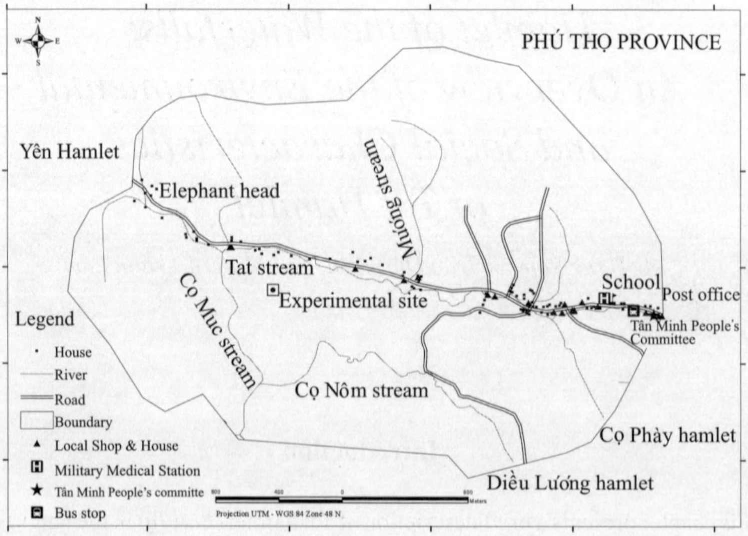

Tat Hamlet is located in the watershed of the Đà River (referred to by the French as the Flueve Noir and known in English as the Black River). The Đà River is the site of the Hòa Bình Dam, Vietnam's largest hydropower project. Flooding of the reservoir in the late 1980s forced relocation of 9,305 households, or an estimated 52,000 people, mostly members of the Mường ethnic minority, from their homes in the valley bottom. Many have resettled on steep slopes above the lake where their swidden fields are a major source of sediments flowing into the reservoir (Vũ Quyết Thắng 1991). Others have migrated into surrounding communities located on higher ground that were not flooded by the reservoir. Several of these households have settled in Tat Hamlet, further intensifying pressure on limited land resources there. Tân Minh village is not classified as a village heavily impacted by resettlement, however, and therefore does not qualify for special assistance under the program established by the government to help local communities overcome problems arising from the in-migration of households forced to move because of the construction of the Hòa Bình dam.

Topography

Tat hamlet is located in the mountain and valley realm of the Northwestern Mountains. The landscape is characterized by sinuous narrow river valleys

Plate 2.1: Tat hamlet viewed from a nearby mountain top (Nguyễn Thanh Lâm)

cutting through steep mountains (Plate 2.1). The total surface area of Tat hamlet is 742 ha, most of which is hills and mountains. Flat land is scarce with less than 20 percent of the land surface having a slope under 25 degrees. Only about 20 hectares in the valley floor are flat enough for construction of bunded wet rice fields. The elevation of the valley floor at Tat hamlet is approximately 360 m asl. Surrounding peaks reach an elevation of 800 to 950 m. The highest peak (Núi Hen) reaches 1,176 m. Slopes of the mountains surrounding the settlement are quite steep (30-60 degrees). Numerous smaller streams flow down from small valleys cut into the slopes surrounding the settlement and join the Tat Stream, which runs through the main valley.

Climate

The climate is monsoonal tropical with a hot wet summer season and a cold dry winter season. The mean temperature is 21.2°C, with an average July high of 25.8°C and an average January low of 14.6°C (Table 2.1). Summer temperatures can reach as high as 40°C and winter temperatures fall as low as 2 or 3°C.

Rainfall averages about 2,195 mm per year, divided into two seasons. The rainy season extends from May until October in which 90 percent of total rainfall occurs (Table 2.2 shows the monthly distribution of rainfall). During this season frequent severe thunderstorms are accompanied by very intense rainfall. During one such storm in May 1999, 40 mm fell in less than 2 hours. The dry season, from November to April, accounts for only 10 percent of total precipitation.

Table 2.1 Monthly mean temperatures in Tat hamlet (2001 to 2004)

	2001	2002	2003	2004	AVERAGE
Jan	16.2	14.1	13.4	14.7	14.6
Feb	15.1	17.4	19.0	15.7	16.8
Mar	19.5	20.1	19.3	18.2	19.3
Apr	25.7	23.7	23.4	21.9	23.7
May	23.9	24.3	25.7	23.9	24.5
Jun	25.7	25.8	25.9	25.4	25.7
Jul	26.0	26.0	26.0	25.2	25.8
Aug	25.3	25.0	25.5	25.6	25.4
Sep	24.4	23.4	23.6	23.9	23.8
Oct	21.9	21.2	21.2	20.6	21.2
Nov	17.6	17.7	18.0	18.8	18.0
Dec	16.5	16.6	16.0	15.0	16.0
Average	21.5	21.3	22.4	20.7	21.2

Source: Data from CARES weather station in Tat hamlet

Table 2.2 Monthly rainfall in Tat hamlet (2001 to 2004)

	2001	2002	2003	2004	AVERAGE
Jan	17.4	43.0	46.8	11.2	29.6
Feb	33.0	20.4	46.8	29.4	32.4
Mar	169.0	9.2	28.2	24.0	57.6
Apr	65.6	45.0	106.4	280.8	124.5
May	379.2	385.2	363.8	267.4	348.9
Jun	387.6	434.6	142.8	258.4	305.9
Jul	479.8	326.6	251.0	474.0	382.9
Aug	395.6	307.0	326.8	326.8	339.1
Sep	328.6	317.4	514.0	270.0	357.5
Oct	276.0	138.4	211.4	2.0	157.0
Nov	3.4	82.6	4.2	55.4	36.4
Dec	12.0	53.4	6.0	20.2	22.9
Total	2547.2	2162.8	2048.2	2019.6	2194.5

Source: Data from CARES weather station in Tat hamlet

Figure 2.2 displays the average monthly mean temperature and rainfall based on data collected by the CARES weather station in Tat hamlet during the 4 years 2001-2004.

Geology

The geological make up of the hamlet is complex. Metamorphic rocks, especially quartzo-feldspartic gneiss, and igneous rocks, especially granite, constitute the

Figure 2.2 Monthly mean temperature and rainfall in Tat hamlet (2001–2004)

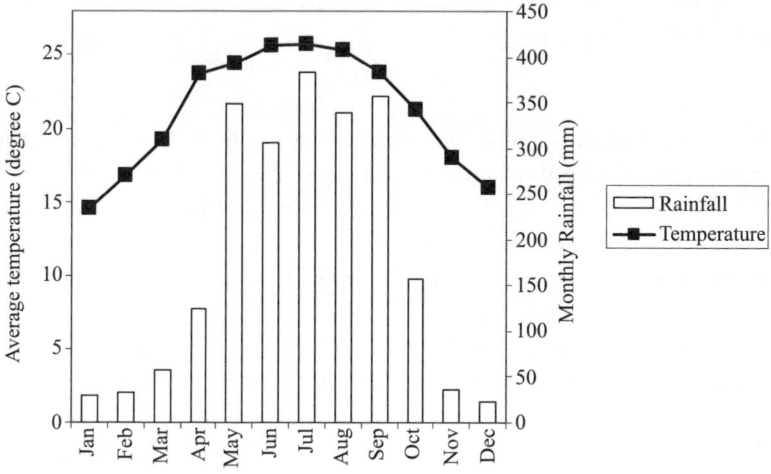

Source: Data from the CARES weather station in Tat hamlet

major rock types as observed on the mountain slopes and in various rock outcrops in swidden fields and paddies. Fresh gneiss appears to be most frequently found as outcrops in paddy fields. Granite is found most frequently in the northern hills. In addition to gneiss, some other metamorphic rocks, including schist and phyllite at different stages of metamorphism, are found. Sedimentary rocks, i.e., sandstone, as floats, are also occasionally observed and large fragments of quartz are found frequently associated with granite and gneiss.

The pre-existing rocks in the area are of argillaceous composition. These are rocks that upon weathering give rise to clay, i.e., sedimentary rocks including shale and sandstone; and metamorphic rocks including slate, phyllite, and schist. Granite found in the northern hills appears to be biotite-hornblende granite. It was probably formed by the intrusion of high temperature magma into the pre-existing rocks at a later stage. These rocks have undergone normal tectonic processes, mainly regional metamorphism, and they also were affected by high temperature from magma intrusion. These processes have metamorphosed the pre-existing rocks into gneissic and meta-pelitic rocks, which have a high aluminum content that tends to give rise to clays. The fact that milky quartz fragments are common in the areas indicates that veins of pegmatic rocks should be found sporadically exposed.

The rocks show differences in degrees of weathering. While the granitic and gneissic rocks are slightly weathered, the meta-pelitic rocks are highly decomposed. Most of the schist decomposes to greenish chalky chloritic clay while weathered phyllite retrogressively changes to reddish brown clay. This indicates that residual soils in the area are high in ferromagnesian content.

Boulders to coarse grain constituents in the residual soils are fragments of highly resistant compositions of the quartzo-feldspartic rocks produced by mechanical processes of degradation.

Soils

Soils in the hamlet are complex and varied. They can be divided into the major categories of upland soils and paddy soils (Table 2.3).

The upland soils are classified as Ferralic Acrisols, which are clayish loamy texture, extremely acidic, low in CEC (6-9 meq/100g of soils), low to medium available nutrient contents of N,P,K. The topsoil in the forest has average to high content of organic carbon, but organic carbon and nutrient contents decrease sharply at lower soil depths (see Table 2.3). There is a great deal of site-specific variability in soils because of differing parent material position in the landscape. Sandstone and schist are common, with areas of quartz and mica-bearing granite. No limestone occurs in the vicinity of the hamlet. Soils are deeply weathered with depths of up to 5 m observed in road cuts. Soil color varies from light grayish-brown to yellowish-brown to reddish-orange, suggesting varying stages in the weathering process and varying contents of iron and aluminum oxides. Upland soils have a fairly high clay content as is revealed by their shiny surfaces and high degree of slickness when wet. These soils are extremely hard when dry, but soften considerably and can be cultivated easily when moist.

Paddy soils in the valleys are of fine silty texture and very dark in color, presumably resulting from a long period of cultivation and deposition of organic matter from upper slopes (see chapters 4 and 15 for further descriptions of paddy soils). The soils are highly acidic (4.1-4.3 of pH_{KCl}), and have high organic carbon in the topsoil (3%), medium available N (4-8 mg/100 g soils), and poor available potassium (9 mg K_2O/100 g soils).

Vegetation cover

The vegetation cover is a complex patchwork of different successional communities that have been highly influenced by human activities (Plate 2.2). Although much of the hamlet was covered by primary forest until the 1960s, today only tiny remnant patches survive on extremely steep and inaccessible peaks (Figure 17.1d is a map of current vegetation cover). Hilltops and ridgelines are covered with secondary forest with a slight degree of canopy differentiation. Most slopes are covered with swiddens or recently fallowed plots covered with grasses, herbs, and scattered patches of bamboo and small trees. Valley crotches and storm drainage courses have largely been left surrounded by woody vegetation and secondary forest, though in some cases these patches have been scorched by fires escaping from the swiddens. Near-slopes and hillocks in

Table 2.3 Physical and chemical properties of upland and paddy soil profiles in Tat hamlet

Soil depth	pH_{KCL}	OC	N	P_2O_5	K_2O	N	P_2O_5	K_2O	CEC	Sand (2–0.05 mm)	Silt (0.05–0.002 mm)	Clay (<0.002 mm)
		(%)				(mg/100 g soils)				(%)		
Upland soils												
0–8	4.12	2.13	0.17	0.04	1.87	8.68	1.12	8.46	9.2	32.6	37.3	30.2
8–27	4.17	1.20	0.09	0.02	2.00	6.44	0.38	4.23	8.1	25.9	38.5	35.6
27–60	4.09	1.15	0.07	0.02	1.91	5.40	0.50	3.76	7.9	29.6	33.7	36.8
60–90	4.23	0.84	0.06	0.02	1.84	5.18	0.13	3.76	7.9	26.8	32.8	40.5
>90	4.34	0.44	0.06	0.01	1.65	2.80	0.13	2.82	6.2	22.0	33.7	44.3
Paddy soils												
0–20	4.08	3.01	0.16	0.05	3.27	8.04	6.38	9.87	8.8	48.9	34.5	19.7
20–24	4.13	2.30	0.12	0.04	3.19	7.48	7.00	9.87	9.0	53.1	32.8	14.2
24–38	4.32	0.80	0.05	0.03	3.29	5.60	2.75	11.75	6.8	44.4	27.7	28.0
38–70	4.34	0.44	0.03	0.02	3.27	4.45	1.00	9.87	6.9	56.6	25.2	19.5

Source: CARES Soil analysis, 1999

Plate 2.2: Swidden landscape in Tat hamlet

non-protected areas surrounding the hamlet have mostly been cleared for the planting of cassava swiddens or tree gardens, with very little fallow land. In a few limited areas, *Imperata cylindrica* grass covers the lower hillocks and isolated patches of Imperata occur even on distant hill slopes. The valley bottom is covered with wet rice fields.

Land use

The hamlet has a very complex pattern of land use reflecting the diverse human activities in its landscape. Table 2.4 presents data on the areas under different types of land use in the hamlet. The data on land use have been derived from two different sources: The first set are based on measurements by CARES researchers using GPS and satellite images in 2004 while the second set are copied from official records of the Tân Minh Village People's Committee in 2002.

It should be pointed out that the official estimate of the land area of the hamlet is only 304 ha compared to the 742 ha that CARES estimated from satellite image interpretation. The official area of 8.4 ha recorded for paddy fields is also much too low. Measurements in 2004 by a CARES team using GPS and tapes revealed an actual paddy area of 17.7 ha. Many reports about Tat hamlet still employ the official figures (that range from 8.2 ha to 9.2 ha), however.

The spatial distribution of different land uses in the hamlet is quite complex as is shown in Figure 2.3. The main land use categories displayed on this map

Table 2.4 Land use in Tat hamlet, area (ha) covered and percentage (%) of total area

LAND USE TYPE	MEASURED BY CARES RESEARCHERS (2004)		OFFICIAL RECORDS OF THE TÂN MINH PEOPLE'S COMMITTEE (2002)	
	AREA (HA)	PERCENTAGE OF TOTAL AREA	AREA (HA)	PERCENTAGE OF TOTAL AREA
Paddy fields	17.7	2.4	8.4	2.8
Swiddens and fallow land	151.0	20.4	141.4	46.5
Ponds	1.3	0.2	0.9	0.3
Homegardens	9.6	1.3	9.6	3.2
Land used for special purposes	3.7	0.5	2.8	0.9
Forest	558.8	75.3	141.2	46.4
Total area	742.1	100.0	304.3	100.0

are paddy fields, swiddens, grass and bush, which is mainly swidden land in the early stages of fallow, young forest, which is swidden land in an intermediate stage of fallow, and old forest, which is mostly fallowed swidden land in an advanced successional stage.

Social and Cultural Characteristics

Almost all of the people living in Tat hamlet belong to the Đà Bắc Tày ethnic minority group. As a minority community, it displays its own distinctive social and cultural characteristics, which are described in the following sections.

Settlement pattern

The hamlet has a lineal settlement pattern with houses built singly or in small clusters for a 3 km stretch along the lower slopes of the east-west oriented valley floor. Running down the center of the valley, and dividing houses on its north side from its south side, is the Mương River, which is paralleled by Provincial Highway No. 433. A few isolated houses have recently been constructed on farmsteads in small side valleys, such as Na Cọ Nôm, and scattered pieces of more or less flat ground in the mountains surrounding the main valley.

Population

Data on the evolution of the population from 1993 to 2005, which are based on records of the Village Family Planning Clinic, are presented in Table 2.5. In 2005,

Figure 2.3 Land use map of Tat hamlet in 2003

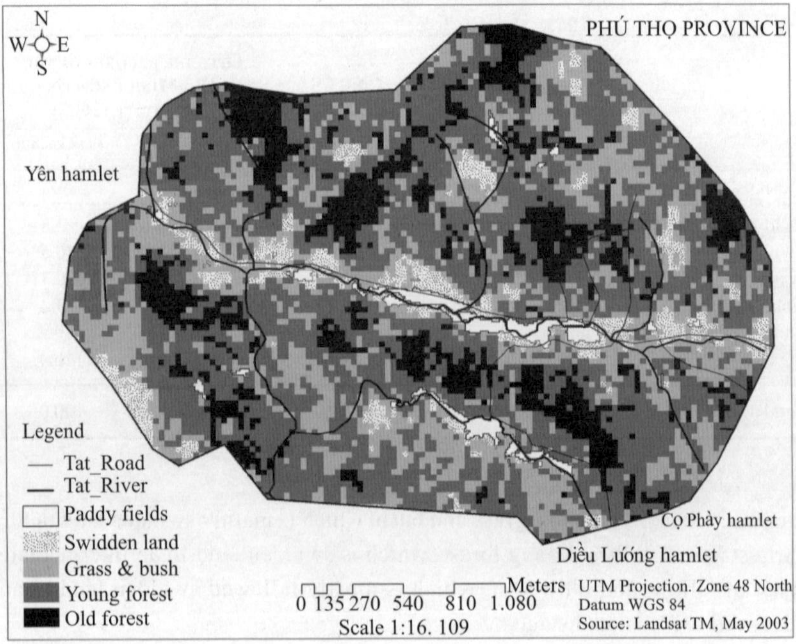

Tat hamlet had a population of 512 persons divided into 111 households compared to 389 persons and 69 households in 1993. In just twelve years, the population of the hamlet had increased by 1.3 times while the number of households had increased by 1.6 times. Formation of new households, even more than increase in number of individuals, places heavy pressure on natural resources to supply materials for building new houses and to provide land for new fields.

In 2005, the mean population density was 69 persons per km^2, compared to 52 persons per km^2 in 1993. No reliable population statistics are available for earlier periods but one informant recalled that there were only seven households in the valley in the mid-1950s when he was a young boy. Assuming seven members per household, the hamlet's population would have been about 50 persons representing a mean density of approximately seven persons per km^2.

Ethnic composition

Although most of the residents of Tat hamlet are Đà Bắc Tày, there are five households (out of a total of 111 households in 2005) that are classified as ethnically Mường or Kinh because one of the spouses belongs to these groups. There are also a number of Kinh, mostly employed as schoolteachers, who

Table 2.5 Changes in the population of Tat hamlet from 1955 to 2005

	1955	1993	1996	1998	2003	2005
No. of people	50	389	402	432	476	512
No. of households	7	69	75	91	107	111
Population density (persons/km2)	7	52	54	58	64	69

Source: Figures for 1955 estimated based on memory of an elderly informant. Figures for all other years are from records of the Tân Minh Village Family Planning Clinic

are temporary residents in the hamlet. They live in quarters provided by the government but are not counted as official members of the community.

The Đà Bắc Tày are a small, geographically isolated group with a population of approximately 17,000 individuals that is found only in Hòa Bình Province, primarily in Đà Bắc district. During the colonial period this population was referred to as the Đà Bắc Thổ (Abadie 1924). In Vietnamese, 'thổ' means 'earth, land, or ground' (Nguyễn Đình Hòa 1966:476). Several older informants in Tat hamlet said that they had referred to themselves as Thổ until the 1960s when they were instructed by government officials to use Tày instead. Similarly, 'Dzao' was substituted for 'Mán' and 'H'mông' for 'Miao' or 'Mèo.' These changes were justified on the grounds that the old ethnonyms had pejorative connotations. People in Tat hamlet do not think that this was true with regard to Thổ and elderly people still employ this word in casual conversation when referring to traditional aspects of their culture. The term is also employed in a somewhat negative sense, however, to refer to individuals that are perceived as being old fashioned or behaving in a feudalistic manner.

Vietnamese ethnologists commonly classify the Đà Bắc Tày as a local subgroup of the Tày, the largest of Vietnam's minority populations. The Tày are concentrated in the provinces of Hà Giang, Tuyên Quang, Cao Bằng, Lạng Sơn, Bắc Kạn, and Thái Nguyên in the central and northeastern parts of the NMR. Although they constitute the largest of Vietnam's more than 50 minority ethnic groups, the Tày are one of the least well known groups from an ethnological standpoint. The Tày were not a major focus of ethnological research during the French colonial era. Only in 1992 was the first relatively comprehensive Vietnamese language ethnography of the Tày published by the Institute of Ethnology in Hanoi (Bế Viết Đặng *et al.* 1992). According to Đặng Nghiêm Vạn (2000: 126), however, there are no scientific grounds for assigning the Đà Bắc Tày to the Tày ethnic group. Instead, Vạn includes them as a 'small local group' within the Thai ethnic group. Some Vietnamese ethnologists (e.g., Chu Thái Sơn and Nguyễn Thị Huyền 1974) believe that the Đà Bắc Tày are closely related to the White Thai but people in Tat hamlet who have visited Mai Châu, a White Thai community to the northwest near the border with Sơn La Province, deny that their cultures are similar and say that they cannot understand the White Thai language.

The historical origins of the Đà Bắc Tày of Tat hamlet are obscure but elderly informants say that the hamlet was settled a little over a hundred years ago when a small group of people belonging to the Xa clan coming from Sơn La province joined with other people already living in the general area to form Tat hamlet. According to one account, Xa Đạo Thành, the leader of the Xa clan, married Lò Thị Lún of nearby Đoàn Kết village. Both clans signed a contract called 'Ka Húa' (Silver-hair paper) sometime at the end of the 19th century. This story is told to explain why most households in Tat hamlet belong to the Xa and Lò clans.

Language

No detailed linguistic studies have been made although there is no question that the people of Tat hamlet speak a language belonging to the T'ai family. Several Thai colleagues from Khon Kaen University, who spoke either Northern Thai or the Isan (Lao) dialect, found that they were able to converse with the villagers with comprehension levels of between 20 and 50 percent.

Formerly, the Tay language was written with an alphabetic script based on Pali but differing from the Thai and Lao scripts. According to one elderly informant (now deceased) who wrote a sample of the script for our team, after the Revolution all written materials in this script were collected by government cadre and taken away. Today no one in the hamlet can write in the old alphabet.

Although most adults in Tat hamlet are reasonably fluent in the Vietnamese language, the local language is still commonly spoken at home and in interactions among neighbors. Teaching in the primary school is in Vietnamese, however, and almost all children develop fluency in that language, which is the lingua franca of trade and government in the mountain region.

Style of dress

In the early 1990s, when we began our research, all men had already adopted the dress style of rural Kinh males, consisting of readymade trousers and shirts, but most women still wore ethnically distinct dress. Older women sometimes wore a home woven white cotton tunic over a dark-colored skirt (Plate 2.3) while most younger women and girls wore an outfit composed of a dark blue or black tubular skirt with a brightly colored woven waist band and a form-fitting long sleeved blouse of pink, green, or white cotton, often worn under a dark blue tunic (Plate 2.4). Most women wore cotton or polyester headscarves. Dark

Plate 2.3: Woman wearing traditional style dress

Plate 2.4: Woman in traditional dress cooking in her kitchen

Plate 2.5: Women in Tat hamlet wearing a mixture of traditional and modern style clothing

blue was the traditional color and was still worn by elderly women but younger women preferred very bright red-orange scarves. In recent years younger women and adolescent girls have increasingly begun to wear Kinh style blouses and trousers. Teenagers of both sexes also now often wear tee shirts and blue jeans (Plate 2.5).

House type

Traditional style houses are the most immediately visible, and in some ways most remarkable component of material culture in Tat hamlet (see Plate 1.2). These houses on stilts are constructed of large hand-hewn timbers extracted from the forest and thatched with palm fronds taken from the tree gardens. The houses are often large, 6 m wide by 12 m in length. The wooden plank floors are raised some 2 m above the ground. Buffalo and cattle are frequently stabled beneath the floor. Walls are made of wide planks. A veranda runs along the front and rear end of the house. Beams are decorated with incised painted patterns. Surprisingly, the people never build their own houses; instead, the frames are usually constructed by itinerant Kinh carpenters from Sơn Tây Province in the Red River Delta or less frequently by Mường specialists from outside the hamlet.

In recent years almost all new houses have been built on the ground in the style of the Kinh, reflecting the near impossibility of finding sufficiently large trees for the pillars needed to support the traditional style of architecture (Plate 2.6). A survey of 42 households in 1998-99 revealed that only 36 percent lived in traditional style houses, while 60 percent lived in Kinh style houses on the ground and 5 percent had modern style masonry houses.

Plate 2.6: House built on ground in Kinh style

Religion

The people of Tat hamlet are animists. They believe in many different spirits that reside in the forest. Most households have small altars inside their houses while some have small roofed spirit houses in their front yards. Different types of altars are found in different houses. Some have altars for the rice spirit (*lao khẩu* in Tày), some for the spirit of the general of hunting (*ma tương san* in Vietnamese). In one house belonging to a retired senior cadre a picture of President Hồ Chí Minh was displayed on an altar hanging on the inside front wall. Some houses also have small thatch-roofed spirit worship houses in their yards where offerings are made on special occasions. Funerals are conducted by a ritual specialist (*thầy mo*) who prays while presenting offerings of food and alcohol to the spirits on a table next to the grave (Plate 2.7). Several tall poles with carved eagles on the tops are set up beside the graves of adult males (Plate 2.8).

Social organization

Until after the August Revolution of 1945, the Đà Bắc Tày of Tat hamlet were subjects of a Mường chiefdom. As commoners, they had to pay tribute and perform labor services for the chief and his nobles. Decision-making was in the hands of officers appointed by the chief and, consequently, local governance institutions were only weakly developed. Although the chiefdom was suppressed after Vietnam gained its independence from the French, the social organization

Plate 2.7: Burial offerings at new grave

Plate 2.8: Ritual poles at a man's grave

of the community is still that of a classical peasantry, with individual households as the dominant institution and very few horizontal ties binding together separate households into a cohesive community.

Individual households are responsible for organization of labor, management and use of resources, and accumulation and allocation of wealth. Multigenerational extended family households are the ideal. These are made up of the aged parents, their unmarried sons and daughters, and one or more married sons and their wives and children. In practice, however, more than 60 percent of households are of the nuclear type, although in some cases a widowed parent lives in the household of the youngest married child. Traditionally, a newly married couple resided for a period ranging from a few months to several years with the bride's parents while accumulating the resources to establish an independent household.

Kinship is patrilineal. There are four exogamous clans: Xa, Lò, Lường and Hà. Marriage between members of the same clan is considered inappropriate, but nevertheless occurs occasionally. Approximately 85 percent of marriages in the hamlet are clan exogamous, however. The clans have no social functions other than regulation of marriage.

Although each household manages its own farming system, much of the agricultural work is done with exchange labor involving relatives and neighbors. Paddy fields are harrowed by several buffalo at once, with their owners working the fields of each member of the labor exchange group in turn. The transplanting is done by teams of as many as one dozen women. Swiddens are planted by large groups of up to 50 or 60 individuals. When a household wants to install a new roof, their neighbors come bringing bundles of palm leaves to help do the job. Meals are provided to helpers when they plant the swiddens and put on roofs but not for work in the paddy fields.

Division of labor and gender relations

Gender, and, to a lesser extent, age are the main differentiating factors in the internal social organization of the hamlet. Ethnicity is important only in relations with the external world. In public affairs there is a well-marked sexual division of labor. Males hold all leading positions in administrative bodies and mass organizations, with the sole exception of the Women's Union, and only men are likely to speak in public meetings.

Production is almost entirely organized on a household basis. Within the household economy there is a clear but not wholly rigid sexual division of labor. Men do the land preparation in the paddy fields. Both men and women carry manure from the houses to the paddies. Men are responsible for clearing and burning the swiddens. Men uproot rice seedlings from the nurseries and help to carry them to the paddy fields. Women transplant seedlings into the paddies

Plate 2.9: Woman threshing rice with her feet

and plant seed in the swiddens using dibble sticks. Weeding is mainly a female responsibility. Rice harvesting is mostly done by women, while men carry the grain from the fields to store in the houses. Threshing is done by the women who until recently trampled the bundles of rice with their feet to release the grain from the heads (Plate 2.9). Husking is done with a mortar and pestle (Plate 2.10). A few households own water-powered rice mills. More recently, many households have begun to take their paddy to be milled at a power-operated mill in the Tan Minh village center.

Both men and women go into the forest to cut bamboo poles and drag them to the road for sale to passing trucks. Illegal logging is done entirely by young men. Bamboo shoots and wild vegetables are mainly collected by women. Women collect most of the firewood in newly burned swiddens or in the secondary forest and carry it to the houses but men bring down an occasional large log. Buffalo are tended by young men or boys but smaller livestock are cared for by women and girls. Most domestic tasks are done by females. Cooking, cleaning of the house and compound, and washing and mending of clothes are done by women and girls. Members of both sexes care for infants although toddlers are most often carried by their older sisters. Men weave baskets for household use and for sale. In the past, when cotton was grown in the swiddens, women

Plate 2.10: Old woman husking rice using mortar and pestle

wove cloth. Only one or two elderly women still practice this craft. Women are responsible for collection of produce from the homegardens and marketing of the fruit. Except at peak labor demand times in harrowing the paddy fields, women have a considerably heavier work load than men. Collection of firewood is an especially heavy burden on women (Plate 2.11).

Socioeconomic differentiation

At the time we began our research, some differentiation among households in terms of material wealth was evident, although, in contrast to the situation during the colonial period, there were no truly wealthy households that were free from the necessity of producing all of their own subsistence needs through their own manual labor. Better-off household had larger houses and more livestock. Given the relatively high equitability in access to agricultural land, the ethos of sharing productive resources with less fortunate kinsmen and neighbors, and the lack of other channels for accumulating wealth, permanent differentiation among

Plate 2.11: Woman carrying firewood

households was inhibited. Most wealth differentiation could be attributed to the different stages in the family cycle occupied by these households. Newly formed households and those of elderly widowed individuals tended to be noticeably poorer than the norm. They were more likely to live in small houses built in Kinh style on the ground. Households with several unmarried teenagers were likely to be relatively prosperous as a result of their large labor force. Families having mainly sons could have their resources depleted by the heavy demands of paying bride price. Four or five large pigs, plus a considerable quantity of rice, alcohol, and other consumption items, were needed to complete a single marriage contract.

In recent years, however, the extent of economic differentiation among households has increased to some extent. There was a gap of 6 times between the per capita cash income of the richest 20 percent and the poorest 20 percent of the 42 households surveyed in Tat hamlet in 1998-99. The Gini coefficient for distribution of income among households, however, was 0.24, which is still remarkably equitable (Lê Trọng Cúc and Rambo: 2001: 150-151).

A number of factors have contributed to increasing differentiation. Growth in population and resultant scarcity of land has made it difficult for newly formed households to obtain paddy fields, forcing them to rely on less productive upland fields and collection of forest products to earn their livelihoods. Allocation of forestlands to households has limited the ability of poorer households to collect resources from formerly open access upland areas. Participation in administrative and political institutions may also confer considerable competitive advantages, with well-connected families deriving benefits from involvement of their members in official roles. Education does not necessarily serve as an effective means of upward mobility, however. One woman who had ensured that all of her children completed high school and obtained jobs as teachers complained of the poverty of her family because of the low salaries paid by the government.

Helping to offset pressures toward differentiation to a limited extent is an ethos of sharing. In the past, hunters who killed wild pig or deer were obliged to share the meat with their neighbors. Young farmers who have not yet accumulated sufficient capital to purchase their own buffalo are able to 'borrow' animals from better-endowed relatives or neighbors. Farmers who do not have sufficient land for their swiddens are able to approach those with larger plots to request the use of some of their extra land.

Administrative organization and public services

Tat hamlet is organized according to the administrative model found everywhere in rural Vietnam. As a sub-unit of Tân Minh administrative village (commune) it is under the authority of the Village People's Council, People's Committee, and Party organization. These organizations, staffed by individuals from all of the hamlets making up Tân Minh village, are responsible for local implementation of national policies. A Hamlet Leader and Vice Leader are responsible for overseeing day-to-day administration of the hamlet and reporting to the Village People's Committee.

A number of mass organizations including the Women's Union, Retired Persons Association, Farmers Association, Veterans Association, and Youth Association, all under the Fatherland Front, are more or less active in the village and recruit members from Tat hamlet. Each of these organizations encourages its members to improve their livelihoods and to manage the forest in approved ways.

The people of Tat hamlet have unusually good access to public services because the Tân Minh village headquarters is located on the eastern edge of the hamlet. In addition to the People's Committee office, there is a tax office, a health station and family planning clinic, a post office, a retail store operated by the provincial trading company (which is also the site of the bus station), a pharmacy, and a newly built headquarters for the road maintenance department. A primary school is located closer to the center of the hamlet.

Transportation and communications

Tat hamlet is located alongside Provincial Highway No. 433. This road runs from the township of Sông Đà next to Hòa Bình Provincial Town to the Đà Bắc district capital of Tu Lý, and from there to Tân Minh and on to Mường Chiềng on the border with Sơn La Province some 50 km to the west of Tat hamlet. A rutted earth jeep trail in 1992, which was only safely usable in the dry season, the highway was hard surfaced in 2000 and is now usable in all weather conditions.

As recently as 1995, motor vehicles were an unusual sight, with only an occasional logging truck or government jeep passing through the hamlet. Now there are three buses every day traveling from Hà Đông in the Red River Delta to Tân Minh. Several trucks hauling produce purchased by dealers from the lowlands pass through the village each day. Motorbike taxis ferry people back and forth to the markets in Cao Sơn and Tu Lý. In 2004, one household in Tat hamlet bought a truck for trading agricultural and forest products with the lowlands.

Beginning in 1999, the infrastructure of Tat hamlet and Tân Minh village was improved with funding provided by Program 135. A public loudspeaker was set up at the hamlet leader's house. A television with a satellite dish was installed at the Tân Minh commune office. One-third of all households have battery-powered transistor radios and many now own television sets (Plate 2.12). The hamlet was linked to the national electricity grid in 2001 and most households now draw electricity from this system, although some households still use cheap mini-hydropower generators which supply enough electricity to light a couple of 40 watt light bulbs.

Economy

In 1992, household production was mostly subsistence-oriented with only minimal involvement in the cash economy. Since then, households have become increasingly involved in the market economy. Farmers report a pressing need to obtain cash, particularly in recent years since the local government began to collect taxes in cash rather than paddy. The need for cash to pay school fees for children is also strong. Informants readily cite prices for different locally produced commodities including livestock, paddy, and fruit.

There is no market in the hamlet or in Tân Minh village. The nearest market is at Cao Sơn, which is about 12 km away over an asphalted road. The closest large daily market is in the district capital at Tu Lý (about 25 kms). Some products, notably bamboo shoots and other forest products, are collected by shopkeepers and sent out to market on buses or logging trucks. Traders from China have even appeared in the hamlet on occasion to buy a variety of wild medicinal plants. Corn, cassava roots, and canna are sold in bulk either to local middlemen or to

Plate 2.12: Television arrives in Tat hamlet

traders from outside the village. Cattle and buffalo are sold to Kinh traders who come to the mountains to purchase animals for use in the delta where traction animals are always in short supply. Government transfer payments and loans from development programs have also become large sources of cash for hamlet households.

There are currently 16 small shops in the hamlet. Their stock is limited to daily use items (e.g., flashlight batteries, cigarettes and matches, detergent, instant noodles, beer, alcohol, and candies). In 1996, a larger better-stocked shop was opened by the provincial trading company and a pharmacy was opened in 1998. Women with fruit to sell simply walk through the hamlet looking for customers for their pineapple or jackfruit. Motorbike vendors come to Tat hamlet periodically to sell pork. Local people can easily purchase chemical fertilizers, pesticides, and other goods not available in the hamlet by ordering them from the bus drivers.

Recent Changes in Tat Hamlet

In recent years, life in Tat hamlet has changed at a quite remarkable speed. In 1992, when we first visited it, Tat was a small, relatively isolated hamlet where motor vehicles were seldom seen, where extensive areas of dense forest could still be found, where growing swidden rice was combined with rice production in small paddy fields, where people lit their homes with kerosene lamps and used torches instead of flashlights when they went out at night, a hamlet that had little cash or much use for it, with a two-day walk to the nearest market. In the intervening 15 years it has been transformed into a very different place. Now a motorbike taxi service shuttles people and goods back and forth between Tat hamlet and a relatively new market 12 km away in Cao Sơn and even to the district seat of Tu Lý, some 25 km distant. One local household has even purchased a truck for trading with the lowlands. Three buses a day leave the nearby village center to go to Hà Đông, a lowland city near Hanoi. Many households have stopped growing swidden rice altogether, instead planting cash crops and using the money thus earned to purchase rice. Quite a few houses have radios or TV sets and light their homes with electricity from the national power grid. Young people often appear in jeans and tee shirts.

Change has been greatly accelerated by two broad processes: Agricultural land area per capita has been decreasing while development assistance has been increasing. Growing population pressure upon the land and other natural resources has been further intensified by changes in government land use policies that have resulted in a rapid reduction in the land available for swidden agriculture. As part of their effort to compensate for this loss in the land area for swidden agriculture, households have extended the cultivation phase of the swidden cycle from two or three years to four or five years and reduced the fallow phase from seven to ten years to only three to five years. This has led to a rapid loss of soil fertility and decline in crop yields in swidden fields, as well as additional need for weeding. Crop cultivation makes a smaller contribution to the household economy and household food needs.

At the same time, a significant and rapid increase in development assistance has made possible a diverse set of activities to increase income from other sources to compensate for drops in subsistence crop production. Thus, a whole range of new opportunities has opened up just at the time the people were in exceptional need of them. The diversification and intensification of other sources of income has taken place at a relatively rapid pace. Government assistance to the hamlet was significantly increased through the 327 and 135 programs, among others. The road through the hamlet, earlier little more than a dry-weather jeep trail, was hard-surfaced, greatly facilitating access to the market. New credit programs gave farmers access to capital for investment. Most people used that credit to expand animal husbandry.

The people of Tat hamlet have carried out many other innovations in their farming systems, some large, some small, some on their own initiative, some with assistance from district extension workers and/or researchers and students from HUA. New fishponds have been dug and others deepened and expanded. New cash crops—including corn, canna, and ginger—have been grown in swidden fields, while the area devoted to dry rice has been significantly reduced. Also, researchers and students from HUA have worked with the farmers to achieve boosts in agricultural productivity in a variety of additional ways. Finally, increased mobility has enabled a number of households to obtain money from off-farm wage labor and yet other households to start small businesses.

At the same time, however, the people have also been exploiting local resources at an unsustainable rate to raise their incomes. Environmental degradation and depletion of resources is not limited to loss of soil fertility. Logging has been banned, but few high-value tree species remain in the forest in and around Tat hamlet and some illegal logging continues to take place. Wild animals, once hunted, have virtually disappeared, including bears, deer, wild boars, wild chickens, monkeys, and large snakes. Certain species of bamboo, medicinal plants, mushrooms, and various other non-timber forest products have become rare because of unsustainable exploitation for sale to traders.

It is obvious that some of the means employed by the people to diversify and increase income sources are not sustainable. But most people now feel a strong need for greater cash incomes. They need money to pay their taxes and to pay school fees. They need money for electricity, and flashlights, and batteries, and plastic ponchos. Now that some people have bought radios and even TV sets and motorbikes, others, of course, now want them too. People want and expect rising incomes. But even with cultivation, animal husbandry, and the sale of non-timber forest products greatly intensified, household income is only several dollars a day.

The educational level in Tat hamlet is still relatively low, but it is rising. New images are flowing into the hamlet on TV screens, generating new desires. New consumer interests—in everything from motorbikes to instant noodles to blue jeans—are emerging. Participation in the market economy is developing, while channels to these markets are expanding. Although agriculture is still the most important element in the household economy, it is becoming just one component in a much larger portfolio of livelihood activities. A few households have undertaken some modest entrepreneurial schemes, with greater or lesser success. A number of households have found some off-farm labor opportunities to supplement their incomes. Already, different households are devising different economic strategies, although all remain strongly subsistence-oriented, and getting enough to eat is still their main concern.

Acknowledgments

This chapter is a much revised and expanded version of Rambo (1998). The section on geology is based on an unpublished report prepared by Patma Vityakon with the assistance of Suwin Busaracome, a geologist in the Faculty of Science of Khon Kaen University (Vityakon 1994).

PART II
STRUCTURE AND FUNCTIONING OF THE COMPOSITE SWIDDEN AGROECOSYSTEM

Introduction

The eight chapters in this part present descriptions of the structure and functioning of the composite swidden agroecosystem and some of its main individual components. The composite swidden agroecosystem of Tat hamlet is a very complex agricultural system involving the interaction of multiple different components that are tied together by flows of energy, materials, and information. The key components are paddy fields, swiddens, fallowed swidden fields, livestock, fishponds, homegardens and tree gardens, and secondary forest. Although the emphasis is on description of biophysical aspects, including biodiversity and species composition, interactions with the people of Tat hamlet are not neglected, although these receive more concentrated attention in Part III on the social system.

Chapter 3 presents an overview of the composite swidden agroecosystem of Tat hamlet. It briefly introduces each of its individual components and examines the roles that these components play in the functioning of the overall system as well as the contributions that they make to the people's livelihoods.

Chapter 4 focuses on the paddy field component of the agroecosystem. It describes the paddy field soils and the classification of fields employed by the farmers and discusses the methods employed by the farmers to maintain soil fertility and manage pests. The paddy fields make an extremely important contribution to the livelihood of the farmers because of their high and relatively stable and sustainable yields.

Chapter 5 describes the vegetation that begins regenerating in swidden fields as soon as they are fallowed. The successional communities are ecologically complex and characterized by high species diversity. In addition to its primary role in regenerating soil fertility to permit the next round of swidden cultivation, the fallow vegetation contains many species that are exploited by the farmers as sources of food, medicine, raw materials, and to sell for cash. The fallow fields also provide an important grazing area for large livestock.

Chapter 6 deals with the livestock component of the agricultural system. Large livestock (buffalo and cattle) make important contributions to the functioning of the overall system. Buffalo provide draft power for harrowing the paddy fields, and both buffalo and cattle are a major source of manure that is used to maintain soil fertility in the paddy fields. In recent years, farmers have increased the size of their herds, especially cattle, which are increasingly raised to sell for cash. Pigs are mostly raised to sell although they are also valued for providing high quality manure used to fertilize fishponds. Chickens and ducks are raised both for household consumption and to sell for cash. Although diseases are a major constraint on chicken production, some farmers are experimenting with using industrial methods of raising them on a larger scale.

Chapter 7 describes the fishpond component of the agricultural system. Fishponds are a useful source of protein for household consumption as well as an occasional source of cash. Productivity is quite low, however, because management is not very intensive.

Chapter 8 is devoted to homegardens and tree gardens. Homegardens have a high species diversity and provide a variety of vegetables and fruit for home consumption as well as occasional sale. Tree gardens provide tea, palm leaves, and other raw materials for construction of houses and sale. Neither homegardens nor tree gardens are intensively managed.

Chapter 9 describes the secondary forest component. Like the fallow swidden fields from which it develops in the natural course of succession, the secondary forest is an important source of wild foods, medicines, and raw materials that are used to meet household needs and to sell for cash. Illegal logging has depleted timber stocks but bamboo stems and shoots continue to be extensively exploited and are a very important source of in-kind and cash income, particularly for poorer households.

Chapter 10 discusses changes in the farming systems of households in Tat during the past twenty years. Over this period, farming systems in Tat hamlet have become more diversified and household economies considerably more differentiated. At the time we began our research in 1992, all households in Tat hamlet still engaged in traditional composite swiddening involving cultivation of both rice swiddens and paddy fields. In recent years, however, although the majority of households continue to employ this system, close to one-third of the households engage in a modified form of composite swiddening in which they cultivate paddy fields together with cassava swiddens, but do not make rice swiddens. And a few households do not have any paddy fields so only cultivate swiddens.

3
The Composite Swidden Agroecosystem of Tat Hamlet

A. Terry Rambo and Nguyễn Thanh Lâm

Introduction

In this chapter, we describe the overall structure of the composite swidden agroecosystem in Tat hamlet. We then briefly describe each of its individual components, including paddy fields, swiddens, fallowed swidden fields, livestock, fishponds, homegardens, tree gardens, and secondary forest, and examine the contributions that they make to the people's livelihoods. Finally, we look at resource flows and other interactions among the different components of the agroecosystem.

The Composite Swidden Agroecosystem in Tat Hamlet

The defining characteristic of composite swiddening in Tat hamlet is that households simultaneously manage both permanent wet rice fields in the valley bottoms and shifting swidden fields on the hill slopes (Rambo 1998). The system is a stable adaptation rather than a transition phase between pure shifting cultivation and pure permanent agriculture. Thus, in the typology of swidden systems, composite swiddening does not fit into Harold Conklin's well-known category of 'partial systems' (Conklin 1957). Cultivation of swiddens by the Đà Bắc Tày is not a gradually vanishing survival of an earlier, more primitive pure swiddening adaptation that is the process of being replaced by more advanced irrigated farming. Nor is swiddening a recent response by traditional wet rice farmers to rapid population growth that has exceeded the carrying capacity of their wet rice fields and forced them to expand their farming onto the forested slopes, as has happened among Northern Thai 'supplementary' swidden farmers (Kunstadter and Chapman 1978: 7). Instead, the farmers in Tat hamlet have practiced both wet rice farming and swidden agriculture together as an integrated system of subsistence for a very long time.

Elderly informants (whom we interviewed in the early 1990s) reported that their parents had told them that they had cultivated both swiddens and paddy fields when they first began to settle the valley at least one hundred years ago. At that time, the entire area was covered by primary forest and there was no scarcity of land on which to make paddy fields in the valley bottoms. One man recalled that in the mid-1950s, when he was a boy, there were only 7 households in the valley so there was plenty of land available for anyone who wanted to exploit it. The area of paddy fields was much smaller than it is now and good forestland was abundant and free for the taking. It would thus have been possible for households to have only cultivated paddy fields or only cleared swiddens, but none are reported to have done so. It may well be that there are survival advantages to subsistence farm households in maintaining a more diversified agroecosystem, as Terry Grandstaff (1988) has suggested is the case with regard to the 'diversified portfolio' of income-generating activities employed by small farmers in Northeast Thailand.

When we began our research in 1992, all households in Tat hamlet engaged in composite swiddening involving both rice swiddens and paddy fields. Today, despite many environmental and social changes, the majority of households continue to employ this system in traditional form, while a substantial minority employs a modified form in which they cultivate paddy fields together with cassava swiddens, but do not have any rice swiddens. Thus, in 1998-99, 61 percent of farming households practiced traditional composite swiddening (paddy fields in combination with rice swiddens), 21 percent engaged in modified composite swiddening, and 17 percent cultivated swiddens but not paddy fields. In 2003, 63 percent of households did traditional composite swiddening, 30 percent did modified composite swiddening, and 7 percent only cultivated swiddens. Thus, the vast majority of households in Tat hamlet manage composite swiddening agroecosystems of either traditional or modified types.

Components of the Composite Swidden Agroecosystem

The key components of this system are paddy fields, swidden fields, livestock, fishponds, homegardens and tree gardens. Fallowed swidden fields in various stages of regeneration and patches of secondary forest are also important system components.

Paddy fields

The paddy fields are built in a series of terraces rising like steps from the stream in the middle of the valley floor (Plate 3.1). Each terrace is quite small, covering

Plate 3.1: Paddy fields in Tat hamlet

an area of 30 to 70 m^2. The height of the almost vertical bunds is one meter or more. The tops of the bunds are very narrow, only 20 or 30 cm wide, making walking on them a treacherous affair. Irrigation water continuously enters the highest terraces from small streams flowing down off the hill slopes. It then spills down from the highest fields to those below and ultimately reaches the lowest fields. Finally, any excess water flows into the large stream running through the floor of the valley.

Because the fields are kept flooded, even when not being actively cultivated, plowing is not usually required. Cultivation is done using a wooden harrow pulled by a single buffalo. Each field is harrowed three times before planting. Farmers from several households exchange their own labor and that of their buffalos to do this task. In larger fields as many as six buffalo may be used at the same time (Plate 3.2). After the final harrowing, the rice seedlings are immediately transplanted by groups of women. The planting pattern is quite casual with little concern shown for placing the seedlings in straight rows.

Manure from the buffalo and cattle, green manure, and chemical fertilizer are all used to fertilize the paddies. Manure is collected from the buffalo and cattle that are stabled at night underneath the house. It is stored in large woven bamboo bins until needed and then carried to the fields in pack baskets by the women and children and with shoulder pole baskets by the men. Farmers

Plate 3.2: Buffalo harrowing paddy fields

say that they generally have adequate quantities of manure and that use of greater amounts would produce excessive vegetative growth with lowered grain production.

In the early 1990s, most farmers grew improved varieties of non-glutinous rice in the spring crop and traditional varieties of glutinous rice in the summer crop. Farmers said that they preferred non-glutinous rice because a small amount of grain cooked up into a large volume of rice. Today, however, traditional varieties have been completely replaced by higher yielding Vietnamese improved varieties (e.g., *Khang dân*) and hybrid varieties from China (*Tạp giao* and *Nhị ưu 838*).

Yields are low in comparison to the lowlands, although in the past ten years they have been gradually increasing. Before 1995, the average yield was 2.5 tons/ha/crop. This increased to 3.25 tons by 1999, and to 3.9 tons in 2004. This change in yields reflects greatly increased use of chemical fertilizers, adoption of high yielding and hybrid varieties, and improved pest management.

Swiddens

Several different types of swiddens are cultivated, including rice, cassava, maize, canna, and ginger swiddens. Most swiddens are essentially monocultural although in recent years some farmers have begun to intercrop maize with canna. Rice swiddens are located on the most fertile available land. Swiddens on less fertile soil are planted with cassava, maize, canna, or, more rarely, ginger. Most households cultivate several different swidden plots simultaneously. One woman interviewed in the spring of 1999 had eight plots: Two of rice, two of cassava, and four of maize.

The system is quite dynamic; when we began our study in 1992 only rice and cassava swiddens were observed. As access to markets improved, maize and canna become important swidden crops. Ginger was important for a short period in the late 1990s but is now cultivated by only a few households because of rapid price fluctuations and problems in protecting this crop from theft.

Rice swiddens

Rice swiddens are generally located on higher slopes at some distance from the settlement area although, in recent years, scarcity of land has caused some households to plant rice on plots with inferior soils on the lower slopes of hills next to the settlement. However, whenever possible, rice swiddens are cleared in relatively mature secondary forest on soils thought to be of higher fertility. Soil fertility is critical for obtaining a high yield because no manure or chemical fertilizer is applied to the rice swiddens. The farmers judge the fertility of swidden soil by sticking a knife into it. If soil sticks to the blade when it is pulled out, they consider the soil suitable for rice. If nothing sticks, it is only good for cassava. Rice swiddens are often planted on extremely steep slopes ranging from 30 to 60 degrees (Plate 3.3).

The rice varieties planted in the swiddens are rather diverse. This helps to ensure a measure of food security because the different varieties ripen at different times so farmers rarely lose the entire crop to pests. We found twenty-two varieties of rice (both non-glutinous and glutinous), but most households plant only five or six varieties at any one time. All are local varieties. Two varieties (*khẩu khinh* and *khẩu chàm lương*) are especially popular because they can tolerate poor soil conditions, have a high yield, and have good eating quality.

During the harvest, seeds for the next year's crop are chosen directly from the fields, bundled up, and taken home to save for the next planting season. In first year fields cleared on very good soil, yields can reach 1.5 tons/ha. In the second year yields fall to about one ton and, by the third year, are only about 0.4 to 0.6 ton/hectare. On poor soil, in the first year the yield is only about 0.8 tons per hectare, while by the second year the yield declines to about 0.1 ton per hectare, or even less. The overall average productivity of all rice swiddens in the hamlet is very low, with yields now averaging only 0.4 tons/ha.

The rice swiddens are almost pure monocultures. For example, one quite large field located about 1 km from the hamlet that we observed in 1993 was planted with four varieties of rice. The only other crop was some sesame planted

Plate 3.3: Women cultivating steep swidden field

scattered in a thin strip along the bottom of the field. Swiddens located at a greater distance from the settlement display somewhat greater crop diversity. Squash and pumpkins, beans, maize, taro, and other vegetables are interplanted with the rice. Most of these secondary crops are grown for consumption on the spot when the farmers stay overnight in their field huts to protect the ripening rice from wild animals but surplus vegetables may be carried back to the village for consumption there.

The species diversity and architecture of the rice swiddens in Tat hamlet is very different from the model of swiddens as very diverse multi-layered polycultural systems that has come to typify swidden agriculture in much of the anthropological literature (Conklin 1957; Geertz 1963). Thus, the monocultural rice swiddens in Tat hamlet do not mimic the architecture of the tropical rainforest with multiple layers of canopy. Instead they resemble fields of tall grass, except that they are much more open with considerable areas of exposed soil (Plate 3.4).

Plate 3.4: Rice swidden in Tat hamlet

Cassava swiddens

Cassava swiddens are commonly sited on the lower slopes of hills near the hamlet where the soil is too infertile to support rice cultivation. During the cooperative period before 1986, when villagers could still clear new swiddens in old-growth forest with fertile soil, cassava would be planted in the final year of the swidden cropping cycle after one or two years of growing rice. At that time, cassava was grown for household consumption, winemaking, pig fodder and for sale to the government through the cooperative system. The villagers also used cassava to pay their taxes, using a certain amount of cassava to substitute for the standard amount of rice. Today, cassava is most often planted in nutrient-depleted soils of swidden fields after one or two crops of maize and canna have been harvested. It is raised both for home use and for sale. When households lack sufficient rice, cassava is eaten as a substitute for rice. The roots are also fed to pigs and the leaves used for carp food. Fresh roots and dried chips are sold to traders from the lowlands.

The villagers call cassava *cọ men*. There are two varieties; one is 'red' or 'banana' cassava (*cọ men denh*). It has a short stalk and the tuber looks like a banana. The villagers prefer growing this variety because of its high yield, ability to grow on poor soil, and because its roots are shallow so that they can be dug up easily. The villagers usually just pull them up by hand. The other variety is 'white' cassava (*cọ men don*). The stalk is tall, and the leaves have a white sheen. When they harvest this variety, the villagers have to use a hoe. The yield of this variety is low, but the roots are tasty and have more tapioca than the red cassava so it is preferred for their own consumption.

When selecting planting material for the next crop, villagers choose cassava stalks with many large leaf buds and a lot of sap. In order to grow cassava, a section of the stalk with at least three leaf buds must be planted three to five cm deep. Selecting planting material for the next crop is very important work, but few households do it carefully. On average, productivity is around ten tons of fresh roots per hectare. All of the leaves and stems are deposited on the fields. No expensive inputs, apart from labor, are made into the cassava swidden. Only a few families apply a small quantity of manure to cassava fields close to their houses and in their homegardens. Cassava is weeded twice in the course of its growth. It is very important to weed the red variety well, because the plants are small and can be easily out-competed by weeds.

Maize and canna swiddens

Maize (corn) was formerly planted as a monoculture but in recent years has been commonly intercropped with canna. Maize is planted in January and February, and harvested in April and May. Canna is planted in February and March, and harvested in November and December.

Maize has been planted in swiddens in Tat hamlet for many years. The villagers used to plant only local varieties of maize but recently some farmers have begun to plant hybrid varieties as well. The local varieties usually yield one small ear per plant, although some have two. The villagers like the local varieties because the husks completely enclose the ears, protecting them from rot due to rainy conditions. There are two varieties of maize: *kle niêu* (sticky corn) and *kle cham* (hard corn). *Kle niêu* has a good consistency and flavor, but *kle cham* is more widely grown because it has a higher yield and is easy to sell. The yield depends on the soil quality, the density of the plants per hectare, and the timing of sowing the seeds. The local varieties yield from 800 to 1,500 kg per hectare.

Canna has only been extensively planted as a cash crop for about ten years in Tat hamlet. Its starchy rhizomes are sold to processors in the lowlands to make a type of clear noodle. Two varieties are grown: Red canna (*mặc đót dênh*) and white canna (*mạc đót don*). These two varieties have similar yields and economic value. Average yields are about seven tons/ha but yields vary greatly from field to field, depending on soil quality, planting density, and the quality of the planting material. Since 2004, production has decreased due to widespread root disease. This has caused many households to stop planting canna in their swiddens.

Ten to fifteen days after the swidden field is burned, the farmers plant maize by making holes with dibble sticks and putting three corn seeds in each hole.

When the plants have reached a height of about 10 to 15 cm, at the three-leaf growth stage, the field is weeded for the first time and the canna planted. Once the canna sprouts emerge, the field is weeded a second time. Only if the weeds begin to smother the crops is a third weeding done.

After the maize is harvested, the canna plants require little additional care. By the end of November or beginning of December the canna plants start flowering and forming the rhizomes, after which they are harvested. Canna usually has a height of 1.5 to 2 m. When the villagers harvest canna, they first cut the stalks down to about 15 cm, after which they dig up the rhizomes, which then have to be cleaned before they are carried to the road to be sold. The villagers usually leave some high-quality plants growing on the upper part of the swidden fields to use as planting material for the next crop. It takes 1.5 to 2 tons of planting material to plant one hectare.

After the canna is harvested, the villagers leave the leaves and stalks of the maize and the canna on the swidden fields to dry for one or two months. Then, after the lunar New Year, they burn these residues to prepare the field for planting another round of crops. If the soil has a high nutrient level, the villagers can grow maize and canna continuously for up to four years. This extended period of cultivation reflects the low rate of soil erosion from maize-canna swiddens. After the maize is harvested, the broad leaves of the canna plants prevent the rains from falling directly onto the field and washing the soil away. This makes it possible to grow maize and canna on areas with relatively steep slopes with less soil degradation than occurs in monocultural rice or cassava swiddens.

Ginger swiddens

People in Tat Hamlet always used to plant small amounts of ginger (*Zingiber officinale*) in their homegardens or swidden fields for home consumption. Then, in 1995, traders from the lowlands came to the hamlet and offered to buy ginger in volume for the relatively high price of 2,000 VND/kg. In response to this new market opportunity, many farmers began to plant ginger as a cash crop. Some households even planted all of their swidden fields with ginger for sale. Soon after that, however, the price of ginger fell to only 700 VND/kg, due to instability in the Japanese market. This sudden drop in the price of ginger caused economic difficulties for those farmers who had allocated much of their land to ginger production. Crop theft, which became a serious problem when ginger prices were high, was another factor causing farmers to cease cultivation of this crop. Now, only a few households continue to plant small areas of ginger swiddens.

Ginger is usually planted in its own separate swiddens but sometimes patches of ginger are planted in the flatter, less stony areas of rice or corn-canna swidden fields. The rhizomes, which grow at shallow depths, can be grown on poor soils if soil texture is light. Ginger grown on newly cleared fields has good vegetative growth but does not necessarily have a high yield. Four different varieties, varying in size, yield and aroma, are now planted. Ginger is planted in February for harvest in November. Ginger yields are highly dependent on the variety planted and the soil quality but average 3.4 tons/ha. No manure, chemical fertilizer, or pesticides are applied to ginger swiddens. Given its aromatic properties, ginger is less subject to attack from rats than cassava or corn.

Livestock

Buffalo, cattle, pigs, and chickens and ducks are the most common livestock species in the hamlet. A few households also keep goats. Almost all households own buffalo, with an average of 2.4 head per household, and over half own cattle as well, with an average of two cattle per household. The population of buffalo has been stable for the past decade but the number of cattle has increased rapidly in recent years. Livestock are allowed to range freely in a valley that is reserved by the cooperative for use as a pasture. After the swiddens are harvested, livestock are allowed to freely graze in the fallowed fields and neighboring areas of secondary forest. Buffalo are used to harrow the paddy fields and to drag logs out of the forest to the road. Cattle are raised mostly as a source of cash. Both buffalo and cattle provide manure used in the paddy fields.

Pigs of the native pot-bellied variety are raised by some households but are said to be vulnerable to disease. Pigs are of great ritual importance because the families of young men need them for bride payments when their sons marry. Pigs are also sold for cash. Some pigs are free ranging during the day while others are kept all of the time in small cages in the homegardens. They are fed cassava roots from the swiddens close to the village. A few households have recently begun to raise hybrid pigs, which require purchased feed.

Almost all households keep some chickens. Chickens are kept to provide meat for feasts and for eggs. Fowl range freely around the houses during the day but are kept in special bamboo pens beside the house at night. Flocks range in size from 5 to 40 birds. Some households have tried to raise larger flocks but have suffered heavy losses from disease outbreaks. A few households have begun experimenting with raising 'industrial' chickens that are kept caged in closed bamboo pens. Some households also raise ducks and Muscovy ducks.

Fishponds

Fishponds are owned by almost three-quarters of the households. The small, shallow ponds are usually located near to the houses of their owners. The ponds are filled with water that flows through bamboo or plastic pipelines from streams or springs on the hill slopes behind the house site. Carp of several species, tilapia, and catfish are the most commonly raised fish. They are fed cassava leaves, weeds, rice bran, and buffalo and pig manure. Management of the ponds is not intensive, however, and production is low, in part because the species used are vulnerable to dying off during the cold winter weather. Fish are mostly consumed by the households but are sometimes also sold for cash.

Homegardens

Most houses are surrounded by small homegardens in which a variety of trees and vegetables are planted. A survey of gardens of 30 households in 2004 found 69 species, most of which had been deliberately planted by the farmers (Nguyễn Thanh Lâm and Nông Hữu Dương 2005). The gardens usually are polycultural with a large number of different species in the same plot. Only a few representatives of each species are usually found in an individual plot (although sometimes, tea is grown in separate monocultural tea gardens). The common plant species in homegardens are local varieties of fruit trees, vegetables and herbs, medicinal plants, tea, and timber trees. Beginning in the mid-1990s, new varieties of apricots, longan, litchi, plum, and persimmon have also been planted. Climbing vines (e.g., cucumber, luffa gourds, and beans) are often grown on bamboo trellises. Some houses also have a small bamboo-fenced enclosure where green vegetables are protected from depredations by the free-ranging pigs and fowl. Green onions are sometimes planted in raised wooden troughs. Marigolds and other ornamental flowers are sometimes also grown in the homegardens.

Vertical stratification of the garden vegetation into several layers was observed in many homegardens in Tat hamlet. The highest layer is composed of the canopies of timber and large fruit trees. Smaller fruit trees and bananas compose a second layer. The under-story is made up of vegetables, tubers, medicinal plants, spices, pineapple, and grass.

Although homegardens occupy only a small area compared to other components of the composite swiddening system, they are a highly efficient use of land. Women usually take care of the gardens, which are located near their houses. Management of the gardens is not very intensive, although some fertilizers, ashes, manure and pesticides are applied to the garden. It is difficult to assess the value of production of such mixed gardens because products are harvested in small quantities on an almost daily basis throughout the year and most products are consumed by the households rather than being sold for cash. The CARES survey of 42 households in 1998-99 found that only one-third of the households having homegardens earned any cash income from them (5,000 to 200,000 VND/year), mainly from the sale of bananas and vegetables.

Thanks to training and capital support provided by extension programs, many new plant species have been introduced into the homegardens such as oranges, persimmons, and improved varieties of litchi. Some households have begun going to the markets in Cao Son or the district town to buy vegetable seeds and fruit seedlings for their gardens. However, attempts to transform mixed species homegardens into specialized fruit orchards have been unsuccessful. Most plum trees were cut down in 2000-2002 due to a lack of market demand and a drop in fruit price to 200 VND/kg. New improved litchis and persimmons have also experienced periodic drops in price. Therefore, most farmers have returned to planting a diversity of species in their homegardens.

Tree gardens

Small patches of planted trees are found on the hillslopes behind many of the houses. Palms (*Livistona cochinchinensis*), timber trees (*Melia azedarach*), candlenut (*Aleurites montana*), and bamboo are the most common tree garden species, usually planted in pure stands. Few of the Aleurites trees bear nuts because of failure to select only female plants. Melia is grown for use at home and for sale. It is often interplanted with the rice plants in the first year rice swiddens. In subsequent years, cassava may be planted between the small trees. Once the trees become larger and their canopy closes, people stop cultivating agricultural crops and the field becomes a tree garden. Melia takes about seven to eight years before it can be cut. Eucalyptus was planted in old cassava swiddens some years ago as part of a reforestation effort supported by PAM (World Food Program), but now that subsidies are no longer provided, farmers show little interest in growing this species. Between 2000 and 2005, planting of bamboo and timber trees in fallowed swiddens in order to create new tree gardens was strongly supported by a government development program. Participating farmers received free seedlings, fertilizer, and payment for their labor.

Fallow swidden fields and secondary forest

There is no clear line between fallowed swiddens and secondary forest. Instead, there is an unbroken successional continuum from a largely open field with very partial cover of grass, weedy annuals, and scattered tree seedlings at the beginning of the fallow to a largely closed forest if the field is left undisturbed for ten or more years. The villagers exploit fallow swiddens and secondary forest in similar ways, using them for grazing their large livestock and collecting wild plants and animals.

Fallow fields

Swidden fields of all types are cultivated for several years, after which they are put into fallow for several more years. Rice swiddens are now usually cropped for two years, after which they may be planted with maize or cassava for one or two more years before being left fallow. Cassava swiddens are cultivated for two years before fallowing. The fallow period has been declining due to land scarcity and is now only three or four years on most sites. However, if the fields are left undisturbed for a longer period, secondary forest will regenerate. Some households manage their swiddens in order to promote forest restoration after fallowing.

In several actively cultivated rice swiddens, Melia seedlings were observed growing scattered among the young rice plants. Some of these plants are volunteers, with seeds probably dispersed by birds, but others were deliberately planted by the farmers. The seeds are dibbled into the field before burning so that the fire will cause germination. Melia and candlenut are also planted in the cassava swiddens. Few other trees were observed in the swiddens and no attempt appears to be made to protect coppicing stumps or to enrich the fallow with more trees than occur naturally.

Following the harvest, swiddens become open access pasture for cattle and buffalo from the hamlet. The effect of grazing on regeneration of secondary forest has not yet been studied but several farmers remarked that efforts to improve management of the fallow period, such as enrichment planting of leguminous trees, would be made difficult by unregulated grazing. If left relatively undisturbed, however, fallowed swiddens gradually undergo a process of succession until they are again covered by forest.

Forest

Virtually no primary forest survives within walking distance of the hamlet although remnant patches survive on the highest mountains along the boundary with Phú Thọ Province to the north of Tat hamlet. The tops of most hills and the channels of water courses are covered with well developed secondary forest. These areas are defined and protected by the cooperative. Individuals who clear swiddens in these protected areas are subject to fines and also may be compelled to plant trees to speed reforestation of the plots.

Forest areas are an important source of wild resources for all households, both for their own consumption and for sale. On average, non-timber forest products provide nearly one-quarter of total household cash income. Bamboo shoots are an especially valuable commodity. They were formerly collected primarily for home consumption, but now, they are mostly sold to outside traders to obtain cash. Exploitation of forest resources is completely unregulated which often results in excessive exploitation. For example, the aromatic root of one kind of forest tree, called in Vietnamese *vù hương*, was traditionally neither used nor sold by the villagers. In 1997, however, Chinese traders came to the hamlet seeking to buy these roots, so many men went to the forest to collect them. Within a period of one month, there were almost no *vù hương* trees remaining in surrounding forest

Plate 3.5: Truck carrying logs to the lowlands

areas. Although cutting of large timber trees is illegal, more than 20 young men, aged 25-35, routinely enter the forest to cut down trees. Groups of two or three men go into the forest together; make a temporary shelter and stay there for four or five days to cut down trees using axes and saws. They then split the logs into thick planks, which are dragged to the hamlet by buffalo. The planks are stored beneath the floors of their stilt houses or in sheds until they can be sold to timber dealers who come to the hamlet from the lowlands and take away the wood, often camouflaged under bamboo or other bulky legal products, in large trucks that move after dark to avoid being stopped at checkpoints (Plate 3.5).

Some men still go hunting for wild animals in the forest areas in Tat hamlet as well as neighboring hamlets. They shoot larger animals with shotguns and use snares and traps to catch smaller mammals and birds (Plate 3.6). The most commonly hunted wild animals are civets, squirrels, wild chickens and other kinds of birds, bamboo rats, mice, wild pigs, and deer. Changes in the abundance of wild game animals over the past 10 years are presented in Table 3.1. Populations of most species have decreased and some species, especially valuable and rare species like tigers, bears, and elephants, have become locally extinct. Civets, monkeys, and deer have also vanished from the forests in the hamlet, but some species, like mice and squirrels, have increased because their natural predators have been eliminated from the local ecosystem.

Wild animal populations have decreased because of excessive hunting and probably also because of habitat loss, in particular the fragmentation of the forest into scattered islands that are separated from one another by cultivated fields and areas of regenerating fallows. Fragmentation has broken the formerly almost

Plate 3.6: Hunter with the trap he used to catch a bird in the forest

Table 3.1 Changes in the abundance of wild game species in Tat hamlet

Species	10 years ago	5 years ago	Now
Small birds	Many	Many	Many
Wild chickens	Many	Some	Few
Civets (*Cầy bay*)	Many	Some	Few
Civets (*Cầy cui*)	Many	Some	Few
Squirrels	Some	Some	Many
Mice	Few	Some	Many
Bamboo rats	Many	Some	Some
Deer (*Hoẵng*)	Many	Few	Few
Wild pigs	Many	Some	Few
Monkeys	Some	Few	Few
Loris	Some	Some	Few
Snakes	Many	Some	Few
Bears	Few	Few	None
Deer (*Nai*)	Many	Few	None
Tigers	None	None	None

Source: Group discussion with farmers in Tat hamlet, 2004

unbroken forest habitat into multiple small patches: Between 1952 and 1995, the number of secondary forest patches within the hamlet increased from 18 to 292 while their mean size fell from 37 ha to 2 ha (Fox *et al.* 2000: 525).

Interactions among Agroecosystem Components

Table 3.2 is a matrix that shows the resource flows and other interactions among all the components making up the composite swidden agroecosystem. The villagers are included in this matrix, although they are usually described as part of the social system, in order to reveal the extent to which they interact with the components of the agroecosystem.

Read horizontally, each row of the matrix shows the resource flows from each individual component to every other component. It also shows the interactions that each component has with each of the other components. For example, in the case of a paddy field, it can be seen that resources flow from it to other paddy fields, to livestock, to wild animals in the forest, and to the villagers. However, the paddy field has no flows to, or other interactions with, swiddens, fallowed swiddens, fishponds, and homegardens and tree gardens. In the case of the livestock component, however, there are flows of resources to, or other interactions with, all of the other system components.

Read vertically, each column in the matrix shows the resources that a single component receives from every other component as well as how it is impacted by each of them. Thus, it can be seen that a paddy field receives resources flows and/or is impacted by the actions of every other component of the agroecosystem.

What is most strikingly revealed by Table 3.2 is the extent to which the composite swidden agroecosystem is a tightly integrated system. Although not every component interacts strongly with every other component, most components display at least some interactions with the majority of the other components. Of course, not all relationships between components are equally important for the functioning of the whole system—some links between components are very strong whereas other links are much weaker—but there are no interactions that are wholly insignificant. Consequently, any change in any individual component is likely to have ramifications for the overall functioning of the system. For example, conversion of swidden fields to secondary forest, which is a major objective of government land use policies, will reduce the flow of water-borne nutrients to the paddy fields and shrink the grazing area for livestock, which can be expected to lead to a decrease in the supply of manure for the paddy fields. Such changes are likely to impact the productivity and sustainability of the paddy field component of the system.

Conclusions

In this chapter, we have described the components that make-up the composite swidden agroecosystem in Tat hamlet and examined some of the key interactions that occur among these individual components. It is clear that households manage a very complex system in which they derive valued resources from many different components. Moreover, the functioning of each component is influenced by other components and it exerts influence on them in turn. Consequently, any change that occurs in one component is likely to have ramifications for the functioning of the entire agroecosystem. Thus, managing such complex and sophisticated agricultural system requires the farmers to balance the requirements of multiple different components, each with its own spatial and temporal aspects and each with its own needed inputs and desired outputs, against the functional integrity of the whole system.

Table 3.2: Resource flows and other interactions among components of the composite swidden agroecosystem

RESOURCE FLOWS/ INTERACTIONS	PADDY FIELD	SWIDDEN FIELD	FALLOWED SWIDDENS	LIVESTOCK	FISHPOND	HOME GARDEN/ TREE GARDEN	FOREST	VILLAGERS
Paddy field →	Run-off water, nutrients			Rice straw, rice bran, weeds, stubble grazed by ducks			Food for wild animals	Grain, winter season vegetables
Swidden field →	Run-off water, sediments, nutrients	Run-off water, sediments, nutrients	Run-off water, sediments, nutrients	Cassava roots and canna rhizomes for pig fodder, stubble grazed by cattle and buffalo during off season	Cassava leaves		Food for wild animals	Food crops, wild vegetables, firewood, cash
Fallowed swiddens →	Run-off water, sediments, nutrients	Run-off water, sediments, nutrients	Run-off water, sediments, nutrients	Grazing area and fodder for buffalo and cattle, bamboo shoots for pigs	Grass and leaves	Seedlings	Food for wild animals	Medicinal plants, wild vegetables, broom grass, bamboo stems and shoots, firewood, cash
Livestock →	Manure Traction for harrowing	Manure Damage to crops	Manure Disturbance of regenerating vegetation	Disease, competition for food resources	Pig manure	Manure Damage to vegetation	Manure Damage to vegetation, buffalo used to haul logs	Meat, cash Traction power

Fishpond →	Run-off water, nutrients			Taro for fodder	Diseases	Run-off water, sediments, nutrients		Fish, cash
Home garden/ tree garden →	Green manure			Fodder		Seed and seedlings Predation on crops by wild animals	Planting material	Fruit, vegetables, spices, medicinal plants, wood, cash
Forest →	Green manure, Predation on crops by wild animals (e.g., insects, rats, birds)	Run-off water, sediments, nutrients, fencing material Predation on crops by wild animals	Run-off water, sediments, nutrients	Grazing area and fodder for buffalo and cattle, bamboo shoots for pigs	Seed and seedlings Predation on crops by wild animals		Run-off water, sediments, nutrients, fencing materials, seedlings	NTFPs, timber, cash Disease (malaria)
Villagers →	Seedlings, chemical fertilizer, pesticides Management	Planting materials Management	Management	Commercial feed, vaccinations Management	Planting material Management	Planting material Management	Seedlings Management	Cooperation (labor exchange, mutual assistance) Competition (land use rights, extraction of NTFPs) Conflict (fights over crop damage by livestock, theft of crops)

4

Paddy Fields

Patma Vityakon

Introduction

The area of paddy fields in Tat hamlet in 2004 was 17.7 ha, which is 2.4 percent of the hamlet's total surface area. In contrast, swiddens and fallowed swiddens and other agricultural land uses covered 162 ha or 21.9 percent of the total surface area. Paddy fields thus accounted for less than eleven percent of all agricultural land in the hamlet. However, despite their small area, the paddy fields play a critical role in the functioning of the composite swidden agroecosystem. This chapter presents information on the location of the paddy fields in hamlet landscape, paddy soils, water supply to the paddy fields, farmer classification of the productive quality of different paddy fields, management of soil fertility, and pest control measures.

Study Methodology

This chapter is primarily based on an investigation conducted at Tat hamlet during 6-13 April 1994. In the course of that study, the author and Trần Đức Viên visited all of the clusters of paddy fields in the hamlet, observed their characteristics, took soil samples, recorded their names and locations on a sketch map, and found out farmer assessments of their quality. Soil profiles were observed from cuts at different levels of terraced paddies. Soil colors were determined using a Munsell soil color chart and pH was measured colorimetrically using a pH quick test kit.

Information about use of manure, fertilizer, and pesticides in the paddy fields in 2004 was obtained from interviews of 26 households using a standardized questionnaire by CARES researchers in February 2005.

Location of the Paddy Fields in the Hamlet Landscape

The location of the various paddy fields in the hamlet is shown on Figure 4.1. Most paddy fields are located in the main valley of the hamlet with its east-west orientation. They are in effect flanked by hills to their north and south directions. They are built in a series of terraces rising from the stream in the valley floor. Paddies at different terrace levels usually have a 0.5-1.5 meter height difference between them. The very narrow (20 cm) paddy bunds are ridges made from piling and packing black top soils to a height of about 10-30 cm.

Although paddy fields are mostly located in the main valley of the hamlet, they are also found in several other areas. One is on the slope of the northern hills (on the north side of the village's main road). There are two such sites of paddy fields. The second location is on the slope of the southern hills at the west end of the hamlet. The paddy fields in both locations are not as productive as those in the main valley due to poorer soils and lack of water. The third location is in an adjacent valley to the south of the main valley. Characteristics of these paddy fields are similar to those in the major valley location. To reach these paddy fields from the hamlet, one must cross the southern hills.

Paddy Soils

The complex geology of the hamlet (see Chapter 2) is reflected in the considerable diversity of soils founding the paddy fields. Paddy soils include both valley and hillside soils, depending on the locations of the fields.

Valley soils

Most of the paddy soils are classified as valley soils. Two layers can be distinguished in their profiles, as observed from cuts at different levels of terraced paddies. The top layer, which is about 50 cm thick (including the thickness of the bund of approximately 20 cm with a range of 10-30 cm), appears black in color with dark brown mottles indicating the frequently inundated condition of the soil. The soil texture is silt loam. In some locations inclusion of coarse sand or gravels was found in the top soils. The top layer ends abruptly at approximately 50 cm as it changes into the brown colored subsoil layer. There is also some brownish-red color material (which appeared greasy) found in the subsoil layer. This is ferrous oxide occurring in ferralitic soils under water-saturated (reduced) conditions. The texture of the subsoil layer is silty clay loam, somewhat heavier than the top layer. There is a lot of gravel in this subsoil layer.

It appears that the black silty top layers are derived from depositional materials from the hillsides and from the stream in the valley floor. Alluvial materials are deposited on the valley floor by water flowing from hillsides and from the stream in the middle of the valley floor.

Figure 4.1 Sketch map of Tat hamlet showing locations of paddy fields in relation to physical structures (1994)

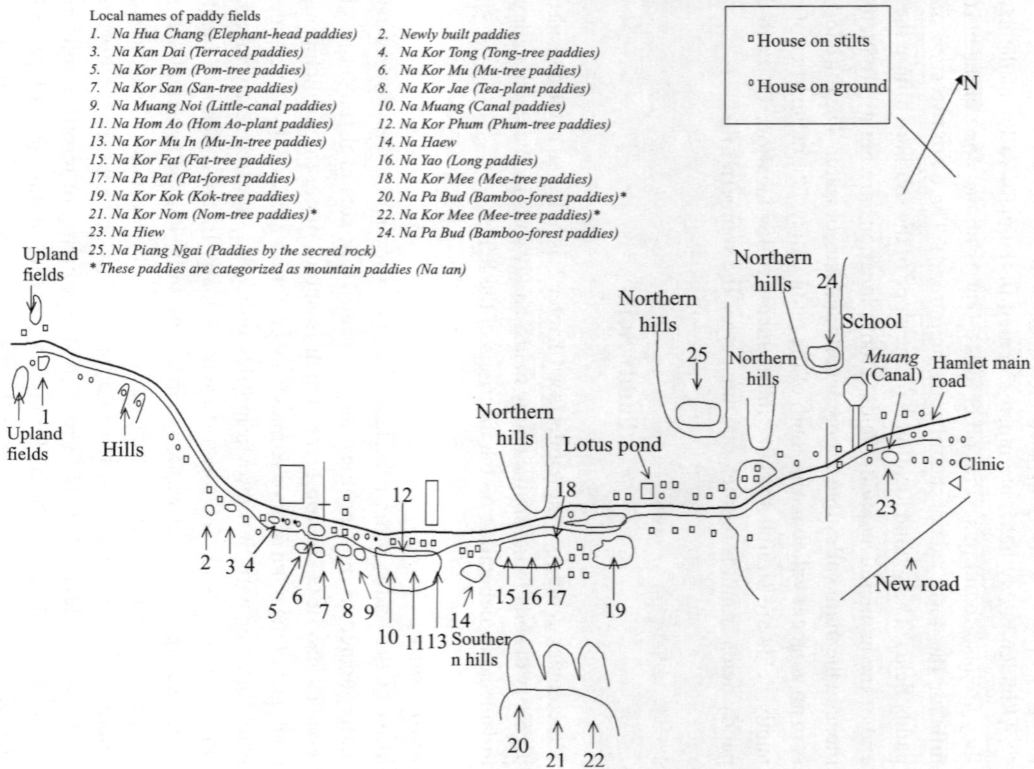

Table 4.1 pH of different layers of the soil profile of *Nà Kor Kok* (Kok-tree paddy field)

DEPTH (CM)	PH	REMARKS
0–20	-	Paddy bund, pH not measured
20–30	6.0	
30– 40	6.0	
50 +	6.4	

The quality of sediments from the southern and northern hills appears to be different. The southern hills give rise to black sediments which are a good growing medium, while the northern hills give rise to brown sediment, which is not as good a growing medium. The soils of the northern and southern hills appear to be derived from the pre-existing argillaceous rocks. However, the color of the soil in the southern hills is blacker than the soils in the northern hills. The outcrops of granite in the northern hills appear to be either fresh or slightly weathered and, hence, not considered to be the parent rock of the soil there.

The subsoil layer is derived from the weathering of the preexisting argillaceous rocks underlying the valley. The coarse grains found in the subsoil are from weathering of quartzo-feldspartic rocks, mainly gneiss found most frequently as outcrops in paddy fields.

Measurements of the pH of the topsoil layer of samples taken from several locations of valley soils revealed values ranging from 5.8 - 6.2. The pH values of different layers of paddy soil of the *Nà Kor Kok* (Kok tree paddy field) are similar (Table 4.1). This is considered to be a good pH range from a soil fertility viewpoint, providing a good balance of nutrients for plants.

The paddy soils are in a puddled condition. The soil is dominated by silt-size particles. In one instance in the *Nà Tan* paddy fields, there was a lot of gravel (1-5 mm in size) in the puddled soil. This gravel is probably from weathering of the quartzo-feldspartic rocks, especially gneiss, found as outcrops in the paddy fields. The depth of the puddled layer varies from 10-30 cm, below which it reaches a hard surface. This corresponds to the depth where the brownish subsoil layer occurs.

Hillside soils

Hillside soils are mainly brownish to brownish-red in color. The color description in terms of hue, value, and chroma (Munsell's soil color chart) of the top layer (0 -10 cm) of this soil, as sampled from the *Nà Pa Bud* ('bamboo forest paddy fields') behind the hamlet school, is 7.5 YR 5/8 (strong brown). A sample of hillside soil from a northern hill has the same color description. Soil texture as determined by the 'feel method' is silty clay loam. The brownish hillside soils are derived from the pre-existing argillaceous rocks. The brownish color of the soil is derived from the weathering of iron-bearing minerals, i.e., ferromagnesian minerals, of the parent rock. Granite, which is found in abundance in the hills as outcrops, should give coarser textured soil than the silty texture found to dominate the soils in the area. There is evidence of coarser-textured hillside soil in one location at the west end of the hamlet. The soil in this location, which is close to the *Nà Hua Chang* ('elephant-head paddy fields'), has a lot of gravel of quartz origin in its matrix (Figure 4.1).

Water Supply for the Paddy Fields

Paddy fields are irrigated mostly with water flowing into them from the surrounding hills in the north and the south. The paddies situated on the north side of the valley stream receive water from the northern hills while those situated on the south side of the stream receive water from the southern hills. Water from the hills flows into the upper-terraced paddies in the main paddy field area through a concrete canal built by the government in 1994. Other fields are supplied through small irrigation canals (*muang*) or bamboo pipelines constructed by individual farmers (Plate 4.1). In recent years some farmers have

Plate 4.1: Bamboo pipeline carrying irrigation water to the paddy fields (Nguyễn Văn Dung)

begun to use plastic pipes in place of bamboo. Water is led to lower-terraced paddies through small channels made in the bunds of upper paddies.

The water supply is ample for all paddies situated in the valley. However, water for hillside paddies is not adequate. Some of these hillside paddies, such as the *Nà Pa Bud* ('bamboo forest paddy fields') were built when there was more water in the past, when the surrounding forest was in better condition. Due to lack of water at a later stage *Nà Pa Bud* had already been abandoned for ten years when the field survey was done in 1994. These paddy fields were given the name *Nà Huea* ('abandoned paddy fields') for this reason. The *Nà Hua Chang* ('elephant head paddy fields'), at the west end of the hamlet, are another cluster of hillside paddy fields which were abandoned due to water shortages.

Water from the southern hills appears to carry with it sediments which are of better quality than those from the northern hills as they are deposited in paddy fields. It can be postulated that water quality as far as dissolved nutrients are concerned is likely better in the water from the southern hills than that from the northern hills.

Farmer Classification of Paddy Fields

The villagers have several systems of paddy field classification based on different criteria. Productivity of the paddy fields is one commonly used classification criterion. Based on this criterion, paddy fields can be divided into two categories which are differentiated by soil type, water supply, and soil fertility management level. Fields that give good yields are classified as 'Good paddy fields' (*nà dee*). Fields that give low yields are classified as 'Bad paddy fields' (*nà hye*).

Good paddy fields

These fields are invariably situated in the valley. They have black soils derived from alluvial sediments from the surrounding mountains and the valley stream. These soils are referred to by the farmers as being 'good soils' (*din dee*). The type of sediments from the surrounding mountains appear to play a major role in determining the quality of the soils in the valley. Alluvial soils derived from the southern hill sediments are black and constitute better soils than those derived from sediments from the northern hills. Sediments from the southern hills are said to have more fertilizing materials (*fun*) in them than those from the northern hills. The northern hills appear to have brown soils while the southern hills appear to have black soils upon examination from a distance. For the same reasons, the paddy fields in the valley situated to the south of the valley stream generally have better soils than those to the north of the stream.

The paddies classified as good are all located in the east side of the valley and at the south side of the stream including *Nà Kor Kok* ('Kok tree paddy fields'), *Nà Kor Mee* ('Mee tree paddy fields'), *Nà Haew* ('former burial ground paddy fields'), *Nà Kor Mint* ('Mint tree paddy fields'), *Nà Hom Ao* ('Ao plant paddy fields'), and *Nà Muang* ('canal paddy fields') (Figure 4.1).

Bad paddy fields

Some of these bad fields are located in the valley and some on the hillside slopes. Their soil is relatively more brownish to reddish in color and is called 'bad soil' (*din hye*) by the farmers. Most of the valley paddy fields classified as bad are located towards the west side of the valley, including *Nà Kor Ja*e ('Tea plant paddy fields'), *Nà Kor San* ('San tree paddy fields'), *Nà Kor Tong* ('Tong tree paddy fields'), and *Nà Kan Dai* ('terraced paddy fields') but some paddy fields on the east side of the valley and the north side of the stream are also classified as bad paddy fields. These are *Nà Pa Bud* ('bamboo forest paddy fields'), and *Nà Ya*o ('long paddy fields'). The bad paddy fields receive water and sediments from the northern hills. The topsoil color is less black with a greater degree of brown-ness than those paddy fields that receive sediments from the southern hills (Table 4.2).

The paddies located on hillside slopes, such as *Nà Pa Bud* ('bamboo forest paddy fields') on the slopes of the northern hills behind the hamlet school, have a strong brown soil color and relatively lower pH than paddy fields such as Nà Kor Kok in the valley (Table 4.2). Nà Pa Bud and similar fields are considered

Table 4.2 Comparative properties of topsoils (0–10 cm below paddy bund) of paddy fields of Tat hamlet

PADDY FIELD NAME	LOCAL CLASSIFICATION	PH	COLOR	TEXTURE
Nà Kor Kok (Kok tree paddies)	Good	6.0	10 YR 3/2 Very dark grayish-brown	Silt loam
Nà Hom Ao (Hom Ao tree paddies)	Good	5.8	10 YR 4/3 Dark brown	Silt loam
Nà Kor Jae (Tea plant paddies)	Bad	5.8	10 YR 4/3 Dark brown	Silt loam
Nà Pa Bud (Bamboo forest paddies)	Bad	5.6	7.5 YR 5/8 Strong brown	Silt loam

Nà Lang ('dry paddy fields') as they do not have adequate water supply. In fact, *Nà Pa Bud* had not been cultivated since the mid-1980s. Another bad hillside paddy area located in the northern hills is called *Nà Piang Ngai* ('paddy fields by the sacred rock'). Soil and water supply characteristics of these fields are similar to those of *Nà Pa Bu*d. During our visit only part of *Nà Piang Ngai* was transplanted and some fields were left idle due to lack of water.

Other bad paddy fields on hillside slopes are those at the west end of the hamlet in the southern hills. These paddies are called *Nà Hua Chang* ('elephant head paddy fields'). The soil is red and there is an inadequate water supply. In 1994, the spring crop was not planted due to water shortage and the farmers returned the fields to the hamlet cooperative.

Farmer Management of Soil Fertility

When we first began to study Tat hamlet, soil fertility in the paddy fields was maintained and improved entirely by application of organic amendments, including cattle manure and green manure. No chemical fertilizers were used at all until the late 1990s but now most farmers apply them (Plate 4.2). This reflects the increased availability of fertilizers due to improved transportation and training of farmers in fertilizer use by extension programs. A CARES survey of 26 households in 2004 found that urea was the most commonly used chemical fertilizer. It was applied by 80 percent of households to the spring crop and by 75 percent of households to the summer crop. NPK combined fertilizer was used by more than half of the households while phosphorus and potassium were applied by only about half of the households. The overall average rate of N:P:K application was 45:33:27 kg/ha in the spring and 38:21:18 kg/ha in the summer. Only 15 percent of households used lime, and then only in the spring season.

Manure was applied by 65 percent of households for both crops. The average amount applied was 3.2 ton/ha in the spring and 2.4 ton/ha in the summer. Green manure is composed of various kinds of tree leaves. Leaves of Melia trees and *Chromolaena odorata,* a common herbaceous weed, are extensively used for green manure. Green manure was applied by 45 percent of households in the spring and 25 percent of households in the summer at an average rate of 600 kg/ha/crop.

Different paddy fields are not treated in the same manner as far as soil fertility management is concerned, and different farmers appear to have different criteria for applying organic amendments. Some farmers apply more organic matter in fields with low fertility. Some apply more on good fields which are on the upper terraces, probably because they know that some of the nutrients will be

Plate 4.2: Woman broadcasting chemical fertilizer (Nguyễn Văn Dung)

later washed down to lower terrace paddies. Some paddy fields located next to the homestead area do not receive much application of organic fertilizers as soils are already enriched by materials washed down from the houses. Some farmers also mentioned distance from the house as a criterion for degree of soil fertility management. The paddies situated far from the homestead usually get lesser applications than those close to the homestead because of difficulties in transporting the heavy organic amendments to distant fields.

Pest Management in the Paddy Fields

Damage by insect pests is the main cause of productivity reduction during the summer season when farmers report a yield gap of up to 30 percent due to pest damage. The most common pest are stem borers (*Chilo simplex But, Sesamia inferens W., Schoenobius incertellus W.*), leaf worms (*Parnara guttata B., Cnaphalocrocis medinalis G.*), stink bugs (*Scotinophora lurida B.*, *Leptocorisa varicomis* F. and *L. acuta Th.*), and thrips (*Haplothrips aculeatus F*).

Traditional methods of pest control include burning chili, buffalo horns, or old tires to drive insects away with the acrid smoke. Farmers sometimes hang dead crabs on poles in the fields to attract stink bugs in order to trap them. Burning of crop residues after the harvest is sometimes done to clean the fields and kill insects. Cattle and pig manure is often mixed with rice husks and burned to kill fungi and grass seeds in the manure. Melia leaves, which are applied as green manure, are also believed to reduce stem borer activities.

In addition to traditional methods, a chemical pesticide called Ofatox is commonly used as a universal pesticide, although it does not effectively control stink bugs, thrips, and stem borers. Ofatox can be purchased at village shops or in the district market. In the mid-1990s, only one pesticide sprayer was available in the whole hamlet but at present there is one sprayer for every two or three households. In 2004, 65 percent of households used chemical pesticides in the summer season but only 35 percent used it in the spring, when pest damage is less severe.

Conclusions

Although paddy fields make up only a small part of the total area of cultivated land, they contribute a large share of the grain produced in Tat hamlet and are thus a critical agroecosystem component in the eyes of the farmers. Not surprisingly, the farmers display a rich body of knowledge regarding the paddy fields and their management. Individual fields have all been assigned names and their specific productive qualities are well known. Productivity depends on soil type, water supply, fertility management, and pest control. The farmers cannot change soil type, which depends on the parent materials, but can to some extent control the supply of water through construction of irrigation canals, and can significantly modify soil fertility by application of organic amendments and chemical fertilizer. Pests are managed using both traditional methods and spraying of chemical pesticides, although success is limited.

5
Fallow Swidden Fields: Floral Composition, Successional Dynamics, and Farmer Management

Mai Văn Thành and Trần Đức Viên

Introduction

Fallow swidden fields are an important subsystem of the composite swidden agroecosystem in Tat hamlet. In this chapter we describe the floral composition and successional dynamics of the vegetation found in fallow swiddens in Tat hamlet, the uses to which the villagers put wild products collected from the fallow fields, and their practices and beliefs regarding systems of fallow management.

Research Design and Methodology

The study was carried out in 2004. It involved identifying the plant species found within the territory of the hamlet and ascertaining the uses of these species by the villagers. Sample fields that had been under fallow for varying lengths of time were selected and the characteristics of vegetation in these fields described and its biomass estimated. Information about farmer knowledge regarding fallow management was also collected.

Identification of flora in the study area

A survey of flora was carried out by walking along four transects with a total length of 12 km together with local key informants, who were deeply knowledgeable about the flora in the study site. All the different flora species encountered were collected and key informants interviewed in order to obtain their local names and an explanation of how these species are used (e.g., as timber, food, medicine). Scientific names of the samples were identified by morphological comparison and from secondary data on flora in Vietnam (Phạm Hoàng Hộ 1993). Botanical phylums were listed in order of vegetation evolution, i.e., from plants that scatter spores [Psilotophyta (Khuyết lá thông), Lycopodiophyta (Thông đất), Polipodiophyta (Dương xỉ)] to plants with seeds [Pinophyta (Thông), Magnoliophyta (Ngọc lan)]. The list of species is in alphabetical order based on phylum, family, genus, and species (except for the Magnoliophyta [Ngọc lan] phylum which is under class order).

Selection of sample fallow fields

A general survey was made of fallow fields in the hamlet, key informants were interviewed regarding land use history, and the conditions of the selected fields were checked to make sure that they met all of the selection criteria. Finally, different fallow fields were selected in the second, fifth, tenth, and twentieth year of fallow. Three plot replications were selected for each fallow period. The fields were carefully selected to be as uniform as possible in terms of soil type, slope gradient, position of the fields, cultivation history, and fallow succession. All the fields selected had been managed using local fallow management practices.

Description of characteristics of the vegetation in different stages of the fallow

Different sized standard plots were used to record the vegetation of fallow fields in different successional stages (Figure 5.1). The standard plot A was 50 x 50 m (0.25 ha) and was applied to forest after 20 years of fallow. All timber trees having a diameter at breast height (DBH) larger than 10 cm were measured: $DBH_{1.3\ m}$, H_{crown} and H_{stem}. Trees shorter than 10 meters in height were directly measured, while trees taller than 10 meters were measured by using the similar triangles method.

The size of standard plot B was 31.5 m x 31.5m (ca. 0.1 ha) and was applied to 4 to 5 year-old fallow plots and 10 year-old fallow plots. Measurement of tree species is similar to that done for the standard plot A. The standard plot C was 10m x 10m and was used to measure the number of species of bushes, secondary timber species, and bamboo. The standard plot D was 2m x 2m and was used to measure grass biomass under the woody canopy.

Estimation of biomass of fallow vegetation

The biomass of vegetation in plots in different fallow stages was measured using methods appropriate to the class of vegetation. For trees, the volume of all the specimens growing inside the A plot was determined by the equation:

$$V_0 = \frac{\Pi * D_{1.3}^2}{4} * H_{crown} * f_{1.3}$$

Where V_o =Volume of timber, $D_{1.3}$ = diameter of timber plant at breast height; H_{crown} = height of the tree; $f_{1.3}$ = coefficient ($f_{1.3}$ = 0.42 for natural forest; $f_{1.3}$ =

Figure 5.1 Diagram comparing the sizes of the A, B, C, and D standard plots

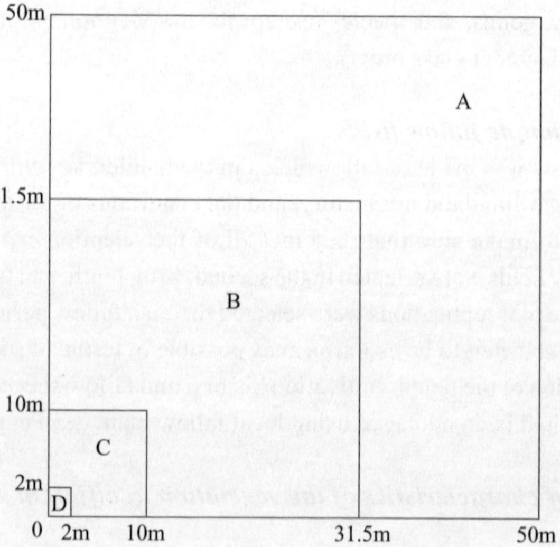

0.45 for planted forest) (Vũ Tiến Hinh and Phạm Ngọc Giao 1998: 60). The tree biomass was estimated from their total volume using the NIRI (Nissho Iwai Research Institute) formula (Ngô Đình Quế *et al.* 2003: 15):

$V_1 = 1.33 * V_o$

$V = V_1 * 1.2$

$M = 0.5 * V = 1.33 * 1.2 * 0.5 \times V_o$

Where: M = Biomass of trees, V_o = Volume of above ground biomass of timber, V_1 = Total volume of above ground biomass of trees (leaves, timber, and stems), and V = Total volume of above and underground biomass of trees.

For bushes, bamboo, and lianas, the average wet weight per plot was directly determined and then converted to dry weight of the specific vegetation (i.e., bushes, bamboo, or vines) per hectare. The biomass of grass was determined by picking, washing and weighing all the specimens growing inside the D-plot (2m x 2m).

Collection of farmer knowledge about fallow management

Key informant interviews were conducted with four knowledgeable farmers and semi-structured interviews were made with six farmers in Tat hamlet to collect information about local fallow management practices. A matrix ranking exercise was done with nine farmers representing men and women in different age groups to gather information about the use of species found in the different fallow stages.

Description of Fallow Flora

The flora of the fallow fields is described in the following sections in terms of diversity, life forms, and the uses made of different species by the villagers.

Diversity of fallow flora

A total of 419 species belonging to 283 Genuses, 113 Families, and 5 Phyla (Psilotophyta, Lycopodiophyta, Polipodiophyta, Pinophyta, and Magnoliophyta) were identified (Appended Table 5.A1). It is evident that the flora in the hamlet is very diverse. Its composition is affected by both a climatic and topographic factors. Variation in the sunlight regimes within the different areas of the hamlet leads to further differentiation. Moreover, the canopy stratification is relatively complex due to a high frequency of human intervention. Almost all ecological groups are represented, including annual and perennial light-demanding pioneer flora; flora with medium light demand; and shade tolerant flora. There is a great variation in the number of species belonging to each family or genus. The Fabaceae family has the largest number of species (28), followed in descending order by Euphorbiaceae (25 species), Poaceae (18 species), Asteraceae (16 species), Rubiaceae (15 species), and Moraceae (13 species) (Appended Table 5.A2).

Life forms

The ecological characteristics of fallow vegetation represent the relationship among species of the fallow community. Table 5.1 displays the five main floral life forms in Tat hamlet: grass, trees, lianas, bushes, and palms, of which the grass species are the most numerous.

Useful plants in the fallow vegetation

Villagers identified 398 out of the 419 species of plants (95%) in the fallow flora as being useful for some purpose. The local and scientific names of these species along with their functions are presented in Appended Table 5.A1. These species are

Table 5.1 Distribution of flora life forms in fallows in Tat hamlet

Type	Number of species	Percentage
Grass	178	42.5
Trees	137	32.7
Lianas	45	10.7
Bushes	50	11.9
Palms	9	2.2
Total	419	100.0

Table 5.2 Useful fallow plant species in Tat hamlet

USES/ FUNCTIONS	NUMBER OF SPECIES	PERCENTAGE
Medicine	225	53.7
Fruit	59	14.1
Timber	39	9.3
Vegetables	44	10.5
Ornamental	8	1.9
Spice	17	4.1
Dye	3	0.7
Chewed with betel	3	0.7
No use identified	21	5.0
Total	419	100.0

categorized according to eight different uses (medicine, fruit, timber, vegetables, ornamentals, spices, dye, and for chewing with betel). As Table 5.2 shows, medicinal plants are the most numerous, followed by fruit, vegetables, and timber.

Successional Dynamics of Fallow Flora

The fallow flora undergoes very important changes in its species composition, life forms, biomass, and presence of useful species over the course of the fallow cycle.

Successional changes in the floral composition of fallow vegetation

The number and type of species within plots changes in the process of succession from one fallow stage to the next (Appended Table 5.A3). The number of tree species increases markedly from the second year fallow to the twentieth year fallow. Also, the economic value of the tree species increases from the second year fallow, where lower value *Styrax tonkinesis* and *Rhus javanica* are found, to the fifth year, where *Neonauclea calycina* and *Saurauia roxburghii* are found, to the tenth year where *Ormosia balansae* is found, and finally to the twentieth year fallow where high value *Elaeocarpus petiolatus* and *Demos chinensis* are found along with several wild fruit trees that have market value, such as *Garcinia cochinchinensis* and wild litchi.

Successional changes in the life forms of fallow vegetation

The floral life forms also change in importance in the course of succession (Table 5.3). Although the species structures of the second and the fifth years of

Table 5.3 Changes in number of species in floral life forms in different fallow stages

LENGTH OF FALLOW (YEARS)	TOTAL NO. OF SPECIES	TYPE				
		GRASS	BUSHES	TREES	LIANAS	BAMBOO
2	62 (100.0)	21 (33.9)	16 (25.8)	19 (30.6)	5 (8.1)	1 (1.6)
5	42 (100.0)	13 (31.0)	7 (16.7)	18 (42.9)	3 (7.1)	1 (2.4)
10	47 (100.0)	13 (27.7)	7 (14.9)	25 (53.2)	2 (4.3)	0 (0.0)
20	76 (100.0)	21 (27.6)	8 (10.5)	42 (55.3)	4 (5.3)	1 (1.3)

fallow are varied, almost all species found are light tolerant plants and the tree species are outnumbered by grass and bush species. The number of species of lianas (vines) decreases between the second year and the tenth year of fallow. However, between the tenth and twentieth year of fallow, the number of lianas increases again as the forest canopy closes. The number of tree species increases from 19 species found in the second year fallow stage to 42 species found in the twenty year fallow stage. In general, the number of tree species increased, while the number of bush, lianas, and grass species decreased, as the length of the fallow increased. However, the number of tree species slightly decreased between the second year fallow stage and the fifth year fallow stage due to human intervention, since the villagers collect certain species for their livelihoods.

Successional changes in vegetation biomass

Overall, the total amount of above ground biomass increased over the duration of the fallow, going from 16.6 tons/ha in year 2 to 63.9 tons/ha after 20 years (Table 5.4). The greatest change occurred for large tree species between 10 and 20 years of fallow. The changes in biomass of small trees, bushes, and grass did not present a clear successional trend, however, due to human interventions that disturbed the natural regeneration pattern, such as planting of bamboo and grazing of livestock.

Table 5.4 Biomass of different life forms in different fallow stages (kg/ha)

	2 YEARS	5 YEARS	10 YEARS	20 YEARS
Grass	16,000	15,000	30,000	19,000
Small trees and bushes	570	342	4,448	1,290
Large trees	-	-	8,131	43,650
Total biomass	16,570	15,342	42,579	63,940

Table 5.5 The number of useful plant species in different fallow stages in Tat hamlet

Fallow stages (years)	Number of useful plant species					
	Medicinal	Fruit	Timber	Vegetables	Ornamental	Spice
2	33	5	7	2	1	1
5	20	2	6	3	1	1
10	22	3	11	2	1	1
20	34	7	20	1	2	0

Changes in the number of useful plants during fallow succession

Table 5.5 displays the change in the number of useful plants found in the different fallow stages. It is notable that the number of useful woody plant species changes rapidly from seven found during the first fallow stage to 20 species found in the last fallow stage. The numbers of other types of useful plants undergo minor changes. The results indicate an ecological succession towards the dominance of woody species in the old fallow.

Utilization of Plants Collected from Fallow Land

The villagers have always depended heavily on products collected from fallow land for both their own subsistence needs and as a source of cash through sales to the market. However, the plants collected and the uses to which they are put has changed over the course of time, with a particularly marked change between the time when land was managed by the cooperative and marketing of wild products was controlled by state enterprises, and the present time, when most land has been returned to household management and a largely free market prevails.

In order to identify changes in the utilization of plants collected from the fallows since the dismantling of the cooperative, a simple matrix analysis was used to identify major fallow products and their functions during the cooperative time and at present. The matrix used a scoring system of 0 to 10, with 0 meaning the plant was not important for that function while 10 means that it was very important for that function. Tables 5.6 and 5.7 show the ways in which the functions of key products have changed from the cooperative period to the present time.

Table 5.6 shows that plant products collected from the fallows were diverse during the cooperative period. The products considered especially important during the cooperative time were bamboo stems, palm fruit and leaves, and

Table 5.6 Rating of relative importance of plants collected from fallows during the cooperative period

USE	PRODUCT							
	BAMBOO SHOOTS	BAMBOO	FIREWOOD	TIMBER	PALM LEAVES	MEDICINAL PLANTS	BROOM GRASS	MUSHROOMS
Food	4	-	-	-	3	-	-	2
Sell for cash	2	-	-	-	-	5	-	5
Medicine	-	-	-	-	-	3	-	-
Fuel	-	3	10	-	-	-	-	-
Housing	-	9	-	6	10	-	-	-
Fencing	-	5	-	4	-	-	-	-
Livestock fodder	6	1	-	-	-	-	10	-
Total	12	18	10	10	13	8	10	7

Note: The higher the number, the greater the perceived importance of the plant type, with "10" having the greatest importance

Table 5.7 Rating of relative importance of plants collected from fallows in the present time

Use	Product							
	Bamboo shoots	Bamboo	Firewood	Timber	Palm leaves	Medicinal plants	Broom grass	Mushrooms
Food	3	-	-	-	-	-	-	2
Sell for cash	10	8	5	4	5	5	5	2
Medicine	-	-	-	-	-	3	-	-
Fuel	-	5	10	-	-	-	-	-
Housing	-	5	-	4	10	-	-	-
Fencing	-	7	-	5	-	-	-	-
Livestock fodder	10	2	-	3	-	-	10	-
Total	23	27	15	16	15	8	15	4

Note: The higher the number, the greater the perceived importance of the plant type, with "10" having the greatest importance

Plate 5.1: Broom grass being dried in the sun after collection from fallowed swiddens

bamboo shoots. During the cooperative period, the fallow products were used mainly for home consumption such as house construction, food, livestock, and firewood. After the cooperative period, however, the relative importance and functions of these products changed (Table 5.7). Bamboo is still the most important product, followed by bamboo shoots, timber, firewood, and broom grass (Plate 5.1). Today, however, many more products are mostly collected for sale in the market, although the villagers continue to use them for livestock fodder, house construction, fuel, fencing, and their own food.

Fallow Management Systems

Farmer strategies for managing fallow land have changed over time in response to increasing population density and changes in government policies. During the cooperative period, population density was still low and people did not face land shortages. Natural fallow and the augmentation of natural fallow regeneration by planting Melia trees and palm trees in the fallows were the most popular systems of fallow management. Recently, villagers have faced greater constraints on land use due to population pressure and the implementation of the government's land allocation program. Several fallow management systems are currently employed in Tat hamlet:

Natural fallow

This is a traditional fallow practice that allows natural vegetation to regenerate without any human intervention. The advantage of this method is that soil fertility is naturally restored without any investment of capital or labor. However, because there is no intervention during the early stages, soil erosion can occur during the early years that the field lays fallow, and a long period of time can be required for the soil in the field to recover. Consequently, this method is only applied to fallow land located a long distance from the settlement area.

Melia planting on the fallow land

This type of agroforestry model has long been practiced in Tat hamlet (Trần Đức Viên 2007). Farmers usually sow *Melia azedarach* seeds in their swiddens before burning the fields because Melia seed only germinates after exposure to high temperatures. Both crops and Melia trees are grown together in the swidden field (Plate 5.2). After several crops are harvested, the field is left fallow while the Melia grow to maturity. Melia grows quickly and its timber is used for domestic purposes. Melia leaf has a bitter taste and is used as green manure and as a method of biological pest control for paddy fields. The harvesting period for Melia usually starts 8 to 10 years after planting. After harvesting of the Melia, the next cycle of swiddening starts. This model is still popular in Tat hamlet.

Bamboo planting on the fallow land

Most households now apply this system on some of their land because planting of bamboo is promoted by various development projects to restore areas considered 'barren land' by the government. These projects provide farmers with subsidized seedlings, fertilizer, and training. Planting of *Luồng* bamboo (*Dendrocalamus brandisii*) is often promoted because it has a high value, is easy to sell, and grows quickly. However, it also has several disadvantages, including being most suitable for growing in moist soils and lowland areas, its high initial cost, its removal of a large quantity of nutrients from the soil when it is harvested, and the high resistance of its leaf litter to decomposition. The latter two factors adversely affect the next swidden cycle.

Styrax planting on the fallow land

The quick growing native tree *Styrax Tonkinesis* is a common pioneer tree in fallow fields in many parts of the northern uplands of Vietnam. It occurs as a volunteer, but can also be planted by the farmers. Although this species can be sold for paper pulp, for timber, and to make matches, it is not popular in Tat hamlet. However, some farmers plant these trees to use for firewood and fencing, as well as to speed up the regeneration of the fallow.

Plate 5.2: Melia planted as successional crop in rice swidden (Trần Mạnh Tường)

Palm planting on the fallow land

This model is relatively popular in Tat hamlet. At the beginning, crops are intercropped with palms for several years. When the palms reach the closed canopy stage, cropping is stopped. Palm leaves are used for roof thatching and for sale. Recently, local farmers can obtain a high cash income from their palms during the off farm season (i.e., from November to January).

Coppicing trees in the swidden period

This method leaves tree stumps with a height of 30-40 cm above the ground when the forest vegetation is cut to open up a new swidden. This results in a reduction of soil erosion and landslides. During the cropping phase, the stumps begin to regenerate buds and shoots, which permits rapid regrowth of secondary forest when the field is fallowed. The regenerating trees are a source of firewood and also encourage quick recovery of soil fertility for the next swidden cultivation cycle.

Farmer Perceptions of Fallow Management Systems

A group discussion was organized in May 2005 to identify the relative value of different fallow management systems used in Tat hamlet in the eyes of the farmers. Seven criteria were identified: (i) generates income; (ii) needs less investment; (iii) offers effective erosion control; (iv) improves soil fertility; (v) requires only simple techniques; (vi) suitable to the farmer's situation; and (vii) minimizes demand for labor. The matrix used a scoring system of 0 to 10, with 0 meaning the value was not important for the system while 10 means that it was very important. Table 5.8 presents the results of group discussion.

The farmers generally favor systems that require less investment, simple technology and are most suitable to their specific situation. Generation of income, improving soil fertility, and controlling erosion are all lesser concerns. Labor demand is not seen as at all important. Consequently, the locally developed models of planting Melia and palms in the fallow are most preferred by the farmers.

Conclusions

The information presented in this chapter shows that fallow land, far from being useless wasteland as it is commonly portrayed in the Vietnamese official mythology of shifting cultivation (Jamieson *et al.* 1998; Rambo 2001), plays vital ecological and economic roles in the continued functioning of the composite swidden agroecosystem in Tat hamlet.

The flora found on fallowed swiddens is highly diverse. The mosaic of fallow fields in various stages of succession provides habitats for species with a wide range of ecological requirements, from early pioneers to those adapted to life under the conditions of mature secondary forest. This habitat diversity contributes to the maintenance of overall floral biodiversity.

The fallow flora is composed of many species having high economic value, especially the timber species and medicinal plants. The products collected from fallow land have always contributed considerably to local livelihoods and local social activities, although the significance of specific products has changed considerably in recent years after the economic reforms. In particular, products from the fallows have become important commodities used by the farmers to earn cash from the market.

The farmers have experimented with a number of innovative agroforestry systems to manage their fallows, including planting fallows with Melia, palms, bamboo or Styrax. Of these, planting of Melia and palm, the two locally developed systems, have been most successful in the eyes of the farmers, and would seem to have the most potential for expansion in Tat hamlet and beyond.

Table 5.8 Farmer perceptions of the value of different fallow management systems

	SYSTEM							
VALUE	NATURAL FALLOW	MELIA PLANTING	BAMBOO PLANTING	STYRAX PLANTING	PALM PLANTING	COPPICING	TOTAL	RANKING
Generates income	3	8	10	4	8	3	33	4
Less investment	10	8	3	8	8	10	47	1
Erosion control	3	6	4	4	8	8	29	6
Improves soil fertility	5	7	2	6	5	7	32	5
Simple technique	10	8	3	8	8	8	45	2
Suitable to situation	7	8	4	7	6	6	37	3
Labor demand	0	6	8	6	7	0	27	7
Total score	38	51	34	43	50	42		
Ranking	6	1	5	3	2	4		
% households using	20	50	90	50	45	30		

Note: The higher the number, the greater the perceived value of the system, with "10" having the highest value.

6
Livestock

Nguyễn Thanh Lâm and Trần Đức Viên

Introduction

This chapter focuses on livestock as an important component of the composite swidden agricultural system in Tat hamlet. It describes the situation of large livestock (buffalo and cattle), small livestock (pigs and goats), and poultry (chickens and ducks) in the hamlet, discusses the way livestock are managed by farmers, looks at farmer perceptions of the value of livestock and major constraints on livestock production, and examines the interactions of livestock with other components of the composite swidden agroecosystem.

Methodology

In 2006, we interviewed 30 randomly selected households using standardized questionnaires in order to obtain information about household social and economic characteristics and the livestock production systems they employed. In addition, we conducted semi-structured interviews and farmer group discussions in order to understand farmers' perceptions of the value of livestock and to identify constraints on livestock production. Information collected in previous field studies in 1998, 2003, and 2004 was also used as were secondary data obtained from the district and village agricultural departments.

To examine changes over time in the situation of livestock, data from two questionnaire surveys of the same sample of households carried out by CARES researchers in 1998-1999 and in 2004 were compared. Forty-two households (out of the 95 households in the hamlet) were randomly selected for interviewing in 1998-1999. In 2004, these same households were re-interviewed using the same questionnaire.

Large Livestock

The current status of buffalo and cattle in the hamlet is shown in Table 6.1. Ninety percent of households own one or more buffalo, with an average of 2.4 animals

Table 6.1 Ownership of buffalo and cattle by households in Tat hamlet (n = 30, percentages inside parentheses)

SPECIES OF LIVESTOCK	HOUSEHOLDS OWNING ANIMALS	MEAN NO. OF ANIMALS/ HOUSEHOLD	HOUSEHOLDS BUYING ANIMALS IN PAST YEAR	HOUSEHOLDS SELLING ANIMALS IN PAST YEAR
Buffalo	27 (90.0)	2.4	5 (18.5)	6 (22.2)
Cattle	19 (63.3)	2.1	11 (57.9)	8 (42.1)

Source: Tat hamlet household survey, March–April, 2006

per household. Cattle are owned by 63 percent of households, all of which also own buffalo, with an average of 2.1 heads per household. The maximum number of buffalo owned by a single household is eight animals, while the maximum number of cattle is 11 animals. These are wealthier households that seek to accumulate large livestock as a means of saving. More households own buffalo than cattle because, in addition to serving as a store of wealth, buffalo can be used for land preparation, to provide manure for the paddy fields, and to drag logs from the forest to the road as part of illegal logging activities, whereas cattle are mostly raised to sell for cash.

The number of large livestock has increased in recent years. In 1998-99 there were 99 buffalo and 83 cattle whereas in 2004 there were 142 buffalo and 154 cattle. According to the hamlet leader, the number of cattle quickly increased once many households were able to obtain credit for raising cattle from a government livestock development project, which provided low interest loans of 3 million VND per participating household to purchase cattle; prior to that buffalo had been the dominant type of large livestock.

Farmers use cattle as an investment vehicle. Whenever they have extra money, they buy additional animals rather than putting their savings in a bank. When money is needed to cover a large expenditure, such as building a new house, the farmers sell the necessary number of cattle. According to the farmers, a mature animal can be sold for 2 or 3 million VND. Each year, many farmers buy calves using credit from the government project to fund these purchases. Of the eleven households that bought cattle in 2006, only three households had to use their own money, while eight households were able to secure loans for purchasing cattle. However, all five households buying buffalo had to use their own funds. As Table 6.1 shows, more than 40 percent of households owning cattle sold animals in 2006, whereas just over 20 percent of households owning buffalo sold animals in the same year, suggesting that cattle are raised mainly for commercial purposes.

Table 6.2 Division of labor in caring for large livestock among household members

	Buffalo (n = 27)		Cattle (n = 19)	
	No. of HHs	%	No. of HHs	%
Wife	3	11.1	2	10.5
Husband	6	22.2	4	21.1
Both spouses	12	44.4	9	47.4
Children	3	11.1	2	10.5
Grandparents	3	11.1	2	10.5

Source: Tat hamlet household survey, March–April, 2006

Table 6.2 shows how the responsibilities for caring for large livestock are allocated among members of the households. There are no rigid cultural rules governing allocation of these duties so that individual households assign tasks based on their specific conditions. The most common allocation is for both spouses to share the labor involved in caring for the large livestock.

Watching the livestock while they engage in free range grazing is a major responsibility for caregivers. Although managed grazing has recently started in the hamlet, most large livestock are still allowed to graze freely in fallow swiddens and certain designated forest areas (especially Suối Cọ Mục, Cọ Nôm, and Suối Mương). Periodically, the caretakers go to the fallow swiddens to care for their livestock grazing there. When heavy storms approach, caretakers bring the animals down to the households for shelter and to feed them (Plate 6.1). During the periods in February and June when the paddy fields are harrowed, buffalo are often kept at home to use for draft power and as a source of manure. Often, they are kept at night under the floors of the houses on stilts (Plate 6.2). During these periods, the caregivers cut and carry grass to feed the buffalo at the houses. In addition, several households have recently begun to graze their livestock on their farmsteads, which are located one or two km from the main settlement area of the hamlet.

Problems in raising large livestock

Farmers identified six main problems affecting current livestock production (Table 6.3). In order of frequency of mention these are lack of capital, Surra disease, insufficient labor, livestock-caused crop destruction, accidental death of livestock (for example, fatal falls down steep hill slopes), and loss of livestock due to theft. Overall, buffalo suffer from more problems than cattle.

Each species has its own problems. For buffalo, the highest risk was identified as Surra disease, while cattle had a higher frequency of being killed in accidents

Plate 6.1: Buffalo being led back to the settlement after grazing in the mountains. A freshly burned swidden field is in the foreground (Nguyễn Thanh Lâm)

Plate 6.2: Livestock are often kept under the houses at night

Table 6.3 Farmer perceptions of problems in raising large livestock in Tat hamlet

	Buffalo (n = 27)		Cattle (n = 19)	
Problem	No. of HHs	%	No. of HHs	%
Lack of capital	1	3.7	0	0.0
Surra disease (*Trypanosomiasis*)	14	51.9	0	0.0
Insufficient labor	1	3.7	1	5.3
Destruction of crops by livestock	4	14.8	5	26.3
Food poisoning by cassava	2	7.4	2	10.5
Accidental death	2	7.4	5	26.3
Theft	1	3.7	1	5.3
None	6	22.2	10	52.6

Source: Tat hamlet household survey, March–April, 2006

Table 6.4 Vaccinations and disease of large livestock in Tat hamlet

Type of livestock owned by household	Households vaccinating animals	Households with diseased animals
Buffalos (n = 27 households)	20 (74.1%)	12 (44.4%)
Cattle (n = 19 households)	11 (5.3%)	0 (0.0%)

Source: Tat hamlet household survey, March–April, 2006

and causing destruction of crops In 2003, two cattle died as a result of falling into a chasm.

Disease is a serious problem for buffalo but not for cattle. In 2006, eleven households had buffalo affected by Surra disease, and one household had a buffalo infected with parasitic worms, but no cattle-owning households reported such problems (Table 6.4).

Surra disease (*Trypanosomiasis* caused by *Trypanosoma evansi*), a previously unknown livestock disease in the hamlet, has recently infected many buffalo. The infected animals exhibit patches of white, chapped skin that spread as the disease progresses. Additional symptoms include loss of appetite and diarrhea, eventually leading to death. Farmers are very concerned about the disease, as several households had had infected buffalo die. As a result, livestock owners keep their animals under close observation and administer medicine when symptoms develop. If the infected buffalo are injected with medicine early enough, their odds of survival are good. Although anti-surra injections cost only 5,000 VND per animal, only 67 percent of buffalo-owners and 37 percent of cattle-owners reported administering the medicine. The farmers said that one reason for this was that the animals were often out in the free-grazing areas when veterinarians came to the hamlet, and thus went untreated.

Buffalo suffer a higher rate of infection than cattle for two reasons: First, cattle appear to have a higher level of natural immunity to the diseases than buffalo. Secondly, cattle spent most of their time grazing in the forest and fallow swiddens away from inhabited areas, while buffalo were kept in the hamlet for land preparation and other domestic uses so were more likely to be exposed to infection.

Both buffalo and cattle are at risk from consuming toxic fodder. In 2005, several buffalo and cattle suffered from food poisoning caused by eating the toxic leaves of a newly introduced high-yielding variety of cassava.

Theft of cattle has recently become a serious problem. In August 2004, two households lost cattle in the forest at the edge of the hamlet. Farmers stated that the authorities in a neighboring commune had arrested a group of cattle thieves in connection with the crime. In 2005, a total of 15 cattle and buffalo were stolen. In addition, the hamlet's women's association lost two cattle in 2006.

Damage caused to swidden crops by free ranging cattle is becoming a more serious problem. This is because the grazing area and the swiddens are located in close proximity to each other. Destruction of crops by livestock causes social tensions within the hamlet. Some cattle have been injured by stones thrown at them by the owners of the swiddens and others have even had their tendons cut with knives.

Small Livestock

According to the hamlet leader, there were 300 pigs in Tat Hamlet in 2006. Only 40 percent of households kept any pigs. Of those households that raise swine, most raise only one or two pigs per year and only a few households raise more than five pigs. There are two breeds of pigs in the hamlet—a hybrid breed and a native (*Mường*) breed. More households raise native pigs than hybrids because they are easier to keep because they are more resistant to disease, require less labor, and do not require special feed. Native pigs are kept at night in pens but during the daytime are allowed to run freely around the houses. However, their productivity is quite low, with weight gains of only 20 to 35 kg per year per head, and the sows produce only three to five piglets per year. Hybrid pigs grow more quickly and produce more manure but require expensive commercial feed. A few households keep both breeds, which they see as an effective way to balance the advantages and disadvantages of each breed. Although native pigs have low productivity, they can eat the surplus food from the households and have a high resistance to disease. Conversely, hybrid pigs have high productivity, but require expensive commercial feed and have very low resistance to disease.

Taro, bamboo shoots, rice bran, cassava roots, and residues from making alcohol are all commonly used for pig food. In addition, several households purchase commercial feed such as 'Cám Cò' (a popular brand of commercial animal feed). Women have most of the responsibility for caring for the pigs. They usually gather food for the pigs when they visit their swiddens or go to the forest to collect bamboo shoots. Several households have built pigpens near their fishponds, so that the manure fertilizes the ponds.

Pigs are sold to district traders for 15,000 VND per kg (pigs weighed between 50-70 kg). In return, the farmers buy piglets weighing 7 to 10 kg for 22,000 VND per kg, which they then fatten before selling. Because there are no butchers in the hamlet, people have to buy meat for their own consumption from a motorcycle vendor for 40,000 VND per kg.

In 2006, only five households reported instances of disease among pigs, usually in the winter months. Farmers said that *Pasteurella* disease *(Pasteurella multocida)* could often be identified by the following symptoms: high temperature, red spots on skin, lack of appetite, and a dark red tinge to the meat.

Villagers do not like to raise goats in the hamlet due to their destructive habit of eating leaves and tree buds. Only two households, located on the outskirts of the hamlet, raise goats, and had been doing so for over 10 years. In 2005, the local Association of Elderly People (one of the mass organizations promoted by the Communist Party) also started to raise five goats. The hamlet leader stated that the local authorities did not promote goat raising because it is widely known that in many countries goats have been responsible for damage to forests.

Poultry

In 2006, there were 1,300 chickens, ducks, and Muscovy ducks in the hamlet. Table 6.5 shows the ownership of different kinds of poultry by households in Tat hamlet. Most households raised at least a few chickens but only a minority raised ducks or Muscovy ducks (Plate 6.3). Numbers of fowl varied but no household had more 40 chickens and 10 ducks.

The majority of the households that raise chickens bred their own chicks whereas all of the households that raise ducks buy small week-old ducklings for 4,000 VND/each and then sell them when adult for 25,000 VND/kg. Poultry are mostly used for home consumption, and only a few households sell their fowl.

Poultry raising has recently undergone significant changes. In particular, until the trend was brought to an abrupt halt by a disease epidemic in 2005, the number of 'industrial' and hybrid chickens (*tam hoàng*), had increased rapidly in recent years. In 2002, the two households with the largest number of chickens kept

Table 6.5 Household ownership of poultry in Tat hamlet (n = 30)

	Chickens	Ducks	Muscovy ducks
No. of households keeping poultry	29 (96.7)	6 (20.0)	4 (13.3)
Number of birds per household	5–40	5–10	2–4
Purchase chicks or ducklings to raise	7 (24.1)	6 (100.0)	4 (100.0)
Raise fowl for home consumption	27 (93.1)	5 (83.3)	4 (100.0)
Raise fowl for sale	2 (6.9)	1 (16.7)	0 (0.0)

Source: Tat hamlet household survey, March–April, 2006

Plate 6.3: Farmer feeding his flock of chickens and ducks (Nguyễn Thanh Lâm)

100 and 200 birds, respectively. In 2004, another farmer raised 400 industrial chickens. These chickens were kept in bamboo pens covered with a palm thatch roof that are placed in the home gardens to take advantage of shade and lower temperatures. Feed consisted mainly of milled maize and rice bran.

There was a serious and deadly epidemic among the poultry in the hamlet in 2005. Some farmers believed that this outbreak was triggered by infected chickens that had been brought into the hamlet from the market. The infected chickens had symptoms that included having white-colored manure, being listless, and drinking a lot of water; after a few days, they died. Some farmers used *Sulfachlopyridazine* for treatment, while others fed their chickens with rice mixed with pounded garlic in order to improve their health. In 2006, farmers were still unable to obtain a sufficient number of breeding chickens as a result of the 2005 epidemic.

Disease is such a serious problem because most chickens are allowed to range freely in gardens around the houses. To escape the epidemic risk, several households relocated their chickens to shelters in their swiddens where they were less likely to be exposed to diseases. Of the six households that relocated their chickens, only one reported that their chickens suffered from disease, whereas the flocks of all 23 households that raised chickens at their houses were affected by disease.

Diversity of Livestock Species Owned by Households

Most households own several different kinds of livestock (Table 6.6). Three-quarters of the households raised three or more species but no single household kept more than five species. Households keeping only one or two kinds of livestock were mostly young families. Only one household did not raise any livestock, although, in the past, this household had kept large numbers of livestock. At the time of the survey, however, this family's income was derived from swidden farming and government subsidy payments from a reforestation program. According to the head of the household, raising livestock was a high-risk venture: Small livestock had great vulnerability to serious epidemics, while large animals are often lost due to theft. Moreover, a reduction of the grazing area has occurred because large areas are now being used for forest plantations and swidden farming.

Ownership of diverse types of livestock helps to ensure that households have access to multiple sources of income. Farmers said that buffalo were useful primarily as draft animals for rice farming, cattle were considered a kind of high-interest savings account, while pigs and poultry were used as short-term income sources. Households showed a tendency to diversify their livestock ownership between 1998 and 2004 (Figure 6.1). This diversification is an alternative way of risk-management that addresses fluctuating market prices, disease outbreaks, and other natural disasters.

Contributions of Livestock to Household Livelihoods

Farmers raise livestock for many reasons. In addition to being an important, and steadily increasing source of cash income (Table 6.7), livestock are kept to provide food for household consumption, to serve as draft animals, to provide

Table 6.6 Number of different species of livestock kept by households in Tat hamlet

	NUMBER OF HOUSEHOLDS	PERCENTAGE (%)
No livestock	1	3.3
One species	1	3.3
Two species	5	16.7
Three species	10	33.3
Four species	7	23.3
Five species	6	20.0
Total	30	100.0

Source: Tat hamlet household survey, March–April, 2006

Figure 6.1 Changes in number of species of livestock owned by households in Tat hamlet between 1998 and 2004

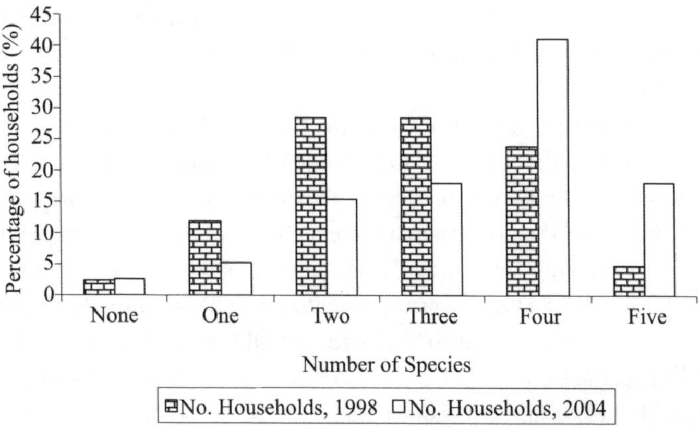

manure for farming, and to make use of otherwise unutilized resources. Pigs and poultry are occasionally eaten by the villagers, especially on festive occasions. A few households raise pigs to utilize the residues from alcohol making and a few households raised ducks every year, from April to July, both to earn cash income and to make use of the paddy fields as a grazing area during times when they are not cultivated with rice. Some farmers keep large livestock to provide manure, which they use as fertilizer in their paddy fields and also sometimes sell to earn cash. Occasionally, traders from Ha Tay Province came to the hamlet to buy manure. The price of 1 m^3 of manure was 50,000 VND.

Farmers recognize that livestock make important contributions to their livelihoods. When asked to rank different agroecosystem components, including livestock, they gave first ranking to paddy fields, followed in descending order of importance by livestock, forests, swiddens, and fishponds. According to

Table 6.7 Main sources of household cash income in 1998 and 2004 in Tat hamlet

SOURCE OF INCOME	MEAN HH CASH INCOME, 1998 (VND)	%	MEAN HH CASH INCOME, 2004 (VND)	%
Crops	753,786	23.6	625,231	6.2
Livestock	705,000	22.1	2,937,769	29.1
Forest	1,000,190	31.3	3,095,128	30.6
Others	734,381	23.0	3,446,498	34.1
Total	3,193,357	100.0	10,104,626	100.0

Source: Tat hamlet household survey, 1998–99 and 2004

the farmers, the paddy fields ensure food security; and reduce fluctuations in productivity while livestock, swiddens, and fishponds are important sources of both cash income and food for subsistence. Forests contribute cash, firewood, and construction materials. Other system components, such as off-farm activities, home gardens and government salaries are perceived as having lower importance.

Livestock are important not just for the direct contributions they make to household livelihoods but also because of the important role they play in maintaining the functioning of the overall agroecosystem through their multiple interactions with other system components. For example, buffalo provide draft power for harrowing the paddy fields, while buffalo, cattle and pigs provide manure for fertilizing the rice crop there. Pigs provide manure that increases the productivity of the fishponds. Of course, not all interactions are positive ones: Buffalo and cattle frequently damage crops in swidden fields and sometimes even in the paddy fields. Large livestock also damage vegetation in the forest.

Livestock also receive inputs from most other system components. Rice straw from the paddy fields is used to feed buffalo and cattle; after the rice harvest they are allowed to graze on the stubble in the paddy fields (Plate 6.4). Cassava roots grown in the swiddens, taro grown in fish ponds and gardens, and bamboo shoots collected from fallow swiddens and forest are used for feeding pigs. Fallow swidden fields and the forest are used for grazing buffalo and cattle. Homegardens are used by chickens and pigs to forage for food and fallow paddy fields provide grazing areas for ducks.

Constraints on Livestock Production

Livestock production in Tat hamlet is constrained by several factors:

Plate 6.4: Buffalo grazing on stubble in the paddy fields after the rice harvest (Nguyễn Thanh Lâm)

1. Local demand for the livestock products is very limited so that surplus production must be sold to outsiders. Thus, livestock production is done mainly for subsistence purposes, although cattle and pig raising are gradually becoming more commercial in nature. In 2005, some households tried to sell pork in the hamlet, but because local demand was low, they were forced to sell surplus meat in neighboring hamlets (See also Chapter 13) as well as to hawkers and small traders from the district town. Thus, product demand fluctuated and prices were unstable due to this high dependency on outsiders.
2. The productivity of native breeds of pigs, cattle, and buffalo is relatively low. A native pig typically gains 20-25 kg per year, while hybrid pigs (F1) can reach 60 to 75 kg in 6 months. Local buffalo and cattle are small in size and sell for only 3 to 5 million VND each, while lowland cattle sell for double that amount due to their larger size.
3. There is high risk in raising new breeds of livestock. Although farmers have tried to raise new livestock breeds in order to increase productivity, chickens and hybrid pigs were often afflicted with serious diseases. The reasons for this included: (i) local knowledge of raising new livestock breeds was still limited, although several training sessions on livestock raising had been carried out in the hamlet; (ii) veterinary services were still limited in the hamlet (the district veterinary officers traveled to the hamlet only twice per year for vaccinations); and (iii) upland conditions were different from lowland conditions, but many, possibly

inappropriate, technical methods from the lowlands were introduced in the uplands.

4. There are increasing conflicts between large livestock raising and swidden farming. This is because the main grazing areas are fallow swidden fields and forest patches, which are not clearly spatially separated from actively cultivated swiddens. Consequently, free ranging animals frequently damage swidden crops so that farmers have to expend considerable extra labor fencing their swiddens to avoid cattle and buffalo damage. In the past, cattle rarely damaged crops because the number of livestock was much smaller.
5. There is insufficient area of pasture in the hamlet. The grazing area for buffalo and cattle has been gradually reduced due to restrictions on swidden farming, which has resulted in a smaller fallow area, the main pasture for livestock in the past. The problem is made worse because many former swidden areas have been converted into forest plantations. This trend has led to a reduction in feeding capacity for large animals, while livestock numbers continued to increase.
6. There is a lack of appropriate slaughtering facilities. Because the hamlet lacks a real slaughterhouse, farmers have to sell their cattle and pigs to outside dealers who transport them to be slaughtered either in the market in Cao Son village (15 km away) or in the district town. Several attempts to establish a slaughterhouse in Tat hamlet failed due to low and irregular supply of animals to process.
7. Institutional support for livestock production is inadequate. Livestock production is exclusively managed by individual households, with each household free to decide the number and kinds of livestock to raise. Only a few households have joined interest groups to share their experience and to help each other in pig raising. There are no local extension or veterinary services to assist farmers with problems. The hamlet authorities have to contact the district extension and veterinary officers in order to receive technical and financial support. The one local veterinary officer in Tan Minh village has poor equipment and lacks appropriate medicines.

Conclusions

Livestock are very important for the farm households in Tat hamlet. Almost all households raise livestock and many have made very large investments in livestock production. Livestock are still mostly used for home consumption, although in recent years more livestock have been raised for the market.

Livestock production has considerably increased in terms of quality and quantity although most farmers still raise low productivity native breeds since new, more productive breeds of livestock (hybrid pigs and chickens) face many problems, including vulnerability to serious diseases.

Farmers highly value the role of livestock in their household economies. In addition, a number of factors exert positive influences on livestock development. These include government policies to promote livestock development, increased wealth of many households which allows them to invest more money in livestock, and a high demand for livestock from the lowland market.

However, livestock development also faces a number of serious constraints, including: (i) the limited local demand for livestock products and the consequent reliance on selling livestock to outsiders; (ii) low productivity of native breeds and risks surrounding the introduction of new hybrid breeds; (iii) conflicts between large livestock and swidden farming; (iv) shortages of pasture for buffalo and cattle; (v) lack of slaughtering facilities; and (vi) lack of institutional support for livestock production.

A number of strategies are employed by farmers to overcome constraints on livestock raising in the composite swiddening agroecosystem. For example, livestock in the hamlet are fed mostly locally produced feed, which reduces dependence on feed purchased from outside sources. The integration of new, improved breeds together with native breeds of animals is an effective strategy to balance the disease vulnerability of the new breeds and the low productivity of the native breeds, as well as to efficiently use locally available feed. Livestock diversification, with individual households raising several different species, may also reduce market output constraints, feed demand, and ease labor pressure. Also, the free ranging grazing of large livestock in the forest and fallow swiddens helps avoid the spread of diseases among livestock, which often occurs among animals kept at houses in the settlement area of the hamlet, although it can also result in damage to crops and regenerating forest vegetation.

7
Fishponds

Nguyễn Thanh Lâm and Trần Đức Viên

Introduction

In this chapter, we describe the fishponds as an important component of the composite swidden agroecosystem in Tat hamlet. We first discuss the history of development of ponds in the hamlet, and then describe their physical characteristics, the species of fish that are raised in them, their stocking density, and the techniques employed by the farmers for managing fish production. Contributions of fishponds to household livelihoods and the perceptions of the farmers regarding the advantages, difficulties, and potentials for development of fishponds in the hamlet are set forth.

Methodology

This chapter is based on information collected using standardized questionnaires from a random sample of respondents from 30 households in the hamlet in March and April of 2006. In addition, semi-structured interviews were conducted with several key informants and group discussions with farmers were held to understand their perceptions regarding opportunities for and constraints on fish production in the hamlet. We have also made use of our direct observations of fishponds made in 1998, 2003, 2004, and 2006.

Fishponds and their Management

In this section we describe the history of development of fishponds in the hamlet, their physical characteristics, their species composition, the sources of fingerlings used to stock them, and examine the techniques employed by farmers to manage their ponds.

The historical development of fishponds in Tat hamlet

Fishponds are a relatively new component of the composite swidden agroecosystem in Tat hamlet. According to older informants, the villagers formerly caught wild fish from the Tat River and there were no ponds in the hamlet. Natural fishing began to decline after the road connecting the hamlet to the district town was constructed in 1969. Ten years later, in 1979 during the cooperative period, the first fishpond was constructed in the middle of the hamlet. After a few years, many new ponds were constructed so that, by 2006, almost three-quarters of households owned one or more ponds (Table 7.1). Other households wanted to have ponds, but were unable to build them because they lacked suitable sites or sufficient space in their house yards or did not have enough labor capacity to do the construction. Consequently, the number of ponds has remained essentially unchanged in recent years.

Physical characteristics of the fishponds

The fishponds are generally small with one-third of the households having very small ponds, with a surface area of less than 100 m^2 and one-third having ponds between 100-600 m^2. Only 10 percent of houses have relatively large ponds (>600 m^2), which can be used for commercial production of fish. The ponds are filled with water that flows through bamboo or plastic pipes from streams or springs on the hill slopes behind the house sites where the ponds are located.

Most fishponds are shallow, with a depth of less than one meter, although some ponds may be as much as two meters deep. Consequently, the ponds have limited water storage capacity, little ability to protect the fish during spells of cold weather, can only sustain low stocking densities, and have low productivity. Attempts to increase the depth of these fishponds are constrained by the shallow bedrock and the threat of cave-ins of the banks due to sandy soil.

All ponds have open-system characteristics, with water continuously flowing in and out. This makes them suitable for raising fish species that primarily eat grass and cassava leaves (e.g., grass carp) but not so favorable for raising species (e.g., tilapia, common carp, and Mrigala carp) that rely on eating food (e.g., phytoplankton, zooplankton, snails and shellfish) that is produced as part

Table 7.1 Ownership of fishponds by households (n = 30)

AREA OF POND (M^2)	NO. OF HOUSEHOLDS	%
No pond	8	26.7
1–100	9	30.0
101–200	4	13.3
201–600	6	20.0
> 600	3	10.0

Source: Tat hamlet household survey, March–April, 2006

of the food chain inside the ponds. This is because the continuous outflow of water carries nutrients out of the ponds, which leads to decreased phytoplankton growth, thus diminishing the supply of natural fish food. Hence, although the density of grass carp in the small fishponds with open-ecosystem characteristics can be quite high if they are supplied with enough grass and cassava leaves, the density of ordinary carp, Mrigala carp, and tilapia is kept at a low level since most naturally-occurring food is lost in runoff water. Increasing the stocking density of these species would require regular supplementary feeding with rice bran and other nutrient-rich feeds.

Fish species

Households raise different mixes of fish species in their ponds. All of the households raise grass carp (*Centopharyngodon idellus*), 87 percent raise Mrigala carp (*Cirrhinus mrigala)*, 50 percent common carp (*Cyprinus carpio*), and 32 percent tilapia (*Oreochromis mosambicus*, *Oreochromis niloticus*). Only one household raises catfish. Many ponds are stocked with several different species, with half of the households raising three species: Near-surface dwelling grass carp, medium-depth dwelling Mrigala carp, and bottom-dwelling common carp.

In an effort to improve fish productivity in ponds in Tat hamlet, a new species of tilapia (*Oreochromis niloticus*), was introduced by CARES researchers in 2000 and soon mostly replaced the local tilapia (*Oreochromis mosambicus*). The local tilapia had several disadvantages, including small size, long length of time needed to reach maturity, and low resistance to cold during the winter season. Conversely, the new variety of tilapia had a larger size, was quick to reach maturity, and could survive cold weather conditions. Some households still have the old variety of tilapia, however, because they failed to clean and lime their ponds before stocking the new species.

Sources of fingerlings to stock the ponds

Farmers buy fingerlings from a variety of sources (Phạm Thị Mai Hương *et al.*, 2005). One household in the hamlet produces fingerlings, which they sell to neighbors, as well as to people from neighboring hamlets. Farmers can also sometimes buy fingerlings from itinerant hawkers, who bring them to the hamlet from the lowlands, and a few households buy fingerlings from friends in other villages.

Stocking density

Table 7.2 shows fish density in the ponds. The majority of households had ponds with very low stocking density of less than two fish per cubic meter of water while less than one fifth of households had high-density fishponds with more

Table 7.2 Stocking density of fishponds in Tat hamlet (n = 22)

DENSITY (NO. OF FISH/M³)	NO. OF HOUSEHOLDS	%
0.1–2.0	12	54.5
2.1–3.0	3	13.6
3.1–4.0	3	13.6
> 4.0	4	18.3
Total	22	100.0

Source: Tat hamlet household survey, March–April, 2006

than four fish per m^3. These households ignored the problems associated with excessive stocking density since their goal was to raise as many individual fish as possible rather than to maximize their growth rate.

Fish raising techniques

Ponds are usually stocked with young fish between March and April, and harvested from November to the Tết lunar New Year in January or February. Some farmers begin to harvest fish only three or four months after stocking. The fish are small and their premature removal results in low productivity at the final harvest.

Methods of raising fish are simple, with most farmers relying on locally available food resources obtained from diverse components of the agroecosystem to feed their fish. The main sources of food, in addition to phytoplankton growing in the ponds, are rice bran from the paddy fields, pig manure from pens near the ponds, taro leaves and banana stems and leaves from the homegardens, cassava leaves from the swiddens, and grass and tree leaves collected from fallowed swiddens and the forest (Plate 7.1). On many occasions while conducting research in the swiddens, we observed women carrying grass and cassava leaves in their shoulder baskets back to their houses to feed their fish. They reported that each time they went to the swiddens, they collected these plants to use for fish food. Several households use the VAC (Garden-Pond-Livestock) system in order to increase production of fish as well as livestock and vegetables. Their ponds are constructed close to the houses so that domestic and kitchen waste drain into the fishpond. The livestock pens and gardens are also situated near the ponds. Taro is planted on the border of the ponds. Nutrient rich eroded materials also flow into the ponds from surrounding house plots and homegardens.

Grass and cassava leaves are the most commonly used feeds but recently pig manure and rice bran have been gaining in popularity to some extent. Rice bran is quite cheap in Tat hamlet (2,500 VND/kg), but only a few households use

Plate 7.1: Farmer feeding his fish with taro leaves (Nguyễn Thanh Lâm)

it for fish food. The aquatic extension officer estimated that 4 kg of rice bran could increase the weight of a carp by 1 kg. Some households feed residues from making alcohol to their fish.

Disease outbreaks are uncommon although there was a serious disease outbreak among grass carp in 2003 that farmers did not know how to control. Some tried to rescue their fish by adding clean water to the ponds to flush out the dirty and sediment laden water and to increase the volume of water in the pond while others tried to move the fish to other ponds. These rescue attempts were rarely successful, however, and many fish died. Most of the affected farmers simply abandoned raising fish for several seasons. Some farmers attributed the cause of the disease (or at least the spread of it) to the recent adoption of high-density fish raising practices by some households in the hamlet.

Contributions of fishponds to household livelihoods

Most of the households owning fishponds (91 percent) raised fish exclusively for home consumption while nine percent raised fish both for home consumption and to sell. Households with ponds catch fish once or twice a week for their own consumption (Plate 7.2). No households engaged in purely commercial production. Only the few households that had large fishponds raised fish for dual purposes. Incomes from selling fish ranged from 100,000 to 4,000,000 VND/yr, with an average of 1.6 million VND per household. Carp sell for the very high price of 25,000 VND/kg in Tat hamlet so villagers rarely ate carp because it was too expensive for them to afford. The owners of large ponds reported that

Plate 7.2: Couple catching fish in their fishpond

local demand for high-quality fish only occurred during traditional holidays or ancestor days because villagers did not have enough money to buy the fish for everyday use.

Farmer Perceptions about Fishponds

In focus group discussions, we asked farmers to identify some of the advantages of raising fish, the current constraints on fish production in Tat hamlet, and the perceived potential for developing fish production in the future.

Advantages of fishponds

When first asked, the farmers ranked cattle as having a higher value than fish. After long discussions and detailed calculations, however, they recognized that this was not necessarily the case. For instance, households, who had large ponds and had fish available for daily consumption, could have much higher total income than households that had only one or two cattle. Finally, all farmers agreed the main advantages of raising fish were as follows:

1. Fishponds provide for daily food needs of their owners.
2. Fish can be fed with cheap locally available feed resources from the forest, swidden, paddy fields, and livestock.

3. Households can receive assistance from the government's fishery development program (training and loans of 1,000,000 VND per household/year).

Constraints on fish production

Some serious constraints on fish raising in Tat hamlet were identified by the farmers:

1. The shallow depth (one meter or less) of the ponds made it difficult to protect fish from the cold during the winter.
2. The ponds had insufficient nutrients because of the continuous throughflow of water. Without vital nutrients, the species that relied on phytoplankton for their food (such as common carp, tilapia, and Mrigala carp) took much longer to grow; some farmers even reported that their fish were still at fingerling size two years after stocking.
3. Some ponds were stocked with a fish density exceeding 4 fish/m^2. Overstocking was particularly a problem with ordinary carp due to their high market value. As a result, the carp could not grow to full size due to insufficient food and space. The farmers were not much concerned about maintaining an appropriate fish density to obtain high productivity, however. Most farmers know about the density problem, but since they usually eat fish before they have fully matured, they do not perceive it as serious. However, a few farmers try to minimize competition for space and food by stocking their ponds with different species of fish that occupy different water depths and others try to build up the level of the rim of the pond in order to increase the water volume in the pond.
4. The feed had poor nutritional quality. Most fish were fed with cassava leaves and grass. This was sufficient for the needs of grass carp but ordinary carp needed supplementary feed, such as rice bran, which was rarely used.
5. The hamlet offered only a limited market for selling fish, and producers were unsure of additional places to sell, potential customers, etc. Moreover, market information was not up-to-date so that if someone had fish for sale in the upper part of the hamlet, potential customers living at the other end of the hamlet did not know.
6. Many ponds were used continuously for several years without drying and cleaning so that toxic elements (H_2S, Al^{3+}, Fe^{2+}) built up in the mud. Many households did not yet use lime for cleaning ponds once per year. As a result, aquatic primary producers (*phytoplankton, phytobenthos*, *azolla* and taro) were slow to grow. This meant that the additional amount of feed contributed by the farmer was insufficient to supplement the lack of these

primary producers, and fish matured slowly with greater susceptibility to disease.

7. Water was sometimes lacking during the dry season, while heavy rains in the summer caused damage to the banks of the ponds and the loss of fish in overflow water.

Potential for developing fish production in the future

The farmers agreed that fishponds had considerable potential for development, because:

1. The amount of capital required for fish investment was not high. For households that already possessed a pond, only a small amount of cash was required to buy fingerlings and locally produced feed.
2. Raising fish had more benefits than pig raising and rice production.
3. Fish suffered less frequent problems with disease than livestock, with only grass carp affected by a serious disease outbreak in 2003.

Fish production was done mainly for subsistence purposes due to the small size of most ponds. Fish were caught once or twice per week, which allowed owners to save cash normally spent on buying food. Even though the new variety of tilapia was raised at a low density, it regenerated quickly in the ponds and met the needs of the farmers for food. Fish could also be a significant source of cash income for households that had large ponds.

Conclusions

Although adopted by farmers in Tat hamlet only relatively recently, fishponds soon became an integral component of the composite swidden agroecosystem and are now highly valued by the villagers. Fish provide high quality protein for home consumption, can be sold for cash, and make use of plentiful locally available feed sources. Techniques for managing the fishponds remain rudimentary, however, and only a few farmers obtain high productivity from their ponds. Considerable potential remains for further intensification of fish production in Tat hamlet.

8
Homegardens and Tree Gardens

Deanna Donovan, Đào Trọng Hưng, and Nguyễn Thanh Lâm

Introduction

Homegardens and tree gardens cover relatively small areas in the landscape of Tat hamlet but make important contributions to the subsistence needs of farmer households and, increasingly, also produce products for sale to earn cash. They also have important ecological functions, serving as seed banks for valuable species, and preserving vegetation cover in the landscape.

This chapter is based on research conducted by Deanna Donovan and Đào Trọng Hưng between 1995 and 1997 and by Đào Trọng Hưng in 1998-99, supplemented by additional information collected by Nguyễn Thanh Lâm in 2004. Fifteen homegardens and five tree gardens were observed in 1998 and 30 gardens were surveyed in 2004.

Homegardens

Many households in Tat hamlet have homegardens with well-defined boundaries, often next to the house itself, and generally carefully fenced to protect them from free-ranging livestock (Plate 8.1). Homegardens are generally well-developed complex agroforestry systems with a large number of fruit, vegetable, herb, spice, and ornamental species (Figure 8.1).

Some forest trees are commonly found in the homegardens including *Melia azedarach, Chukrasia tabularis,* and *Canarium album;* of which *Melia azedarach* is the most popular. However, there are only a few forest trees found in each garden and they are often grown as the fence of the gardens. Forest trees in the gardens supply wood, firewood and occasionally some cash income for the households.

Fruit trees are generally grown in the vicinity of the house on flatter land but may also be found on surrounding slopes. Common species included jackfruit (*Artocarpus heterophyllus*), guava (*Psidium guajava*), papaya (*Carica papaya*), banana (*Musa sp.*), lime (*Citrus aurantifolia*), orange (*Citrus sinensis*), pomelo

Plate 8.1: Homegarden protected from livestock by a split bamboo fence (Nguyễn Thanh Lâm)

Figure 8.1 Map and profile of a typical homegarden in Tat hamlet in 1999

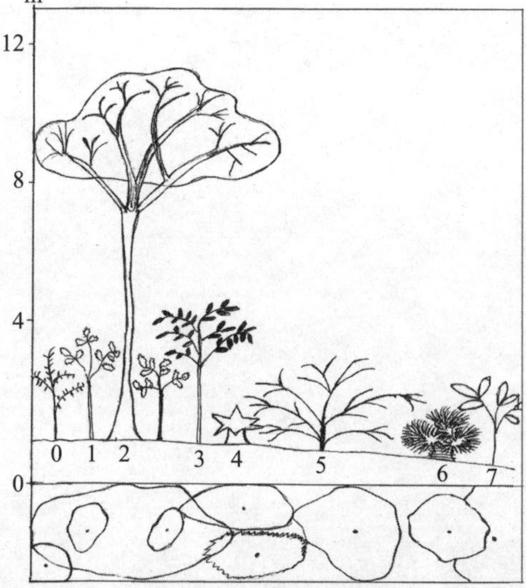

Code: 0=*Manihot esculenta;* 1=*Camelia sinensis*; 2=*Melia azedarach* (12m); 3=*Camelia sinensis*; 4=*Annanas comosus*; 5=*prunus* sp.; 6=*Livistona cochinchinensis*; 7=*Canna orientalis.*

(*Citrus grandis*), lychee (*Litchi chinensis*), jujube (*Zizyphus mauritiana*), plum (*Prunus salicina*) and peach (*Prunus persica*).

Herbs and spice plants used in cooking or for medicinal purposes are often found growing under the shade of the fruit trees. Common species include ginger (*Zingiber zerumbel*), turmeric (*Curcuma sp.*), cardamom (*Amomum sp.*), and galangas (*Kaemferia galanga*).

Some annual vegetables are grown in more open and shade-free parts of the homegardens, while many households also maintain separate vegetable plots adjacent to their houses (Plate 8.2). Popular vegetable species cultivated include chili (*Capsicum annuum*), garlic (*Allium sativum*), pumpkin (*Cucurbita pepo*), wax gourd (*Benincasa hispida*), eggplant (*Solanum album and Solanum melongena*), hyacinth bean (*Dolichos lablab*), mung bean (*Vigna aurea*), taro (*Colocasia* sp.), sesame (*Sesamum indicum*), buckwheat (*Fagopyrum cymosum*), spring onions (*Allium fistulosum*), and bamboo (*Sinarundinaria sp.* and *Neohouzanea dullosa*).

Some 77 percent of species found in homegardens are food plants (vegetables, fruits, herbs and spices); 29 percent are medicinal plants, 27 percent are used as stimulants (e.g., tea, betel, tobacco) and 13 percent are ornamentals. Ranked by frequency the most common species encountered are tea (*Camelia sinensis*), followed by banana (*Musa* sp.) and plum (*Prunus salicina*).

Plate 8.2: Woman watering household vegetable plot (Nông Hữu Dương)

Although the homegardens provide many useful products for home consumption, most are not intensively managed. They often are overgrown with weeds and have relatively low productivity; possibly because the heavy labor demands of raising their primary field crops leaves the farmers with insufficient time and energy to manage their gardens more intensively. The area available for homegardens in the settlement area is also very limited so that expansion of this system component is very constrained by the supply of suitable land.

Tree gardens

Tree gardens are in one sense an extension of the homegardens. Most tree gardens were established many years ago and have been inherited from generation to generation. After the large forest trees were cut down to build houses, some smaller woody species such as melia and candlenut (*Aleurites montana*) were planted to replace them. The villagers also grow tea, palms for roof thatching, and bamboo for house building and making household goods. The bamboo shoots are also used for consumption and sale. In general, if tree gardens are protected from the free ranging cattle and buffalo, they grow well with little management. If species such as *Ormosia balansae*, *Quercus*, *Garcinia planconii*, *Psychotria rubra*, wild bananas, and ferns grow well, then the forest can regenerate more quickly.

The tree gardens generally have a multistory vegetation community. The lower story may include volunteer seedlings or wild seedlings transplanted from the forest by the farmers. *Livistona* palms are common as are forest tree species such as *Desmos cinensis*, *Viburnum lutescens*, *Psychotria rubra*, *Ixora coccinea*, *Rhodomyrtus tomentosa*, *Maesa balansae*, *Melastoma malabathrica*, *Litsea cubeba*, and *Tetracera sarmetosa*. *Macaranga denticurata* and *Mallotus apelta*, *Phylantus emblica*, *Saccharum spontaneaum*, *Evodia lepta*, *Rhus javanicus*, *Aralia armata*, and various ferns and vines can also be found in the understory (Figure 8.2).

Tea gardens, bamboo gardens, and palm gardens are specialized types of monocultural tree gardens. Some households plant bamboo gardens for the production of bamboo shoots and stems. Farmers favor *Sinarundinaria sp.* and *Neohouzanea dullosa*. *Neohouzanea dullosa* is also often grown interspersed with *Livistona* palms in old well-established palm gardens. In the 1990s, the 327 and PAM programs provided eucalyptus, tea, and a few fruit trees for planting in the tree gardens. After 2000, the 661 Program (the successor reforestation program to the 327 Program) introduced a number of tree species into the hamlet including *Chukrasia tabularis, Dendrocalamus membranaceus, Canarium*

Figure 8.2 Map and profile of a typical tree garden in Tat hamlet in 1999

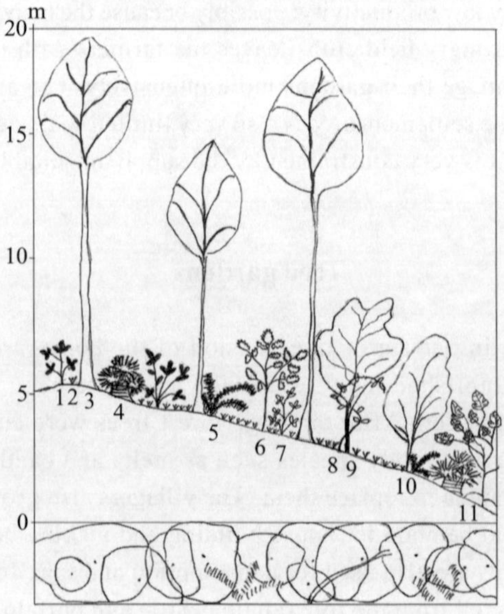

Code: 1=*Melastoma malabathrica*; 2=*Oplismenus compositus*; 3=*Eucalyptus camadulensis*; 4=*Livistona cochinchinensis*; 5=Fern tree; 6=*Viburnum lutescen*; 7=*Eupatorium odoratum*; 8=*Cissampelos pareira*; 9=*Antidesma ghaesembila*; 10=*Apozo mierocalix*; 11=*Rubus alceaefolius*

album, Manglietia insignis and *Dracuntomelon duperreanum*. These trees were often planted mixed together but *Dendrocalamus membranaceus* was considered as the most important species. It is a type of bamboo, which grows quickly, especially on moist soil (near watercourses), and generates income from bamboo shoots and stalks after about 4 years. Moreover, it is a perennial that can be harvested many times. Therefore, more and more households have planted this species.

In the same period, a local extension project encouraged farmers to plant several species of fruit trees, including mango (*Mangifera indica*), persimmon (*Diospyros kaki*), and oranges (*Citrus sinensis*) in their tree gardens. However, attempts to convert subsistence-oriented mixed tree gardens into commercial fruit orchards have proven largely unsuccessful, largely because of the small and unstable market for their produce. Most plum trees were cut down in 2000-2002 after the price for the fruit fell to 200 VND/kg. New improved litchi (*Litchi chinensis*), although highly productive, are difficult to sell because of the small size of the market.

Tree gardens offer a good example of household woodlot management. Trees are grown for a variety of products both for home consumption and cash sales. Leaves of the palm *Livistona cochinchinensis* are used for roofing material, various bamboo species provide edible shoots as well as raw material for construction, basketry, and for sale, and herbaceous plants often have medicinal uses. The most popular tree species planted are *Aleurites montana*, grown for the sale of its oil-producing nuts, *Melia azedarach* for pole timber, *Dendrocalamus mebranaceus* for bamboo shoots and building materials, and *Eucalyptus camaldulensis* for pulpwood and timber.

Only older, more established households own tree gardens. All households want to obtain "red book" certificates to ensure their tenure and their right to use the land for a long period of time. They also want the red book certificates so that they can obtain loans to plant more trees. At present, all tree gardens have been allocated to households but they have not yet received their certificates and there is also no map showing the locations and areas of tree gardens. In fact, some owners do not even know the exact size of their tree gardens.

Biodiversity in homegardens and tree gardens

Plant species diversity in homegardens and tree gardens is quite high. Species in the homegardens are mostly domesticated ones but, because many of the tree gardens are next to fallow lands and natural forests, many volunteer wild forest species become established there. This considerably increases the species diversity of the tree gardens. Appended Table 8.A1 presents a list of the domesticated and semi-domesticated species found in homegardens and tree gardens in Tat hamlet. In the survey made in 1998-99 of fifteen gardens, 82 species were identified; while in the 2004 survey of 30 gardens, 65 species were observed. In 2004, 19 species not recorded in 1998-99 were observed but 37 species listed in 1998-99 were not found in 2004. This does not necessarily mean, of course, that these species had become extinct in the five-year period between the surveys. Sampling error is the more likely explanation, given the rarity of many of these species.

Conclusions

Although visitors from the lowlands often comment that the homegardens in Tat hamlet are small, undeveloped, and poorly managed, these small plots, along with the tree gardens, do provide many useful products needed by the villagers while requiring only minimal investments of labor and capital. The products from the homegardens and the tree gardens help reduce the impact of villagers on natural forest resources in the hamlet. They also provide habitats for a quite diverse assemblage of species, some fully domesticated and common, but others only semi-domesticated or wild, and sometimes now relatively rare in the natural forest. Moreover, the homegardens and trees gardens contribute to maintaining tree cover in the hamlet landscape and thus play an important role in managing natural resources more sustainably.

Home gardens and tree gardens are examples of perennial systems that minimize soil disturbance, erosion and runoff hazards through the use of perennial vegetation. Tree gardens are often used as buffer strips on sloping land adjacent to watercourses to ameliorate the affects of soil disturbance of swidden cultivation of annual or biennial crops on neighboring land. Rotational harvesting of crops in this system also serves to minimize the environmental impacts.

These gardens can be thought of as being manmade or reconstructed forests that yield many of the same useful products found in natural forests. They also have a structure that to some extent mimics that of the natural forest. From the standpoint of hydrology as well, they may function much like natural forest, although this in part depends on soil conditions and the level of human activity therein. Biologically, however, they display lower species diversity than the natural forest, although the tree gardens do supply many useful forest products to the households.

9
Secondary Forests

Deanna Donovan, Đào Trọng Hưng, Trịnh Thị Tiến, Trần Chí Trung, and Stephen J. Leisz

Introduction

In this chapter we describe the ecological characteristics of secondary forests in Tat hamlet and analyze their environmental functions within the composite swidden agricultural system. We also examine the contributions that wild products collected from the forest make to the livelihoods of the villagers.

Methodology

Three separate surveys of the secondary forest community were carried-out between 1995 and 1999. An initial botanical survey was made along transects across the hamlet in order to identify and distinguish different vegetation types and to record their general architectural structure, species composition and location on the landscape.

A second survey collected information on the categories of vegetation that were distinguishable in the remotely sensed images (see Chapter 17). We sought to detail the distinctive characteristics of the different vegetation types within the hamlet and to assess soil and site conditions. The total area of the hamlet was divided into quadrats and sample plots were identified in each of the six categories of distinct vegetation types. A team of six people sampled in 36 randomly situated square plots within the hamlet. Plot size was determined by vegetation type: tree and shrub vegetation were sampled in 20 m x 20 m plots whereas the more uniform grass communities were sampled in 5 m x 5 m plots. At each site an examination of ground cover, litter, and soil composition was also done.

A third survey was carried-out in 1998-99 to examine the main vegetation communities found in this area. We established three ecological transects to survey the dominant ecological communities and to establish a permanent baseline for monitoring changes in vegetation cover over time in the areas most heavily influenced by human intervention. These transects were established in an approximately north-south direction perpendicular to the main road, which passes through the hamlet in an east-west direction. Along each transect, sample plots were laid-out parallel to the transect line in order to gather information about species composition, vegetation structure and canopy cover. Sample plot sizes differed according to community type. In communities with predominantly large trees the plot size was 10 m x 20 m. In plots covered by shrubs and grasses, a 5 m x 10 m plot was surveyed. In predominantly grass communities, plot size was 4 m x 5 m. For each transect, we recorded the date of survey, longitude, latitude, elevation and length.

In 2002, information on the collection and utilization of wild products from secondary forest was gathered by means of a questionnaire survey of a stratified random sample of 60 households. The households were classified into three categories of better-off, medium, and worse-off households. Standardized questionnaires were used to collect information on demographic characteristics, farming activities, participation in collection and utilization of forest products, and the sources of income (Trịnh Thị Tiện 2003).

The Ecological Setting of the Secondary Forests

With a tropical monsoon climate that is very favorable to plant growth, the natural vegetation cover of Tat hamlet is understandably lush. The number of woody ferns (e.g., *Cyathea polophylla*) in the remaining high forest provides evidence that this area was formerly covered by dense tropical seasonal broadleaf evergreen forest. The current vegetation structure and composition, however, reflect more than a century of human intervention, perhaps longer. The pattern of vegetation on the landscape reflects the impacts of a broad spectrum of human activities, including logging, swidden cultivation, non-timber forest products exploitation, and livestock grazing. Even the few remaining patches of tall, dense evergreen forest, surviving mainly in steep, rocky areas unsuitable for cultivation, show the effects of selective logging. Relatively young regenerating forest and shrub and grass communities in various successional stages cover most of the landscape. Ecological transects show that the fragmentation of forest into different successional types reflects variation in topography, soil, microclimate and drainage factors, as well as prior land use. The landscape reflects not only the history of economic activities, but also changes in the social and political situation in the community. Most significant among these are recent institutional changes, especially the allocation of leasehold rights to agricultural land and forestland (Sikor and Đào Minh Trường 2002). New conditions of land tenure and market development may be particularly influential in altering patterns of land use and resource exploitation.

Types of Secondary Forest

There are two main types of secondary forest in Tat hamlet: The secondary evergreen broad leaf forest of medium to tall trees community and the bamboo mixed with woody trees and shrubs community. These types, which reflect human intervention and often the effects of repeated exploitation of the vegetation or soil resources, make different contributions to the livelihoods of the villagers.

Secondary evergreen broad leaf forest of medium to tall trees community

The evergreen broadleaf forests found in this area are almost entirely secondary forests that have been extensively modified by the influence of people over several generations. These plant communities are found at elevations of between 300 to 800 m. Examples with the oldest specimens exist in steep, rocky terrain where access is difficult (Plate 9.1). Such areas, although often subject to selective logging, have not been cultivated; consequently, the soil contains a relatively high percentage of humus. High humidity and ample rainfall create conditions favorable to plant growth and the diversity of plant species is high. Appended Table 9.A1, which was compiled by Lê Trần Chấn, is a checklist of plant species found in the closed forest.

These forests often have complex multi-storey structures. Figures 9.1 and 9.2 show architectural profiles and canopy cover maps of two sample plots.

Plate 9.1: Mature secondary forest on hill top (Nông Hữu Dương)

Figure 9.1 Map and profile of a secondary forest community. Slope: 35°

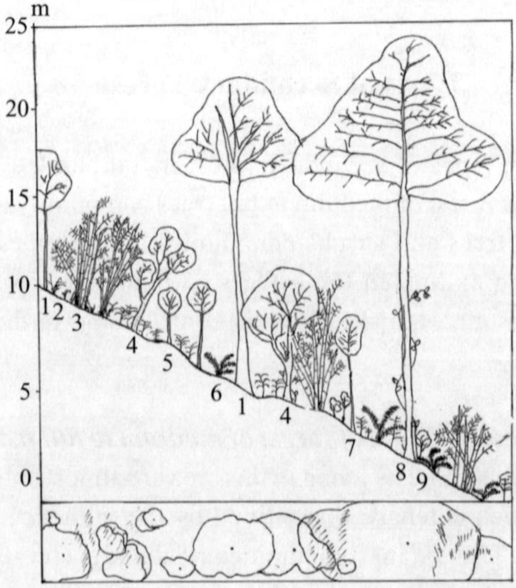

1. *Ormosia balansae*; 2. *Carex*; 3. *Neohouzeaua dullooa*; 4. *Castatanopsis*; 5. *Lithocarpus balansae*; 6. Fern trees; 7. *Heteropanax*; 8. *Garcinia*; 9. *Pueraria thomsonii.*

Four stories can often be delineated: two of woody trees, one of shrubs and one of grasses and ferns:

- Storey I: Large trees 25-30 m tall and 30-35 cm in diameter. Species include: *Aphanamixis polystachya* (*Meliaceae*), *Acanthopanax trifoliatus* (*Meliaceae*), *Heteropanax fragrans* (*Araliaceae*), *Garcinia planchonii* (*Clusiaceae*), *Nephelium cuspidatus* (*Sapindaceae*), *Canarium album*, *C. tramdenum* (*Burceraceae*), *Dipterocarpus retusus* (*Dipterocarpaceae*), *Elaeocarpus bamansae* (*Elaeocarpaceae*).
- Storey II: Medium trees 8-25 m tall and 25-30 cm in diameter. Species include *Pithecellobium clypearia*, *Rachidendron chevalieri*, *Ormosia balansae* (*Fabaceae*), *Garcinia cowa* (*Cluciaceae*) *Engelhartia spicata* (*Juglandaceae*), *Aglaia gigantean*, *Cipadessa baccifera* (*Meliaceae*).
- Storey III: Shrubs 2-8 m tall. Species composition is complex but includes: *Ostodes paniculata*, *Antidesma fordii*, *Breynia angustifolia*, *Bridelia hermandii*, *Antidesma ghaesembilla* (*Euphorbiaceae*), *Ficus variegata* (*Moraceae*), *Ervatamia bonii* (*Apocynaceae*), *Heynia trijuga* (*Meliaceae*), *Viburnum lutescens* (*Caprifoliaceae*), *Ixora coccinea* (*Rubiaceae*), and *Sterculia hymenocalyx* (*Sterculiaceae*).

Figure 9.2 Map and profile of the high secondary forest. Slope: >40°

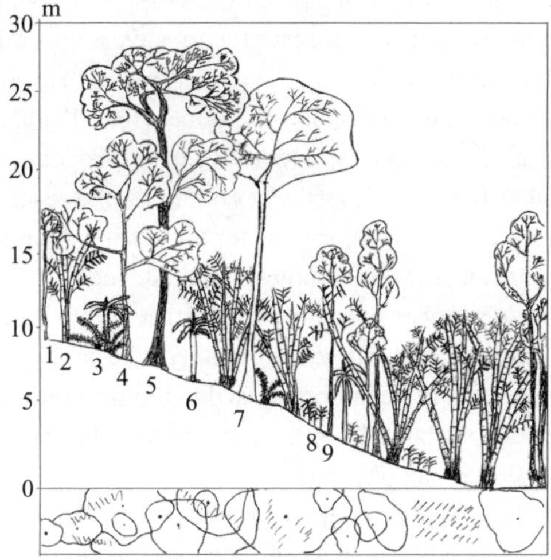

1. *Ormosia balansae*; 2. *Neohouzeaua dullooa*; 3. Fern; 4. *Lithocarpus balansea*; 5. *Castanopsis sp.* (25m); 6. *Musa sp.*; 7. *Aleurites montana* (20m); 8. *Carex sp.*; 9. *Mangalietia sp.*

- Storey IV: Grasses and ferns. Species composition is rich and includes: *Selaginella monosposa* (*Selaginellaceae*), *Cyclosorus baviensis* (*Thelypteridaceae*), *Pteris vitta* (*Pteridaceae*), *Lygodium flexuosum* (*Lygodiaceae*), *Musa coccinea* (*Musaceae*), *Setaria palmifolia*, *Microstegium vagans*, *Themeda caudate*, *Pogonatherun crinitum*, *Paspalum scrobiculatum*, *Micanthus japonicus* and *Thysanolaena maxima* (*Poaceae*).

Few examples of high forest remain in the hamlet. The majority of high forest areas that do persist are in steep inaccessible areas. Remaining large trees include *Garcinia sp.*, *Ormosia*, *Castanopsis* and *Lithocarpus*. The understory contains different species of bamboo as well as shrubs such as *Pyschotria sp.*, *Antidesma sp.*, and *Ixora cochinensis*, vines of the families *Wendlandia, Convolvulaceae, Smilaceae* and *Dioscoreaceae*, and the climbing fern *Lygodium sp.*

Land covered with high forests has not been allocated to individual households, but is instead considered a communal asset for the hamlet to use and manage. The old growth forest at Pu Co Hiên and the bamboo forest along Yên Stream, are examples of such areas that have been retained under communal control and that can be exploited by any resident of the hamlet. Villagers obtain timber, fuelwood, and mushrooms there. Species valued for their timber include *Lithocarpus*, *Castanopsis*, *Ormosia balansae*, *Syzygium* and *Litsea citrata*.

Logging, mostly illicit, still occurs in old-growth secondary forest. According to forestry staff and the villagers, forest officers gave no permission to exploit timber in the Tat hamlet area between 1996 and 1998. District and village forest officers visited the hamlet to confiscate any timber from illegal exploitation and fine the miscreants. Consequently, according to informants, villagers rarely risked cutting timber for sale. Nevertheless, in 1998, the researchers observed skid trails on the upper slopes and frames for pit-sawing deep in the forest that indicate that logging had not been completely halted. Today people in Tat hamlet have permission to log timber only for the construction of their new homes or the repair of old ones. Reportedly villagers only take faulty trees for making walls and floors. The quantity of wood needed to build a new three-room single floor traditional house is about 4-5 m^3, which, at the time of our surveys, cost VND 100,000-120,000 in tax payment.

Regardless of the efficacy of government enforcement, logging in Tat Hamlet is now considerably restricted because the supply of large timber trees is essentially exhausted. The only timber remaining is either of poor quality or in difficult to reach terrain. In addition, much of the remaining degraded forest has been allocated to individual households for protection and management. At the same time forest protection by the state and local authorities has increased.

Edible mushrooms are found mostly in the old growth forests. Women in the hamlet could name at least eight different types of mushrooms, all but one of which reportedly grows on rotten damp timber. Mainly women and children collect them in the rainy season, May through September. As the area of forest has decreased so has the availability of the mushrooms so people spend less time collecting them. Now they are collected mainly when encountered by chance.

In 1997-98, villagers collected *Cinnamomum parthenoxylon* for distillation of oil from its wood and roots. Known as Safrol, this oil is used in the production of cosmetics and beverages. In 1998 traders were paying VND 350/kg of unproccessed plant materials. The extraction of the roots of these trees involved extensive destruction of the surrounding forest area. The extraction of Safrol has recently been prohibited.

Bamboo mixed with woody trees and shrubs community

This early successional community is characterized by two bamboo species that typically follow swidden cultivation (Figure 9.3). Appended Table 9.A2, which was compiled by Lê Trần Chấn, presents a checklist of plant species

Figure 9.3 Map and profile of a shrub and grass community in a fallowed swidden field. Slope:35°

1. *Imperata cylindrica*; 2. *Thysanolaena latifolia (Average height: 3m)*; 3. *Eupatorium odoratum* (Average height: 3.5m); 4. *Macaganca denticulata*; 5. *Cissampelos pareira*; 6. *Saccharum arundinaceum*; 7. *Neohouzeaua dullooa* (5m); 8. *Livistona cochinchinensis*; 9. *Trema orientalis*

found in this community. The dominant bamboo species are *Schizotachium funghomi* (*Neohouzeanua sp.*) (*nứa*) and *Dendrocalamus hamiltonii* (*giang*). *Schizotachium* prefers damp river areas but can also be found on slopes. It forms a stable, long-term community with a continuous low dense canopy. *Dendrocalamus* flourishes in fallowed swiddens. It is small in size with a soft stalk. The canes do not grow erect but lean over one another, creating a very dense canopy cover, which apart from any shading limitations, is difficult if not impossible for other plant species to penetrate. Regeneration of forest tree species under this cover is very difficult and therefore this plant community tends to persist for a long time. It can, however, be destroyed by cutting and burning.

The dominant trees that do emerge in this community include the fast growing, pioneering species *Trema angustifolia*, *Mallotus cochinensis*, *M. apelta* and *Macaranga denticulata*. In addition to these rather common species, scattered individuals of the following species are also found: *Rhus chinensis*,

Aralia armata, *Peltophorum tonkenensis*, *Styrax tonkinensis*, *Litsea citrata*, *Ficus fulva*, *Zanthoxylum accanthopodium*, and *Alangium sinensis*. The tall grasses *Sacccharum spontaneum* and *Thysanolaena latifolia*, are also found scattered in this community. In damper areas, ferns, especially *Lygodium flexuosum*, may also be found

The mixed bamboo, tree and shrub community performs an important economic role as a sort of wild tree garden. Bamboo is the dominant plant type and dominates the architecture of the community. There are many species of bamboo, which is a plant with multiple uses. Of the eleven species identified as most useful to the population in Tat hamlet, nine are a source of food, seven are used in construction, eight are used in weaving mats and baskets and making cordage, and three are used as firewood. Some are used to provide household utensils and containers as well as water pipes. Several species are cut for sale. At the current rate of exploitation bamboo will not be able to regenerate sufficiently to meet projected future demand. Already villagers must go to the forestlands of neighboring villages to collect bamboo.

Collection and Utilization of Forest Products

All households in the hamlet engage in some collection of wild forest plants, both for their own use and to sell for cash, while less than one-third of households also engaged in hunting of wild animals (Table 9.1). In recent years, collecting non-timber forest products (NTFPs) to sell for cash has become an increasingly important part of household economies, but the main objective remains household consumption. Households collect forest products for food, medicine, fodder for livestock, green manure, fuel, construction materials, and to generate cash income:

Food

At least 32 wild species of wild vegetables, fruits and roots are collected from the forest and consumed as food, drinks, and flavorings. Although villagers can now buy vegetables from vendors, all of the households still consume many wild foods. Food from the forest helps households to meet their daily needs while minimizing expenditure of cash to purchase food. It is particularly important in the periods of food shortages between harvests, but even during the period of peak workload in agriculture, green vegetables are usually collected from the forest when farmers travel to and from the fields. Food from the forest is especially important to the farmers when they stay in the temporary shelters while they tend their swidden fields and during this period they totally rely upon wild vegetables for fresh foods.

Table 9.1: Collection and utilization of non-timber forest products by households (n=60)

Type of product	Number and share of households collecting		Percentage of households collecting NTFPs for different purposes		
	N	%	For sale only	For use only	Both for sale and use
Fuelwood	60	100.0	0.0	66.7	33.3
Bamboo stems	53	88.3	24.5	37.7	37.7
Bamboo shoots	58	96.6	0.0	17.2	82.8
Mushrooms	25	41.6	24.0	64.0	12.0
Wild vegetables	60	100.0	0.0	100.0	0.0
Medicinal plants	59	98.3	55.1	0.0	44.9
Wild animals	17	28.3	0.0	64.7	35.3
Broom grass	52	86.6	82.7	7.7	9.6
Pig food	43	71.7	0.0	100.0	0.0
Buffalo fodder	35	58.3	0.0	100.0	0.0
Cattle fodder	33	55.0	0.0	100.0	0.0
Fish food	31	51.7	0.0	100.0	0.0

Source: Household survey, 2002

Bamboo shoots are eaten throughout the year. They can be used in many ways including fresh-boiled, dried, or preserved in salt (Plate 9.2). Bitter bamboo shoots are a favorite food of almost all households. During the harvesting season from February to early May, they are usually collected at least twice a week, sometimes four or five times, with a total quantity of 10 kg to 25 kg of shoots for a whole week. In the peak period (March) bitter bamboo shoots are consumed at every meal. Shoots of other bamboo species (e.g., *Neohouzeaua dullooa*) are consumed when fresh and also dried to eat during times of shortages or to sometimes sell for cash when prices are high.

Wild yams are the most common staple food collected from the forest. Traditionally, yams played a very important role in providing carbohydrates to the households, especially during rice shortages. Recently, however, populations of wild yams have declined so that farmers no longer make special trips just to collect them. Instead, yams are collected only when they are encountered by chance when the villagers are in the forest for other reasons. Several species of yams including *Dioscorea persimilis* (*Củ từ*) and *Dioscorea chingii* (*Củ mài*) are

Plate 9.2: Women boiling bamboo shoots collected from the forest

commonly used in soups and sometimes steamed or boiled. *Dioscorea cirrhosa* (*Củ nâu*) is less common than before. It is only used when they have a serious shortage of food.

Wild fruits are also collected and consumed in many ways. *Garcinia mutiflora* is used to make sour soup, while forest litchis, figs, and bananas are eaten as dessert after meals, and banana flowers are eaten as a favorite vegetable, especially in the dry season when other vegetables are less available.

Wild forest animals are sometimes hunted or trapped for their meat. Most animals that are caught are smaller animals such as rats, squirrels, and wild chickens which are usually consumed by the households. Bigger animals like wild pigs, deer, and monkeys are now very rare. One hunter reported that in 2001 he trapped two monkeys. The meat is usually processed into jelly that can be sold for about one million VND.

Medicines

Villagers use many wild forest plants as medicines, both in prevention and treatment of common ailments. In an earlier study of forest plant collection, 22 species of medical plants were identified (Ireson and Ireson 1996: Appendix 5). An elderly woman, who was well known for her abilities as a curer, said that she used more than 200 different medicines, most of which she could collect from the forest (Ireson and Ireson 1996: 12). Medicinal plants are especially important for the treatment of postpartum ailments amongst new mothers. A study by Miller (2001) reported that Tay women follow a number of traditional practices after they give birth, including drinking a mixture of medicinal plants for at least one month to restore their strength for nursing their babies.

Fodder

The leaves of several forest plants are used as feed for buffalo, cattle, pigs, and fish. Fodder is often collected by women on their way home from working in their swidden fields. Men sometimes also gather wild banana leaves to use for fish food on their way home from their swiddens.

Buffalo and cattle are allowed to roam freely in the forests where they graze on wild plants. Cattle usually stay by themselves in the forest for long periods, and are only fed salt sometimes when their owners train them to find their way home. Buffalo are kept at the houses at least twice a year at the time of harrowing the paddy fields. Some households also keep their cattle at home at night in order to collect manure for crops. During the day, these animals are allowed to graze freely in the forest around the house.

Pigs used to be allowed to roam freely around the house to forage for food but, in recent years, they are more commonly kept in pigsties. They are fed banana stems, cassava, and a mash made of vegetables like *Colocasia esculenta, Alccasia macrorrhiza, Musa coccinea,* etc., which are mainly collected from the forest. All households that keep pigs reported that they depended mostly on forest food. Only some of them sometimes collect pig feed from their homegardens. On the average, households spent about one hour a day to collect about 14kg of wild vegetables for pigs and fish. No cultivated vegetables were used as pig food.

Green manure

Wild forest plants are used in the paddy fields as green manure. Melia leaves and leaves and stalks of *Chromolaena odorata*, an early successional weed in fallow areas, are commonly used. Farmers usually gather green manure twice a year, once in July and again in December. The green manure is then mixed with manure from animals to make compost before applying it in the fields.

Fuel

Firewood is the only source of energy used for household cooking and heating. It is gathered from forests in the form of fallen branches and dead trees as well as from unburned trunks and branches in newly cleared swidden fields. Women and children collect most of the firewood although men sometimes carry bigger logs to their homes when they are returning from the fields or forests (Plate 9.3).

The amount of firewood used by each family is estimated to be between 20 and 45kg/day, with the amount used reflecting the number of pigs kept by the household. Pigs are fattened by being fed a cooked mash made from a mixture of cassava roots, rice bran, and wild vegetables from the forest, which increases the daily demands for firewood. Thus, better-off households, which tend to keep more pigs, have a higher demand for firewood than poorer households.

Plate 9.3: Couple returning to house with firewood collected in the forest

Construction materials

Non-timber forest products are important sources of materials used for construction of housing and implements and making of handicrafts. Bamboo, palms, and rattan are the most heavily used materials.

Bamboo stems are used to construct houses, fences, and a great variety of implements. Houses are sometimes made entirely from bamboo stems, as is the furniture inside them. The larger culms are used for the piles, stilts and the major framework. Smaller sized pieces are used for floors, windows and doorframes. The bamboo can be slit into slats for weaving into mats for walls. Houses with walls made from split bamboo are quite common in Tat hamlet. This is especially true for the poor, who cannot afford to use wood, but even wealthier households generally make their kitchen walls from bamboo because it is cooler. Bamboo stems are commonly used for building fences and walls along the home garden and agriculture fields.

Bamboo is also used to make furniture, fences, cages, mats, farming implements, ladders and blinds. Pipes for irrigation are also fabricated from bamboo. Rattan is used for binding, basketry, weaving, and traps.

Cash income

Forest products that are extracted for sale include medicinal plants, broom grass, bamboo products, ferns, and, sometimes, firewood (Table 9.2). Bamboo stems and shoots are considered as the most valuable products in Tat hamlet, followed by medicinal plants and broom grass. Some food species collected from the forest are sold if there is a surplus or if there is a favorable market or if the household has an urgent need for cash for other expenses. Because of its low price, only some, mostly poorer, households collect firewood for sale.

Table 9.2: The most commonly traded non-timber forest products in Tat hamlet

SCIENTIFIC NAME	TYPE OF PRODUCT	COLLECTION TIME	USES
Neohouzeaua dullooa	Edible bamboo shoots	May–Sept	Food
Dendrocalamus patellaris	Bamboo stems	All year	Cords/string
Cibotium barometz	Fern root	All year	Medicine
Smilax glabra	Medicinal plant	All year	Medicine
Adenosma glutinosum	Medicinal plant	May–Aug	Medicine
Thysanoleana maxima	Broom grass	Dec–Mar	Brooms
Heminthostachys sp.	Fern fibers	All year	Hats/bags
Phrynium sp.	Green leaves	Dec–Mar	Wrapping for cakes
Alpinia sp.	Galanga roots	All-year	Seasoning/ flavorings

Source: Household survey, 2002

Contributions of Forest Products to Household Income

The sale of non-timber forest products makes a very important contribution to household incomes, both in-kind income and cash income (Table 9.3). On average, NTFPs contribute more than one-third of the total in-kind and cash income of households. But the share differs according to the socio-economic status of the households. The proportion for poor households is much higher (53%) compared with medium (38%) and well-off (29%) households. Thus, NTFPs are more important for the subsistence of the poorer groups than the other two groups. This is because poor households have fewer productive assets in the form of land and capital, so they have to rely on collecting NTFPs to obtain food and other daily consumption items as well as income. Thus, the forest is very important in providing a "safety net" for poor villagers.

The cash income from this source was compared to the total cash income of the households to determine the importance of NTFPs in the cash economy (Table 9.3). On the average, NTFPs contribute about 42 percent of total household cash income, but this proportion varies with the economic level of the households. NTFPs made up 53 percent of the total cash income of poor households, compared to 46 percent for medium households, and 33 percent for well-off households. In absolute terms, however, the total amount of income from NTFPs among the different groups of households does not differ very much. The poorer group has lower total income, so the portion of income contributed by NTFPs represents a greater share of their total income.

Table 9.3: Contribution of NTFPs to household income in Tat hamlet

Household type	Income from NTFPs (000 VND)			Total income (000 VND)		Share of income from NTFPs	
	In-kind	Cash	Total	Cash	Total	Cash as % of total cash income	In-kind + cash as % of total income
Poor	34,499	29,809	64,308	55,918	121,175	53.3	53.1
Medium	27,756	35,687	63,443	77,374	167,350	46.1	37.9
Well-off	26,648	34,350	60,998	104,845	207,392	32.8	29.4
Mean	28,648	33,282	59,930	79,379	165,107	41.9	36.3

Source: Household survey, 2002

Conclusions

An understanding of the ecological basis of villager livelihoods in Tat hamlet requires that we acknowledge the importance of secondary forests in the composite swidden agricultural system. The farmers actively manage a mosaic of plots representing the full spectrum of the various successional stages of forest regeneration. This complex indigenous land-use system both maximizes the stability of food production and the percent of the landscape dominated by secondary forests. Changes in the agricultural system as farmers switch from traditional subsistence farming to cash crops will have important implications for the stability of agriculture, biodiversity, cultural preservation, carbon sequestration, watershed hydrology and other variables. It is important to acknowledge and pay attention to how land use is changing and to invest in research on methods aimed at maintaining the benefits associated with swidden fallows while increasing their productivity and soil-sustaining properties.

It is certainly a misnomer to describe this form of land cover as "land not yet in use" (*đất chưa sử dụng*), as Vietnamese government land use planners commonly refer to fallowed swiddens and areas of regenerating secondary forest in early successional stages. This categorization belies the complex management of the landscape undertaken by the composite swidden farmer. The secondary forest, in all successional stages, plays an important role in the life of the people of Tat hamlet, who obtain a wide variety of useful products from the forest. In addition to house construction material, forests provide firewood, bamboo shoots, rattan, mushrooms, wild greens, fruits and tubers, medicinal plants, as well as fodder and green manure. Today, as in the past, these products, whether consumed or sold, ensure the survival of the households whose agricultural production is insufficient to meet their subsistence needs.

10
Recent Changes in the Composite Swiddening Farming System in Tat Hamlet

Nguyễn Thanh Lâm, Aran Patanothai, and A. Terry Rambo

Introduction

Recent changes in environmental and economic conditions have brought about many important changes in the composite swidden farming system in Tat hamlet. In the past, all households in the hamlet employed composite swiddening as a means to achieve food security. In recent years, however, the households have begun to diversify their farming systems. Currently, not all households in Tat hamlet practice composite swiddening, and for those who do, the components are not exactly the same as those in the earlier days (Rambo 1998). In-depth understanding of these changes and their causes and consequences are needed to assess the sustainability of the system and its potential for intensification. This chapter describes the key changes that have occurred in farming systems in Tat hamlet over the past fifteen years, and identifies some of the factors influencing these changes.

Methods

This chapter is based on analysis of data collected in several surveys of households in Tat Hamlet over the years. Secondary data from Rambo (1998) and Rambo and Trần Đức Viên (2001a) were used to describe the farming system in Tat hamlet before the mid-1990s. Data collected in 1998-1999 by the Center for Agricultural Research and Environmental Studies (CARES) on 42 randomly selected households (out of 95 households) in Tat Hamlet were used to characterize farming systems at the end of the 1990s. These data included agricultural and non-agricultural activities, household income and expenditure, debt, farm size and composition, tenure, production of all commodities, and age and sex of all household members. Finally, in 2003, an additional survey was made of these same households, initially employing semi-structured interviews, followed by use of formal questionnaires.

Changes in Tat Hamlet and its Farming Systems from 1988 to 2003

Before the establishment of the cooperative in the mid-1950s, there were only a few households in the hamlet, and all of them engaged in traditional composite swiddening (Rambo and Trần Đức Viên 2001a, Chapter 11 this volume). During the cooperative time, composite swiddening continued, but farming was managed by the cooperative with households contributing labor and receiving payment according to number of work points they accumulated. Thus, there was little economic differentiation among households. The household management period began in the mid-1980s following the issuance of Directive 100, which returned control over agricultural production to the individual households. The land, though still belonging to the state, was allocated to individual households with long-term use and inheritance rights. It was during this period that households began to become differentiated, and their farming systems also gradually changed. This study examined these changes during the period from 1988, which is the early stage of the household management period, to 2003, when the final field surveys were conducted.

To facilitate the identification of changes, the period 1988-2003 was divided into three stages: Stage 1 covers the years 1988-1994, Stage 2 from 1995-1999 and Stage 3 from 2000-2003. Table 10.1 shows major characteristics of Tat hamlet and its farming systems in each of the stages.

Stage 1—1988-1994

Toward the end of this stage in 1993, when the first statistics were collected, Tat hamlet had a population of 389 people with a population density of 52 persons per km^2. It had 69 households, all of which were Đà Bắc Tày except for three Mường households that had recently migrated from the area flooded by the Hòa Bình reservoir.

The farming system at the beginning of Stage I was subsistence-oriented composite swiddening. Farming system differentiation began when the paddies and agricultural tools of the cooperative were divided among the individual households as per the Resolution 10-NQ/TW (April 4, 1988) on renewal of household responsibility in agricultural management. The hamlet headman recalled that the cooperative lands were divided among individual households according to family size at a rate of 350 m^2 per person. Land use at that time

Table 10.1 Major characteristics of Tat hamlet and its farming systems in 1988–1994, 1995–1999, and 2000–2003

CATEGORY	STAGE 1 (1988–1994)	STAGE 2 (1995–1999)	STAGE 3 (2000–2003)
Demography			
Population (persons)	389 (1993)	432 (1998)	476 (2003)
No. of households	69 (1994)	91 (1998)	107 (2003)
Density (persons/km^2)	52	58	64
Migration	3 hh. moved in	2 hh. moved out	None
Policy			
Forest land use	Controlled by government and the local authorities	Land allocation to households by 327 Program	Land allocation to households by 01 and 02 decrees NQ/TW
Agriculture	Allocation of paddy land to households. Development assistance provided by Program on Fixed Cultivation and Sedentarization, Agricultural taxes	Extension program for paddy, no state credit, agricultural taxes	Extension programs, credits for cattle and fish raising, no agricultural taxes
Transportation and Infrastructure			
Road condition	Rough dry season dirt road	Improved all season gravel road	Hard surface road
Bus to Hanoi	None	One trip in every two days from 1997	Several trips daily
Trucks passing	Few each week	Daily	Several per day
Travel to market	Walking, motorcycle	Bus, motorcycle	Bus, motorcycle
Electricity	Few mini-hydropower electric generators	Mini-hydropower electric generators commonly used	National electricity service installed in 2001
Marketing			
Village shop	Only 1–2 shops belong to the cooperative	5–6 private shops, early stage of marketing network	15 private shops, well developed marketing network
Agricultural products	Small amounts of cassava and maize	Cash crops especially ginger; high price of ginger in the early stage and then dropped abruptly	A broad range of cash crops; crop prices gradually decreased

Table 10.1 Continued

Category	Stage 1 (1988–1994)	Stage 2 (1995–1999)	Stage 3 (2000–2003)
Forest products	Timber	Timber & NTFPs	NTFPs
Ratio of cash/ total in-kind income	n.a.	0.4	0.4
Household types			
Composite with rice swidden	95%	60%	61%
Modified composite without rice swidden	0	21%	29%
Swidden only	5%	17%	7%
No agricultural land	0	2%	3%
Paddy fields			
Area/hh (m^2)	770–1,750	300–2,770	200–2,770
Yield (ton/ha/crop)	2.5	3.2	3.8
Chemical Fertilizer	Rarely used	Commonly used	Increasingly used
Rice varieties	Local and a few improved varieties	Mostly improved varieties (CR203, C70, Q5)	Entirely improved or new hybrid varieties
Swidden			
Total swidden area/ hh (m^2)	12,500	8,800	6,400
Rice swidden area (m^2/hh)	10,000	5,550	3,310
Yield (ton/ha/crop)	0.59	0.60	0.43
% households planting	100%	77%	70%
Cassava swidden area (m^2/hh)	2,000–10,000	2,000–5,000	1,000–2,000
% households planting	100%	100%	97.5%
Maize swidden area (m^2/hh)	n.a.	100–3,500	300–6,000
% households planting	100%	75%	58%
Varieties	Local	Improved and local	Hybrid

Table 10.1 Continued

	STAGE 1 (1988–1994)	STAGE 2 (1995–1999)	STAGE 3 (2000–2003)
Canna swidden area (m²/hh)	n.a.	1,180	500–5,000
% households planting	n.a.	80%	75%
Ginger (m²/hh)	n.a.	475	624
% households planting	Only a few	100% (1995–96) but reduce to 35% (1998).	12.5%
Varieties	Local	High yielding	High yielding
Forest			
Natural forest	Scattered patches of degraded primary forest, extensive area of dense secondary forest	Secondary forest abundant	No forest with big trees; grass, bamboo, bushes abundant
Tree plantation	Palm, Melia, Styrax	Eucalyptus, Acacia, bamboo	High value bamboo on allocated land, supported by government
Fallow swidden fields			
Fallow length (yr)	5–8	4–5	3–5
Fallow vegetation	Small trees, bush and bamboo	Bush, bamboo, tall grass, Imperata	Bamboo, tall grass, Imperata and bush; pressure on grazing land (conflict)
Home gardens			
Area (m²)	100–400	100–400	100–400
% households	55.6 %	57.5 %	57.5%
Varieties	Local varieties with low economic value	New fruit varieties planted by several households	Mango, litchi, longan, persimmon in several households
Fishponds			
Area (m²)	50–2,000	50–2,000	50–2,000
% households	78.8 %	62.5 %	62.5 %
Purpose	Home consumption	Home consumption	Home consumption & sale, more investment (credit)
Fish species	Carps, tilapia	Carps, tilapia	Old & hybrid

Table 10.1 Continued

	STAGE 1 (1988–1994)	STAGE 2 (1995–1999)	STAGE 3 (2000–2003)
Livestock			
Buffalo and cattle (no./hh)	1	2	2.5
Ratio of buffalo to cattle	1:1	0.83:1	0.54:1
Pigs (no./hh)	1	1	0.5
% households having cattle or buffalo	90	73	63
Breeds of cattle and pigs	Local	Local and improved varieties	Improved varieties in majority

was governed by two systems. Agricultural lands were managed under the customary system in which land uses were governed by local social organizations and traditional norms. On forest lands, however, laws to protect the forest were in principle enforced by state and village administrative authorities, although in practice, local farmers could still freely exploit forest land to construct new paddy fields and clear swiddens. Many households converted some upland areas at the foot of the hill slopes into paddy fields. One farmer recalled that his family constructed 1,500 m^2 of paddy fields in addition to the 1,350 m^2 they received from the cooperative. Average paddy field area per household at that time was only 1,400 m^2 with a yield of 2.5 tons per ha per crop, providing a family with only 650 kg of unhusked paddy rice annually (Rambo 1998).

The area of other components of the composite swiddening system also expanded quickly in this stage. An elderly farmer recalled that each household normally planted a rice swidden with an area equivalent to 1.2 to 2 ha, and harvested 600 kg to 1.5 tons of unhusked rice. The survey in 1993 showed that nine randomly selected households had total swidden areas of from 6,500 to 23,400 m^2, of which rice swiddens had an average area of 10,000 m^2 per household, followed by cassava swiddens occupying areas of 2,000 to 10,000 m^2. The average yield of the rice swiddens was 590 kg/ha (Rambo 1998).

Homegardens and fishponds were mainly used for home consumption. About half of the households had mixed homegardens with native fruit and vegetable species with areas from 100 to 400 m^2 (Rambo 1998). Most households had small fishponds of 100 m^2, and only a few households had large fishponds of 700-2,000 m^2. Several carp species and tilapia were commonly fed grass and cassava leaves.

Livestock raising was still based on traditional practices with free grazing and native breeds. The numbers of buffalo and cattle were about equal, although the value of buffalo was higher than cattle because buffalo could be used for both land preparation and logging. Native pigs freely roamed around the houses, and their feed supply was inadequate. The number of poultry varied among households and only few chickens survived periodic disease epidemics. No veterinary service was available in this period.

Some small patches of degraded primary forest and large areas of secondary forest still covered the mountain slopes surrounding the hamlet. Cash income from forest products was mostly from timber, and more rarely from hunting and collection of non-timber forest products (bamboo shoots and broom grass). However, the government had implemented several fixed cultivation and sedentarization programs to encourage abandonment of swiddening. As a result, more than half of the households had converted up to 2.5 ha of their swidden areas to tree gardens of palm, tea, *Melia azedarach*, and *Styrax tonkinensis* for household and commercial purposes.

Stage 2—1995-1999

By 1998, the number of households had increased to 91 with a population of 432 persons (58 persons per km^2). Two households, however, left the hamlet to settle in a new economic zone established by the government.

During Stage 2, the infrastructure of the hamlet improved significantly. The road connecting the hamlet to the district town was upgraded into an all weather gravel surfaced road. The number of local private shops increased to five, but the supply of commercial goods was still limited. A canna root processing plant was built at the center of the village, and several households had their own rice mills. In 1997, the hamlet bus station was opened with a bus service to Hanoi once in every two days. Access to information of the local people was also improved, both from training for agricultural technology transfer and from radio and television.

To prevent further forest degradation, the government launched the 327 Program during this period. Under this program, forestland was allocated to individual households to look after, and each household annually received 47,000 VND per ha in return. Clearing of swiddens was forbidden on this land (Rambo and Trần Đức Viên 2001a). The hamlet authorities delimited some relatively small areas of forestland where swiddening and grazing were permitted. These measures restricted the area available for swiddening, making the long rotation cycle no longer possible. Consequently, the fallow period was reduced to 4-5 years.

Variation in paddy field areas among households became more pronounced as many households opened up new fields, while others had to share paddies with their children when they married and established independent households. During Stage 2, the paddy field area per household ranged from 300 to 2,770 m^2. Productivity, however, increased to 3 tons per ha per crop with the use of new rice varieties and chemical fertilizers in addition to compost, green manure, and ashes.

Swidden cultivation began to shift from purely subsistence to semi-subsistence. The rice swidden area per household gradually declined to 5,550 m^2, while swidden areas for canna and ginger, which were new cash crops, increased. The average yield of the rice swiddens was 0.6 ton per ha per crop. At the beginning of this stage, ginger price was high and all households planted the crop. The price of ginger dropped drastically in later years and the number of households planting ginger declined to 35 percent in 1998. All households still planted cassava, but only a part of the harvest was sold while a part was kept to use for animal feed. The area of maize declined to some extent, as did the percentage of households planting maize, but new varieties were increasingly used.

Due to the scarcity of cultivated land nearby the settlement, five households moved their houses to establish new farmsteads in the largest swidden area 2 km away from the residential area of the hamlet. There they engaged in typical composite swiddening with paddy fields in the valley bottom and swidden fields on the surrounding hillsides. One owner said that the reasons he opened the new farm were that he could take better care of his crops, and could secure land ownership for his descendants.

Although the areas of homegardens and fishponds during Stage 2 remained the same as in Stage 1, their management was intensified. New fruit species such as longan, litchi, plum and persimmon were planted in the gardens. Several households had small vegetable (mustard and onion) gardens of 20-100 m^2 protected by bamboo fences. Fishponds were dug deeper and fish were raised at a higher density with additional feed provided. However, most products from gardens and fishponds were still used only for home consumption.

The number of livestock in Stage 2 also remained more or less the same as in Stage 1, but there were changes in species composition and breeds. The ratio of buffalo to cattle declined to 0.83:1 as the use of buffalo for logging was diminishing due to forest degradation and its allocation to individual households to protect. Several households had two or three large animals, and one had eight head. Hybrid pigs were introduced with new production practices recommended by extension officers. About 8-10 chickens or other poultry were commonly observed in a household. During this stage, the village veterinary service was being established and each hamlet had a quota to send one local farmer for veterinary training. However, the service facilities were rather poor, and, as one informant recalled, animal diseases could still easily spread throughout the whole hamlet.

During this stage, the market-oriented economy began to have a significant impact on household livelihoods. Firstly, the farmers had experienced drastic drops of cash crop prices, first in ginger, followed by cassava and canna. Secondly, non-timber forest products and fallow vegetation became a major source of cash income of the households. As a consequence, people spent more time collecting forest products so that their swidden areas gradually declined. Moreover, after several years spent improving their homegardens, several households faced losses on their long-term investment due to price falls. For instance, the price of apricots dropped drastically from 2,000 VND per kg when the trees were planted to 200 VND per kg when the trees started to produce fruits. Farmers responded by cutting down their fruit trees or leaving them unattended. Several households turned to doing business as an alternative source of extra income.

Reallocation of household labor forces was also observed in this period. In the past, family labor was freely expended in the rice and cassava swiddens. But in this period, a part of labor force of both men and women was engaging in business and off-farm activities, so that, the amounts of labor spent on swidden cultivation substantially decreased. Women took care of cash crops, rice transplanting, raising pigs and collecting medicinal plants and bamboo shoots in the forest and broom grass in fallow fields. Men took care of land preparation for swiddens and paddy fields, logging, and collecting bamboo. Cattle and buffalo were tended by children and elderly people when they got back home from working in their swiddens.

Stage 3—2000-2003

The population of Tat hamlet showed continued increase with the number of households rising to 107 and the population to 476 (64 persons km^2) in 2003. The infrastructure was further improved with significant contributions from the 135 Program. The road was paved, and several trucks passed through the hamlet each day. The bus service to the lowlands increased to four routes operating on a daily basis. The number of shops in the hamlet increased to fifteen, offering a much greater variety of merchandise. The national electricity service was also extended to most households in the hamlet.

By the beginning of this stage, the supply of land that could be made into new paddies had been nearly exhausted. Only few households were able to construct small paddy areas of 200-300 m^2. However, the yield of paddy rice increased to 3-4 tons per ha per crop with the use of hybrid rice and improved practices recommended by the extension workers.

Pressure on swidden land became critical so that a few households even cut down small areas of their allocated forest to plant swidden rice without permission from the district authorities. These households suffered from food shortage and could not find suitable land for rice growing in the restricted area designated by the authorities for swiddening. Recognizing the seriousness of this problem, the district authorities then allowed the villagers to implement an agroforestry model in their forest land in which upland rice was intercropped with bamboo (*Dendrocalamus mebranaceus*). Farmers were supported by the extension of bamboo seedlings and fertilizers.

Total swidden area also greatly declined with the rice swidden area per household decreasing to 1,000–5,000 m^2. The yield of swidden rice decreased to 0.43 ton per ha, largely because of increased damage by rats and insects. The market greatly influenced the kinds of cash crops planted. Although most households still planted cassava, the area per household decreased to 1,000-2,000 m^2 as it was mainly used for livestock. A few farmers still planted ginger around their houses to maintain the stock. Local varieties of maize were replaced by high yielding hybrid varieties with the area per household ranging from 300-6,000 m^2. Canna area generally increased with the area per household ranging from 500-5,000 m^2. However, none of the households grew only a single crop in a large area. Apparently, farmers had used crop diversification as a strategy to cope with price fluctuation.

Homegardens continued to be developed with more planting of several new fruit species. However, these were generally cared for by elderly people since the younger generation paid little attention to their homegardens due to lack of time.

A new credit program for cattle raising was initiated. According to the hamlet headman, 30 households obtained loans. After three years (1999-2001), the number of cattle had doubled and several households had raised the size of their herds to 20 heads. The proportion of buffalo to cattle further declined to 0.5 : 1.

Poultry and pig raising had shifted more toward commercial purposes. Two households had raised 50-300 chickens in one year, but then had to quit because of diseases. Integration of crops, livestock, and fish was also implemented by the farmers. For instance, a number of households raised ducks in their fishponds and also released them in paddy fields after rice harvesting. In additions, pig stables were built on fishpond bunds, directly supplying manure to increase the productivity of the ponds. The government also provided credit to improve fish rearing. Eight households that had sufficiently large fishponds received loans of 1,000,000 VND each, and were trained in fish raising practices. Among them, five were successful, but three lost all of their fish because of disease.

Changes in Household Level Farming Systems from 1988 to 2003

At the beginning of Stage 1 in 1988, all households presumably practiced traditional composite swiddening (except for three Mường families, who had recently migrated from the site of Hòa Bình dam. These households had no paddy fields because they were outsiders so were not allocated land from the former cooperative, thus, could not practice the system). Farming system differentiation began at the time when the households were allocated land, and continued after that. As data from the same set of 42 households were available for 1998-99 and 2003, we were able to examine the changes that occurred between these two points.

Household farming systems in 1998–1999

The 1998-99 survey data for 42 households were used to classify these households into different types based on the structure of their farming systems. Four types of households were recognized: Type 1 practiced the traditional composite swiddening consisting of paddy fields, rice swiddens, and cash crop swiddens. Type 2 engaged in modified composite swiddening consisting of paddy fields and cash crop swiddens but no rice swiddens. Type 3 practiced only swiddening including both rice swiddens and cash crop swiddens but no paddy fields. Type 4 had no agricultural land, thus, had neither paddy fields nor swiddens. In 1998-99, 25 households (59.5 %) were Type 1, nine (21.4 %) were Type 2, seven (16.7 %) were Type 3 and one (2.4 %) was Type 4. All of these households also had other components of farming systems, i.e., homegardens, tree plantations, livestock and fishponds. Characteristics of households in Types 1-3 are shown in Table 10.2.

Changes in household farming systems between 1998-99 and 2003

Type 1 households that practiced traditional composite swiddening with rice swiddens in 1998-99 were either original Đà Bắc Tày residents of the hamlet, who had been allocated paddy fields by the cooperative, or newly established families who received paddy fields from their parents. Some households also constructed new paddy fields when they could find suitable lands. However, paddy field area per capita was still small (averaging 226 m^2). Thus, these households had to produce much of their rice in swiddens, and 55 percent of the total swidden area was devoted to rice (Table 10.2). These households were allotted less forestland than those of Types 2 and 3.

Type 2 households, who practiced modified composite swiddening without rice swiddens, were households headed by government officers and teachers who had salaries, and merchants and elderly people. All were Đà Bắc Tày. These

Table 10.2 Characteristics of households of Types 1, 2 and 3 in 1998

CATEGORY		TYPE 1[a]	TYPE 2[a]	TYPE 3[a]
Number of households		25	9	7
Age of household head (yr)	Range	20–78	25–69	25–50
	Mean	38	39	33
Family size (no. of members)	Range	3–10	2–10	3–5
	Mean	5	6	4
Paddy area per capita (m^2)	Mean[b]	226	317	0
Total swidden area per capita (m^2)	Mean[b]	1,974	572	2,934
Rice swidden area per capita (m^2)	Mean[b]	1,092	0	2,252
Rice production per capita (kg)	Range	80–617	5–500	121–506
	Mean[b]	229	175	158
Cassava swidden area per capita (m^2)	Mean[b]	467	278	396
	(% of hh)	100.0	100.0	100.0
Maize swidden area per capita (m^2)	Mean[b]	140	111	200
	(% of hh)	72.0	67.0	86.0
Canna swidden area per capita (m^2)	Mean[b]	229	119	47
	(% of hh)	92.0	67.0	43.0
Ginger swidden area per capita (m^2)	Mean[b]	46	65	39
	(% of hh)	32.0	33.0	57.0
Homegarden area per capita (m^2)	Mean[b]	26	13	39
	(% of hh)	60.0	67.0	57.0
Tree garden area per capita (m^2)	Mean[b]	135	9	478
	(% of hh)	24.0	44.4	28.6
Fish pond area per capita (m^2)	Mean[b]	33	54	104
	(% of hh)	36.0	100.0	29.0
Forest land area per capita (m^2)	Mean[b]	3,466	9,770	6,458
	(% of hh)	80.0	67.0	71.0
Buffalo and cattle (head/capita)	Mean[b]	0.4	0.6	0.1
	(% of hh)	80.0	78.0	43.0
Pigs (animals/capita)	Mean[b]	3.7	3.4	2.5
	(% of hh)	100.0	100.0	100.0

Notes.

a: Type 1 = true composite swiddening with rice swidden. Type 2 = modified composite swiddening without rice swidden. Type 3 = swiddening only

b: Mean of households having this land category or type of livestock

households had more paddy field area and less swidden area per capita than those of Types 1 and 3 (Table 10.2). These households used to plant swidden rice in mid-1990s but gave up this practice and devoted all of their swidden areas to cash crops. Labor shortage was given as the reason for the change as most adult members worked full time as government officers, while some were too old to work in the swidden. This type of household earned their income from salary, animal husbandry, and collection of non-timber forest products (NTFPs).

Type 3 households that practiced only swiddening were newly established families that did not receive any paddy field from their parents or households that migrated in from other places. Young couples originating from families with only a small area of paddy fields normally did not receive any paddy land when they established separate households. The newly resettled households were Mường and Kinh who were not allotted paddy fields by the cooperative because they were considered outsiders. These households had no fishponds and only poorly developed homegardens, and had to depend on rice swiddens and NTFPs for their living.

The Type 4 household that had no agricultural land is a Kinh family that had migrated to Tat hamlet after the construction of Hòa Bình dam. This family owned a shop in the hamlet and earned its income from trading and animal raising. In 1998, it had 10 cattle, 3 pigs, and 30 chickens. This household was also allotted a small plot (3,000 m^2) of forest to look after.

By comparing survey data collected from the same households in 1998-99 and 2003 we were able to trace changes in their farming system during a 5-year period. Figure 10.1 presents the changes in type of farming system from 1998 to 2003 of these households. Overall, the number of Type 1 households (households that practiced composite swiddening with rice swiddens) and Type 4 households (no agricultural land) in 2003 remained the same as in 1998, but the number of Type 2 households (modified composite swiddening without rice swiddens) increased from 9 to 12, and that of Type 3 households (swiddens only) decreased from 7 to 3.

Of the 25 households of Type 1 that had practiced composite swiddening with rice swiddens in 1998-99, most continued to do so in 2003 but more than one third shifted to Type 2. One half of Type 2 households continued to engage in modified composite swiddening, but the other half became Type 1. On the other hand, more than half of the Type 3 households (swidden only) obtained paddy fields and shifted to Type 1 or Type 2, while the remainder still only cultivated swiddens.

In examining the changes in farming systems at the household level during 1998-2003, we also looked at the numbers of households who increased, maintained constant or decreased their involvement with individual farming

Figure 10.1 Changes in type of individual households between 1998–99 and 2003

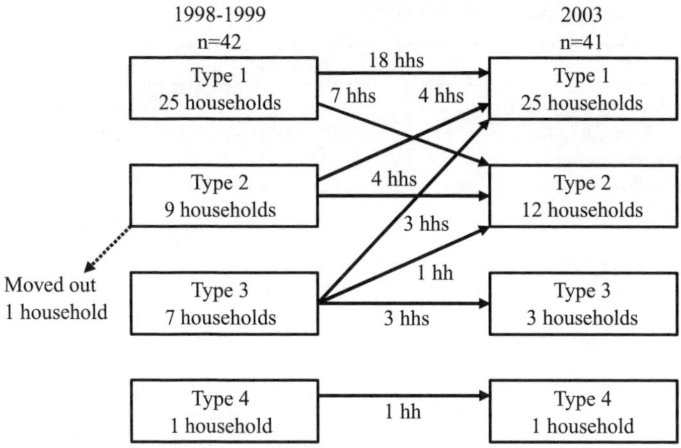

Note: Type 1 = Composite swiddening with rice swidden
Type 2 = Modified composite swiddening without rice swidden
Type 3 = Swiddening only
Type 4 = No agricultural land

system components during this period. These are presented in Tables 10.3 and 10.4 for crop and livestock components, respectively.

Among the 40 households interviewed in both 1998-99 and 2003 (i.e., excluding the household with no agricultural land and the one that moved out), 33 (82.5 %) planted paddy rice in 1998-99 and the number increased to 37 (92.5%) in 2003. Each household has two to five paddy fields at various locations surrounding the settlement area. A few households have newly constructed paddies about 2 km from the hamlet. Paddy field area per household shows less change than for the areas of other annual crops. The average paddy field area per household decreased slightly from 1,237 m^2 in 1998 to 1,170 m^2 in 2003 (Table 10.3). The average paddy area per capita, however, was slightly higher in 2003 than in 1998-99 as several households obtained additional paddies and the numbers of persons in some households were less in 2003 due to the separation of their children into independent households or death. During this period, more than half of the sampled households maintained the same paddy areas, one-third had more, but 15 percent had less. The latter were those that gave some paddy land to their children who separated out to form new households. Yield of paddy rice, however, increased from 3.2 to 3.8 tons/ha/crop due to the increased use of hybrid varieties and chemical fertilizers.

In 1998-99, 40 households practiced swiddening but in 2003 the number had declined to 39, while the average area both per household and per capita dropped

Table 10.3 Changes in crop components of farming systems of 40 households in Tat hamlet from 1998–99 to 2003

CATEGORY	PADDY FIELDS	SWIDDEN
Households planting in 1998–99	33 (82.5 %)	40 (100 %)
Households planting in 2003	37 (92.5 %)	39 (97.5 %)
Households with increased area	13 (32.5 %)	13 (32.5 %)
Households with the same area	21 (52.5 %)	1 (2.5 %)
Households with reduced area	6 (15.0 %)	26 (65 %)
Average area/household in 1998–99 (m^2)	1,237	8,842
Average area/household in 2003 (m^2)	1,170	6,398
Average area/person in 1998–99 (m^2)	255	2,142
Average area/person in 2003 (m^2)	274	1,492

CATEGORY	RICE SWIDDEN	CASSAVA	MAIZE	CANNA	GINGER
Households planting in 1998–99	32 (80.0 %)	40 (100 %)	30 (75.0 %)	32 (80.0 %)	14 (35.0 %)
Households planting in 2003	28 (70.0 %)	39 (97.5 %)	23 (57.5 %)	30 (75.0 %)	5 (12.5 %)
Households with increased area	9 (22.5 %)	12 (30.0 %)	12 (30.0 %)	18 (45.0 %)	5 (12.5 %)
Households with the same area	7 (17.5 %)	17 (42.5 %)	7 (17.5 %)	10 (25.0 %)	14 (35.0 %)
Households with reduced area	24 (60.0 %)	11 (27.5 %)	21 (52.5 %)	12 (30.0 %)	21 (52.5 %)
Average area/household in 1998–99 (m^2)	5,550 (54.4%)	1,951 (19.1%)	1,051 (10.3%)	1,181 (11.6%)	475 (4.7%)
Average area/household in 2003 (m^2)	3,310 (36.8%)	1,793 (19.9%)	1,139 (12.6%)	2,140 (23.8%)	624 (6.9%)
Average area/person in 1998–99 (m^2)	1,345 (59.4%)	418 (18.5%)	196 (8.7%)	223 (9.8%)	83 (3.7%)
Average area/person in 2003 (m^2)	772 (38.9%)	379 (19.1%)	220 (11.1%)	470 (23.7%)	142 (7.2%)

sharply. Although the majority of households had smaller swidden areas than in 1998-99, 13 households had larger swidden areas.

Reduction in swidden areas could be primarily accounted for by the reduction in the size of rice swiddens. In 1998-99, the average area of rice swidden per household was 5,550 m^2, equivalent to 54 percent of the household's total swidden area. The area declined to 3,310 m^2 or 37 percent of the household swidden area in 2003 (Table 10.3). The number of households cultivating rice swiddens also declined from 32 in 1998-99 to 28 in 2003. Differences among households were also observed for the changes in areas of rice swidden during this period. Most of the households had reduced swidden rice areas but nine had increased areas and seven had the same areas as in 1998-99.

Cash crop swiddens, on the other hand, remained more or less the same from 1998-99 to 2003, both in terms of number of the households cultivating them and the average area per household. In fact, most households (39 of 40)

Table 10.4 Changes in livestock ownership by households in Tat hamlet between 1998–99 and 2003 (n=41)

Category	Buffalo		Cattle		Pigs	
	1998–99	2003	1998–1999	2003	1998–1999	2003
Total no. of animals	39	37	47	69	102	48
Households with no animals	19	18	24	19	11	20
Households with animals	22	23	17	22	30	21
Households with increased no.	12		15		9	
Households with the same no.	17		17		10	
Households with animals	5		4		4	
Households without animals	12		13		6	
Households with reduced no.	12		9		22	

cultivated cash crop swiddens in both years. However, the areas of individual crops grown by a household as well as the numbers of households that grew a particular crop varied from year to year, responding to the fluctuation in prices of these crops. Cassava was less affected by price fluctuations because this crop was grown both for sale and for animal feed, and could be left in the field for variable lengths of time. Cassava was planted by almost all households throughout this period. Although the numbers of households planting maize and canna declined somewhat, both areas per household and per capita increased for both crops. Ginger faced a problem of serious price reduction, and the number of households growing this crop dropped greatly. Again, there were variations among households in area changes of individual crops.

The total number of buffalo remained more or less the same from 1998-99 to 2003, but the number of cattle substantially increased (Table 10.4). The number of pigs was greatly reduced. The increase in the number of cattle was the result of the government promotion program that provided loans for farmers to purchase livestock. However, the frequency of conflicts caused by livestock damaging crops in actively cultivated swiddens greatly increased. Similar to the crop components, changes in numbers of individual animals of all kinds varied among households. Some households increased animal numbers; some maintained the same numbers and some reduced the numbers of animals. Apparently, each household follows its own strategy in making its living.

Discussion

Over the 15 years from 1988 to 2003, farming systems in Tat hamlet have undergone major changes. At the beginning of the study period, households were essentially undifferentiated. All households engaged in traditional composite swiddening, as the cooperative lands had just been allocated to individual households for their own management, with each family receiving both paddy and swidden areas on a per capita basis. Components of the farming system included paddy, rice swidden, cash crop swidden, homegarden, fishpond and livestock. These were managed mainly for subsistence, and local varieties or breeds and traditional practices were used. After the management of land was returned to individual households, farming system differentiation began. By the end of this period, household farming systems had become more diversified and substantially differentiated. Farming activities were oriented more toward commercial production, although a decreasing part, particularly rice production, was still for subsistence. As the length of the fallow period was shortened and soil fertility declined, the contribution of swiddening to household income decreased in later years. On the other hand, livestock, NTFPs, and off-farm activities played an increasing role as sources of household income.

At the household level there was great variation in the ways that the farming system changed. Thus, for all farming system components, there were households with increased, the same, and decreased involvement with any particular component. With additional sources of income from non-farm activities, individual households became more differentiated both in terms of activities and wealth.

These changes were the consequence of several factors acting both individually and in combination. Important factors include population increase, government policies on management of agriculture and forest land, changes in the macro-economic environment, improved infrastructure and communication, improved access to market, government development programs and services, changes in family structure and ages of family members that influenced the adaptive strategies of individual households, and environmental degradation.

Population growth has affected food security, which is a primary concern of all households. Local farmers increased the productivity of paddy rice by employing high yielding varieties and chemical fertilizers. However, even with increased productivity, the rice harvest from their paddy fields was not enough to maintain food security for most farmers due to small area of paddy fields per capita (169 m^2). Therefore, many households still had to plant rice swiddens to obtain sufficient rice to meet their consumption needs. As paddy fields are a productive and reliable source of rice for consumption, suitable areas were made into new paddy fields wherever they were found. However, because of hilly topography, there was not much additional land that could be converted to paddy fields. As a result, the total paddy field area was only slightly increased. With increasing population and new households being formed, paddy area per household decreased in later years. Consequently, households have had to find alternative sources of income in order to purchase needed rice and other commodities.

The most important government policy affecting agriculture was the change from the collective production system to the individual responsibility system governed by Directive 100 and Resolution 10-NQ/TW which returned management of agricultural production to the hands of individual households, with lands allocated to them for long-term use. This allowed the households to make their own management decisions and provided a strong incentive for them to make changes to improve production.

The government policy on forest protection had significant implications for swiddening in Tat hamlet. In implementing this policy, plots of forestland were allocated to individual households to look after and clearing of swiddens on them was strictly prohibited. Swiddening was only permitted in some relatively small areas designated by local authorities so that the area for swiddening was insufficient for a long fallow period. The consequence was a reduction of the length of fallow period from 7-10 years to 4-5 years. Such a short fallow period is insufficient to allow for full recovery of soil fertility. Hence, yields of swidden rice have declined. Several villagers complained about the difficulty of finding sufficiently fertile land to grow swidden rice. Such a condition made farmers reluctant to grow swidden rice as the crop demands high labor input. Improved productivity of paddy rice and availability of other income sources provided alternatives to several households to achieve rice sufficiency. Consequently, planting of swidden rice declined substantially in later years.

Progress in the overall development of Vietnam has resulted in a great change in the macro-economic environment of the country, which has impacted Tat hamlet in several aspects. Infrastructure and communications were greatly improved, access to market became easier, marketing opportunities were broadened and income sources were diversified. Modernization has also created greater demand on the part of the villagers for consumer goods and material comforts. As a consequence, the livelihood of the villagers became more and more dependent on cash. To earn more cash income, farmers turned toward commercial production, which is subject to price fluctuations of agricultural commodities. This was evident in the fluctuation of prices of swidden cash crops, particularly the drastic drop of ginger price. Changes in areas of these crops occurred in the response to market demand and prices. Improved transportation also created greater market opportunity for NTFPs, and these had become a major source of cash income of several households in Tat hamlet.

Government development programs and services also contributed significantly to the improvement and diversification of household farming systems. The use of high yielding and hybrid varieties, chemical fertilizers and recommended cultural practices for crops and of new breeds and improved management for animals was partly the results of training for technology transfer and provision of agricultural services. Loans provided to farmers under the livestock promotion program resulted in a substantial increase in number of cattle in the hamlet. Implementation of an agroforestry system also was the result of the support provided by the extension service. These programs and services provided households with more alternatives to improve production and diversify their income sources.

Differentiation among households reflected adoption of different strategies by individual households to cope with all these changes. In addition to skill, preference and available resources, changes in family structure during the family development cycle also had a great influence on the choice of household adaptive strategy. A young couple with a newly established household has limited land and labor so would follow a different strategy from a large well-developed household with more labor and land. Similarly, the household of a government officer with a monthly salary might choose different farming activities from a household without any regular source of cash income. Differentiation in subsistence production and income earning activities had also resulted in a wide variation in economic status of households in Tat hamlet.

The carrying capacity of composite swiddening for a given area is considerably higher than that of pure swiddening. This is because the paddy fields produce about one half of the grain needed to meet the local population's consumption requirements so that the per capita area needed for rice swiddens is only half that of a pure swiddening system. Additionally, the high diversity of components in the composite swiddening system reduces vulnerability to risks of crop loss due to weather or pests and provides some buffering against fluctuations in market prices of different crops.

Environmental degradation, especially the decline in the quality of land available for swiddening, has exerted a particularly strong force for change in the farming system. Because of growing land scarcity, farmers have intensified their cultivation of existing swiddens, extending the period of cultivation from 2-3 years to 4-5 years and reducing the fallow period from 5-8 years to 4-5 years. Consequently, as is discussed in detail in Nguyễn Thanh Lâm (2004), the fertility of the swidden land is declining and the productivity of swidden rice and cash crops falling accordingly. Many farmers have given up trying to grow swidden rice because they cannot find plots with sufficiently fertile soil to make the effort worthwhile. Others, unable to find sufficiently fertile land in the areas where swiddening is permitted, have illegally cleared plots in the forestland allocated to their households to protect.

Conclusions

Composite swiddening has successfully met the subsistence needs of the people of Tat hamlet for more than a century. It has supported a much higher population density than would have been possible with pure swiddening. The system has also displayed great flexibility by allowing farmers to modify their production strategies in response to new constraints and opportunities, most notably those associated with increased integration into the market economy. In recent years, however, the long-term viability of composite swiddening has come under threat, largely as a consequence of environmental degradation. Population growth, increased need to grow cash crops, and, especially, government restrictions on clearing of new swiddens on forest land have forced farmers to drastically shorten the length of the fallow period, to the extent that there is no longer sufficient time to fully restore the fertility of the soil. Unless ways can be found to accelerate the regeneration of fallow plots within a time frame of only 4-5 years, yields of the swiddens will continue to decline, threatening the future ability of the farmers of Tat hamlet to adequately meet their survival needs.

Acknowledgments

This chapter is a revised version of an article by Nguyễn Thanh Lâm, Aran Patanothai, and A. Terry Rambo that originally was published in *Southeast Asian Studies (Kyoto),* 42 (3) 2004, pp. 273-293. It is reprinted with permission of the publisher.

This research was funded by a Ford Foundation grant to the Center for Agricultural Research and Ecological Studies (CARES) of Hanoi Agricultural University. We would like to thank Patma Vityakon, Viriya Limpinuntana, Trần Đức Viên, and Stephen J. Leisz for their assistance during this research. We are also grateful to Xa Văn Lan and all the people of Tat hamlet who have cooperated with us during the field surveys. Finally, this chapter is based on research completed as a partial fulfillment for the Ph.D. requirements for the first author at Khon Kaen University, Khon Kaen, Thailand.

PART III
THE SOCIAL SYSTEM OF TAT HAMLET

Introduction

The four chapters in this part describe several key elements of the social system of Tat hamlet as they have been changing over time. Taken together, they provide an overview of the adaptive strategies employed by the people of Tat hamlet and how and why they have been modifying their composite swidden agricultural system and associated institutions and practices to adapt to changing circumstances. The outcome of these changes is the transformation of the hamlet from a largely subsistence-oriented system to one that is increasingly market-oriented. Each of the four chapters in this part deals with a different aspect of the social system:

Chapter 11 describes the social organization of Tat hamlet. It shows how the social institutions for managing land and forest resources have evolved over the past 100 years and suggests how these specific historical experiences, especially the nature of the several different administrative systems under which the hamlet has functioned as a dependent entity, have resulted in a relative absence of indigenous institutions for managing natural resources.

Chapter 12 explores the ways in which these composite swidden farmers have made changes in the use of their productive resources (land, labor, and capital) to adapt to changing economic circumstances, especially the shift from a managed economy to a largely free-market one. Different groups of households are shown to be employing different strategies for diversifying their sources of income as a means of ensuring their livelihoods. These strategies include not just changes in farming systems but also efforts to generate more off-farm income

Chapter 13 focuses on changes over time in the market itself as these affect the livelihood activities of the farmers in Tat hamlet. Although agriculture is still largely focused on food security, with households consuming much of what they produce, involvement in the market has increased considerably in recent years. Households now strive to earn cash income from the sale, mostly to lowland markets, of crops, livestock, and non-timber forest products. A growing share of households also earns income from various off-farm activities. This chapter describes the relationship between the market and farm household products, an enumeration of the various actors in market transactions, where these transactions take place, how prices are set and methods of payment.

Chapter 14 examines gender roles and some of the ways they have been changing as part of the larger transformation of the way of life in Tat hamlet. It shows how changes in livelihood strategies, particularly the pursuit of new sources of income and changing emphases within the traditional composite swidden system, have been affecting the gender-based division of labor.

Emphasis is placed on gender differentiation in the division of labor and in household decision-making, especially as these relate to the composite swidden system and the management of natural resources, including land and forests.

Taken together, these chapters reveal that an astonishing transformation of Tat hamlet has taken place in less than twenty years. In that short period, the hamlet has shifted from a system based on collective management of land and agricultural production to one reliant on management by individual households, and from an almost wholly subsistence-oriented economy to one that is now increasingly market-oriented. This transformation has affected everything from cropping patterns to gender roles to clothing styles to the use of certain natural resources.

Changes in the composite swidden system have been an integral part of this transformation. Major trends observed include a shortened fallow period and a decline in swidden areas, with some farmers abandoning rice swidden cultivation altogether. Cash crops have become an important part of the swidden system. At the same time, paddy field areas have been expanded wherever possible. There has been an increase in on-farm and off-farm supplemental activities, including diversification and expansion of various forms of self-employment as well as an increase in wage labor. In recent years there has also been a very substantial increase in the percentage of households using credit, and much of the money borrowed has been used to finance a dramatic expansion of the livestock component of the farm system. Farmers in Tat hamlet are shown to be evolving several different livelihood strategies in response to new opportunities and new challenges of their continuously changing environment.

11
Social Organization and the Management of Natural Resources in Tat Hamlet

A. Terry Rambo and Trần Đức Viên

Introduction

This chapter presents a review of the historical evolution of social institutions for managing natural resources in Tat hamlet. It describes the social organization of the hamlet during the colonial era and the Resistance War against the French, the time of collectivization under the cooperative system, and the current period of household-based management that began in the late 1980s with the implementation of economic reforms and adoption of the market system. It then describes the various institutions that influence contemporary management of natural resources.

Methodology

This chapter is based on a wide range of information collected between 1992 and 2001. Historical information on life during the feudal and cooperative periods was obtained from informal interviews conducted by the authors with elderly informants. Only relatively elderly people in Tat hamlet had had direct experience of the feudal regime and their memories were often not very clear about this period which came to an end in the early 1950s. Our informants often disagreed about specific details but they generally shared a common image of the overall form of the feudal regime. Memories about the cooperative period were much clearer and there was unanimity about the poverty and harshness of the life during that time.

Information about contemporary social organization was derived from observations made during our multiple visits to the hamlet, which included frequent extended stays in the houses of the villagers, as well as analysis of data on social behavior and attitudes that were collected as part of the CARES survey of 42 households in 1998-99.

The Importance of Community Social Organization in Management of Resources

In recent years the concept of 'community-based resource management' has gained great popularity among governments, NGOs, and international development assistance agencies as a solution to the problems of environmental degradation (Ford Foundation, 1997). Devolving power to manage natural resources from central government agencies to local communities has become the latest in a long series of magic bullets employed by development planners working in Southeast Asia's uplands.

As Agrawal and Gibson (1999) have pointed out, however, remarkably little attention has been paid to the question of how the internal social organization of villages constrains their capabilities for effectively managing natural resources that are entrusted to them. The prevalent assumption seems to be that rural villages are inherently endowed with the institutional capacity to organize successful collective action to use resources in an equitable and sustainable manner. Social solidarity, that most problematic of social science concepts, is simply assumed to exist, providing the necessary basis for cooperative management of community resources. But, as has been repeatedly revealed by empirical investigations of rural villages, social organization is not a constant. Some rural communities may approximate the ideal model, but others are characterized by high levels of inter-household competition, pervasive distrust among members of unrelated households, and a near total absence of community solidarity. Social capital, that elusive concept defined by Francis Fukuyama (1999: 16) as '... a set of informal values or norms shared among members of a group that permits cooperation among them,' is often in short supply. Indeed, it can be argued that in many such villages, social capital is in shorter supply than financial capital. In such situations, the villagers, whatever their individual intentions and desires, lack the institutional capacity to manage resources for the common good.

The existence of social capital is as problematic in villages in Vietnam's Northern Mountain Region (NMR) (Jamieson *et al.* 1998) as it is elsewhere in the developing world. Although it is commonly assumed by outsiders that all ethnic minority communities there are tightly organized entities displaying a high level of social solidarity, empirical investigations show that there is a considerable range of variation in the extent of their social cohesion (Lê Trọng Cúc and Rambo 2001). Many communities, including those of the H'Mông and Dao, display great cohesion and ability to collectively manage resources (Khổng Diễn 1996; Phạm Công Hoan 1995) as does the Black Thai community in Sơn La Province studied by Thomas Sikor and Đào Minh Trường (2000). Other communities are much less cohesive and have very limited endogenous capacity to collectively manage resources. Tat hamlet is definitely in the latter category. Given the paucity of community studies in the NMR, it is impossible to say how many communities are similar to Tat hamlet. Our impression, however, is that it represents a relatively extreme, although not unique, case of the scarcity of social capital and very limited capability for collective action.

Historical Changes in the Social Organization of Tat Hamlet

The social organization of Tat hamlet is not 'traditional' in the sense often attributed by ethnologists to indigenous peoples in upland Southeast Asia (McCaskill and Kampe 1997). In contrast to other ethnic groups in the Northwestern Mountains such as the H'Mông and Dao, the Đà Bắc Tày of Tat hamlet have always been members of a dependent 'peasant' society rather than belonging to an autonomous 'tribal' society. The hamlet has always been subject to powerful external forces that have dictated the form of its internal structure. Undoubtedly, this historical experience has greatly influenced the character of local social institutions and limited the emergence of communal capacity for resource management.

During the more than one hundred years since it was first settled, Tat hamlet has passed through four historical periods, each characterized by a different pattern of social organization. These periods are the feudal period, the Resistance War period, the cooperative period, and the current period in which the household is the key unit for management of land and natural resources. In each of these periods the balance of power between the individual households, local community governance institutions, and extra-local governance institutions has varied with consequent differences in the way in which resources have been managed.

The feudal period

From 1885 until 1945 Tat hamlet was part of the French protectorate of Tonkin, which was one of the five component 'states' of French Indochina. The French did not impose a uniform administrative system in Tonkin. In the Red River Delta, the administrative system of the Vietnamese state was largely left in place although at the provincial and district levels of government the Vietnamese mandarins worked under the direct supervision of French civilian administrators. In the northwestern mountains, however, active military resistance to French occupation continued into the twentieth century and French administrative presence did not extend beyond the main towns. A more indirect system of rule, working through powerful native chiefs, was established in areas that had been semi-autonomous Thái and Mường principalities in the pre-colonial period

(Rambo 1997a). Tat hamlet was no exception and, until the end of the French colonial period, it was a tributary unit in a larger feudal social formation ruled by Mường overlords. This regime was referred to as the *lang đạo* system. In this pyramidal administrative system, the present day districts of Đà Bắc and Mai Châu formed the single feudal domain (*châu*) of Mai Đà. The hereditary ruler of Mai Đà, who bore the title of *Tri Châu*, was Đinh Công Quyền.

The *châu* of Mai Đà was made up of several *làng*. The *làng* was an area equivalent in size to a modern district. Under the *làng* were several *mường*, each equivalent in size to a modern village (*xã*) or group of villages (*cụm xã*), and beneath each *mường* were several *xóm* or hamlets.

Tat hamlet was a tributary *xóm* of the *làng* ruled by a *lang đạo* named Đinh Công Phú who was a member of the Mường ethnic group. He controlled his domain from his 'capital' in Bến Diem on the south bank of the Đà River near Chợ Bồ (the administrative center of Đà Bắc district until it was submerged by the Hoa Binh reservoir and the government offices relocated to Tu Lý). Different informants used different titles to refer to the Mường overlord including *quan lang*, *lang đạo*, *thổ lang*, and *quan hội*. *Quan hội* is the most commonly used term, however, and will be employed in the rest of this account. This terminology distinguishes the Đà Bắc Tày from the main body of Tày in the north who refer to their hereditary leaders as *thổ ti*. The *thổ ti* were reportedly the descendants of Vietnamese mandarins who had intermarried with Tày women (Abadie, 1924).

Đinh Công Phú had inherited his position from his father, who had also served as *quan hội* for the French colonial authorities. The first son of the lord's first wife normally inherited the title. Đinh Công Phú is remembered as loud-voiced and a very strong man who could fight with a big buffalo. He had a violent temper and would beat any commoner who offended him. One 82 year-old man in Tat hamlet still has a scar on his leg from where the *quan hội* struck him with a stick in a fit of anger. Another informant reported that his 17 year-old elder brother was killed by Đinh Công Phú. Commoners, referred to as *dân đen* (black people), had to bow and hold their hands together in front of their faces when encountering the *quan hội*. They had to preface every sentence with the respectful particle *dạ* when addressing him. The *quan hội* was a member of the Mường ethnic group but also spoke Kinh and Tày. He lived in a very large house on stilts with six or seven rooms. He also had a brick house inherited from his father. He had several wives (some informants said six wives, others seven, one said nine) of Kinh, Mường, Tày, and Dao origin. Each wife had her own house separated from the lord's house by about 500 meters. The *quan hội* could force any pretty girl who caught his fancy to become his wife without paying bride price or having a formal wedding. No women from Tat hamlet or neighboring hamlets were taken as 'wives' by Đinh Công Phú.

The *quan hội* ruled his domain through appointed subordinate officials. Living in his house were four administrative officials who carried out his instructions. In descending order of power they were the *Cai Nhất*, *Cai Nhì*, *Cai Ba* and *Cai Tư*. These officials used conscript soldiers to carry instructions to lower-ranking officials in the villages and hamlets telling them how many subjects to send to perform corvée labor for the *quan hội*. Two soldiers were recruited from every hamlet by the local officials. Sometimes the *quan hội* would order a strong young man who came to his attention to join his force. The soldiers had a poor life but if they gained favor with the *quan hội* they might be given buffalo or cattle. Those who performed their duties well might be selected to become one of the *cai*.

The next lower level of administration of Đinh Công Phú's feudal domain was the *mường* (a unit resembling a modern commune that included several hamlets or *xóm*). It was headed by the *đạo mường* who was appointed by the *quan hội*. This was a non-hereditary position. Tat hamlet was under the authority of Mr. Ơm who lived in Diều Nội. The *đạo mường* was assisted by an official called *ông xã*. Each year five people from Tat hamlet had to do corvée labor for three days for the *đạo mường*. Informants are unclear as to what functions the *đạo mường* performed within the feudal system.

The administrative unit beneath the *mường* was the *xóm* or hamlet. Each hamlet was headed by a *đạo xóm*. The *đạo xóm* was selected by the residents of the hamlet and confirmed in his position by the *quan hội* who sent an official (one of the four *cai*) from his headquarters to announce the appointment of a new *đạo xóm* in a public meeting. This post was normally inherited by the eldest son of the former *đạo xóm*. If, however, he was not considered a suitable choice by the villagers, someone else could be appointed. The *đạo xóm* was expected to speak well, have prestige in the eyes of the people, and have the means to feed visiting nobles. A *đạo xóm* who failed to do his job properly could be replaced. The *đạo xóm* of Tat hamlet was quite wealthy with one hectare of paddy land and many buffalo. One person from every house in the hamlet had to work for one day each year cultivating his field. The *đạo xóm* killed a buffalo and gave a feast to reward these workers. On the anniversary days of their parent's deaths, the villagers had to present gifts to the *đạo xóm* and the *đạo mường*. The *đạo xóm* had an assistant called *ông xã* whose main responsibility was to go to the houses of people selected by the *đạo xóm* to do corvée to inform them to report for work. The *ông xã* took his meals every day at the house of the *đạo xóm*.

The main responsibility of the *đạo xóm* was to ensure that the commoners paid tribute to the *quan hội* in the form of corvée labor. The *quan hội* sent a *cai* or a soldier to the hamlet to tell the *đạo xóm* how many workers were required to report for duty in the next corvée cycle. The *đạo xóm* selected who would go from among the commoners. In theory each male 15 years of age and older had to perform one period of corvée each year but in practice many villagers only went a few times in their lives. Widows whose children were old enough to stay alone also had to do corvée. The *đạo xóm* was exempt from corvée and the *ông xã* only had to perform corvée if the hamlet was short of its quota of workers. If a person were too old or sick to work he had to pay for a substitute. If a person failed to report for corvée the *quan hội* would send soldiers to take away any valuables from his house. The *đạo xóm* would also be punished by the *quan hội* if he failed to provide enough workers for the corvée.

There were two forms of corvée: *làm phu* and *làm phiên*. *Làm phu* involved spending one month doing heavy labor for the *quan hội* in his capital village. The workers had to clear and cultivate his paddy fields and swiddens, collect firewood and bamboo for his house, do construction work, and watch his herds of buffalo. Workers were called to do *làm phu* only as labor was needed. *Làm phiên* involved a 20-day tour working as a servant in the houses of the *quan hội* and his wives. At any one time there were 10 people performing *làm phiên*. According to one informant the people doing *làm phiên* had to bring their own rice to eat. Those doing *làm phu* did not bring rice with them but had to forage for food in the forest because they received no support from the *quan hội*. Other informants recall that the *quan hội* provided the workers with food but it was of very poor quality, just three small tea cups of banana roots mixed with a little rice. If, at the end of their tour their replacements had not arrived they had to remain in service until they did. In any single year villagers had to perform either *làm phu* or *làm phiên*, but not both.

The commoners in Tat hamlet did not have to pay the *quan hội* taxes on their harvest in rice. This probably reflected the remoteness of their hamlet, which made transporting rice to his granary uneconomic. This is in accord with one generalized ethnological description of the colonial Tày social system, which reports that commoners in settlements located close to the residence of the noble were required to cultivate special fields in their village that were reserved for support of his household whereas subjects living in more distant settlements were required to come to the noble's village to perform corvée service instead (Viện dân tộc học 1978). Registered adult males were supposed to pay an annual head tax of eight silver piastres but many people evaded this tax. People were also required to pay other kinds of tribute in kind to their overlord. For example, successful hunters were supposed to present the *quan hội* with a choice cut of any large wild animals that they trapped or shot with their crossbows (commoners were not allowed to own firearms which were exclusively reserved to the nobles). They also had to present a share of any wild honey they collected in the forest to his wives. They were not required, however, to give a share of fish to the nobles.

During each Tet Lunar New Year, the *đạo xóm* of Tat hamlet collected three liters of alcohol, one kg of rice, and a sticky rice cake (*bánh chưng*) from each household and took these along with valuable gifts of forest products such as squirrels, porcupines, and bear gall bladders to present to the *quan hội*. He would be invited together with all of the other hamlet leaders from the domain to eat together with the *quan hội*. If the gifts he brought pleased the *quan hội* he might be rewarded with the higher title of *Đạo Nhật*.

The feudal regime was concerned primarily with extracting labor power from the village. It did not attempt to regulate land use or tax agricultural production. Exploitation of forest resources was also largely unregulated. Villagers could freely fell large timber trees for construction of houses. Management of land was in the hands of individual households. Whichever household first cleared a low-lying area and constructed a wet rice field there was acknowledged by all others to have permanent rights to that plot. Households could clear swiddens wherever they chose to in the forest. They had control over the plot only as long as they planted crops on it, but retained no special rights to it after it was fallowed. Wild plants and animals could be collected freely by anyone in the forest. Because of the remoteness of the hamlet from large population centers and the lack of road access, outsiders were not interested in exploiting its resources and the colonial state showed no interest in its affairs. There was no nearby market so people who needed to buy or sell anything had to walk across the steep mountains for one whole day to the market in Phú Thọ province. They usually sold roots collected from the forest, mainly the *củ nâu* tuber (a type of wild yam) used as a brown colored dye stuff and the *rễ đỏ* bark that was chewed with betel. Thirty kg of wild yams bought from 40 to 60 kg of rice. Population density was very low and unused land and forest abundant, so competition for resources among villagers was low and the need for regulatory institutions minimal.

The Resistance War period

In 1945, the Japanese seized control of Vietnam from the French and Japanese troops occupied the main strategic points in the mountain region. Informants, in nearby Ênh hamlet recall that Japanese soldiers ordered them to plant kenaf in their paddy fields. The period of Japanese control was too brief to have any major direct effects on the lives of people in Tat hamlet although, by shattering the colonial administrative system, it opened the way for the Viet Minh-led forces of national liberation to establish their own control of the mountains.

The French colonial regime operating through the feudal *lang đạo* system was a parasitic system that employed force to extract labor and tribute from the commoners of Tat hamlet but provided no real services in return. It is thus hardly surprising that the people in Tat hamlet sided with the Viet Minh during the long war of resistance against the French (1947-54). The whole of Đà Bắc district became a resistance base area. The *quan hội* fled to Hòa Bình town where he served as an advisor to the short-lived Mường Autonomous Zone established by the French as part of their divide and rule strategy of setting the ethnic minorities against the Kinh. His son, Đinh Công Đọc, however, joined the Việt Minh and served as a combatant in the resistance forces in Sơn La as did the son of the *đạo xóm* of Tat hamlet.

Việt Minh cadre organized a guerrilla force in Tat hamlet. Although no major battles were fought there, the people suffered badly during the long war. In 1945 a French column passed through the hamlet. Local people were forced to carry rice for the French soldiers. Two young men who fainted from hunger and fatigue on the long trek were thrown into a stream and left to die. After they regained consciousness they asked a Mường man for directions to return to their hamlet. In 1947, French soldiers returned and burned the hamlet. They killed two men, one a Đà Bắc Tày, the other a Kinh. The Tày was Ông Đào Đạc who was the *đạo mường* serving under Đinh Công Phú. Because he was riding a horse and carrying a gun the French mistook him for a Việt Minh cadre. They stabbed him and cut out his liver. The inhabitants then evacuated to Suối Mương and Cọ Nôm away from the main trail.

In 1948 another French column passed through Tat hamlet and ambushed a high-ranking Việt Minh cadre. Afraid that the French forces would attack the area again the people built small houses scattered in the forest where the elderly and children stayed. They also dispersed their rice granaries in the forest. Theft of rice became a problem and it was during this period that they abandoned their former custom of storing rice in separate granaries and began keeping it inside their houses, a practice that continues today. The greatest fear was of aircraft. The hamlet was never bombed but people recall seeing observation planes (which they referred to as 'old woman planes') flying overhead. They say that Tu Lý and Phú Thọ were bombed in 1951 and 1952.

After the final victory of the revolutionary forces in 1954, the large rice field belonging to the *đạo xóm* was confiscated and distributed to other families. During the land reform of 1956 the former *đạo xóm* was punished by being sent to live in the neighboring Dao hamlet of Yên for one year. His son and daughter-in-law accompanied him. The son, who had served in the resistance forces, was later rehabilitated and became a high level cadre in the district.

The cooperative period

The cooperative period was divided into two phases, a period from 1957 to 1959 when the villagers were encouraged to join a mutual aid team (*tổ đổi công*), and a period from 1959 until the mid-1980s when Tat hamlet was a full fledged cooperative (*hợp tác xã*).

The mutual aid team in Tat hamlet was organized in 1957 when cadre from the district and the hamlet began mobilizing the villagers to participate. People who had no links with the former feudal and colonial regime were picked by the cadre to serve as leaders of the team and confirmed in office by an election in which villagers raised their hands. All 17 households then in Tat hamlet joined the mutual aid team.

The mutual aid team had a leader (*tổ trưởng*), vice leader (*phó tổ trưởng*) and secretary (*thư ký*). The team leader was responsible for overseeing productive activities and trying to improve living conditions in the hamlet. The vice-leader was responsible for overseeing daily productive activity, particularly in the paddy fields. The secretary recorded all labor contributions by members, taking into account their health and physical capacity and whether the work done was light or heavy to ensure that contributions were counted with equitability.

Before each planting season, each household requested whatever assistance it needed from the team. The leader held a meeting to allocate labor, deciding which household was to be helped first, which helped last. Poor households and households lacking a sufficient number of workers were given preference. The leader then allocated work accordingly with the unanimous consent of the team. The team then worked on the land of each household in turn. People who could not work hard or who were incapable of work had to pay paddy rice or corn to the team in return for its assistance. The team leader decided on the amount to be paid after discussing this with all team members.

At this time Tat hamlet had about 12 ha of paddy fields. Only local varieties of rice were grown and yields were low. Approximately half of the households suffered from food shortages. They had to clear swiddens and dig wild roots in the forest to get sufficient food. Better-off households were encouraged by the mutual aid team to make interest-free loans of grain to households suffering from hunger.

The forest was still very dense with many large trees. Wild animals were abundant and troops of monkeys destroyed corn in the more distant fields. No one was responsible for managing the forest, which was considered an 'open access resource,' and everyone was free to cut trees and collect forest products, hunt wild animals, and clear swiddens as they pleased.

At the beginning of 1959 the mutual aid team was converted into a cooperative (*hợp tác xã*). All paddy fields, buffalo and cattle, and plows and harrows were collectivized. Initially only 20 out of the approximately 25 households then living in the hamlet joined the cooperative. Households owning several head of livestock were reluctant to join and have their animals become common property but after a mobilization campaign all joined the cooperative.

A cadre proposed the names of seven individuals to become leaders of the cooperative and from this list the members elected five: the chairman, vice-chairman, accountant, and two members of the administrative committee. Initially, the cooperative had a single production brigade (*đội sản xuất*), which was led by the former leader of the mutual aid team. Later it was divided into two brigades based on the location of member's houses but livestock were borrowed back and forth between the brigades depending on need. The two brigades contributed 20 percent of the rice and corn each produced to a common cooperative fund that was used to purchase additional livestock and agricultural implements and pay for meetings. In 1967 the cooperative was reorganized into three production brigades but later reverted to having two brigades.

Paddy fields were worked collectively, cattle and buffalo managed by the cooperative, and large agricultural tools treated as common property. Buffalo and cattle were allowed to range freely. Individuals responsible for herding them received one labor point per animal per day. Usually one herder looked after ten animals. Ten labor points per animal were awarded on days when the buffalo were used for harrowing. Members of the youth association enthusiastically collected manure because they received three kg of paddy rice for every 100 kg of manure.

The harvest was divided among cooperative members according to the number of work points they had accumulated. Individuals who worked hard were supposed to receive more points than those who did not. Cheating to get extra work points was widespread. Informants talk with some glee and no evident disapproval of women who would split the rattan ties used to bind the bundles of seedlings into two in order to claim that they had transplanted a greater number of bundles than they actually had. Although these individuals may have seen themselves as tricking the cooperative they were in reality taking food away from their neighbors. Of course, such trickery was hardly a problem unique to the cooperative in Tat hamlet; cheating was also a problem in cooperatives in Kinh hamlets in the Red River Delta but there social institutions evolved to control this problem. For example, several households would be assigned responsibility for caring for a single buffalo with each family in turn feeding the animal for a set period of time (Bouahom *et al.* 1993: 121). Given the time and energy needed to collect fodder in the densely populated delta, it was expected that each household would try to economize by short weighting the food ration it provided. This was avoided, however, by the practice of all of the households gathering at the stable at feeding time to watch the fodder being weighed.

The leader of the production brigade recorded each individual's labor points on a daily basis and submitted the record to the cooperative's accountant after each harvest. The accountant divided the total harvest by the total number of points accumulated by all cooperative members to determine the value of one point. The value was very low, with one point valued at approximately 30 to 50 grams of paddy rice. One informant recalls that the household with the maximum number of points received only 300 kg of paddy rice for one crop, which would be only enough to feed a household of six or seven people for a couple of months.

In the attempt to increase rice production, labor was mobilized to build new terraced fields but many were subsequently abandoned because of lack of sufficient irrigation water. The cooperative also directed people to plant tree gardens of *Livingstonia* palms to provide thatching for house roofs.

Because many households lacked food for four to six months a year they relied on forest resources to make up the short fall. Many people collected wild yams (*củ nâu*) in the forest to sell in Yên Lãng market in Phú Thọ. A load of yams sold for only 1 dong, but 1 dong would purchase 20 kg of rice. At that time cash was used only to buy rice and clothing. No one bought kerosene but instead everyone used resin collected from forest trees for lighting.

Use of the forest was unregulated and individuals could exploit it as they chose. To produce additional food every household cleared swiddens in the forest and kept all of the produce for their own use with no contribution to the cooperative. Beginning in the late 1950s, elimination of shifting cultivation had become a national policy objective, but in Tat hamlet people continued to quietly engage in the practice and no one made any serious effort to control or stop them. In 1978 government cadre announced a new program in which anyone who wanted to clear swiddens was permitted to do so but in turn was required to plant Styrax trees after harvesting the rice, in an agroforestry system resembling the *Taungya* system of colonial Burma. Later, the same system was applied to planting of tea in the swiddens. For this work, the state paid people 192 dong/ha or about 35 kg of rice/ha (the official price of rice at that time was 45 dong/10 kg). Not surprisingly, the villagers began clearing more swiddens, destroying a large area of forest in the process, since they gained the double benefit of having a larger rice harvest and receiving cash from the government for planting trees.

Known by the villagers as the 'regime of the gong' (*chế độ kẻng*) after the large brass gong that was beaten to call people to collective work in the fields, the time when the hamlet was managed as a cooperative is remembered today with mixed feelings. Some informants recall that it was enjoyable to work together in large groups with young men and women taking advantage of the many opportunities for flirting and courtship that this offered. Other informants, mainly older women, blame the spread of gambling among men on the numerous times when they sat around with nothing else to occupy them while waiting for instructions before starting work. But all informants agree that productivity was low and hunger a constant problem during the cooperative period. No one expresses any desire to return to those difficult days.

The household management period

During the latter part of the 1980s, the management role of the cooperative was gradually downgraded as the collectivized economy was replaced by the household economy. The collective labor regime was abandoned and rights to paddy fields assigned to individual households. Initially, households worked the land and retained whatever share of the harvest exceeded a fixed amount that they had contracted to supply to the cooperative. Later, the contract system was replaced by the current system in which households retain all of their harvest from the paddies except the relatively small share they had to pay to the state in the form of agricultural land taxes until this tax was abolished a few years ago. Households are also supposed to make annual contributions to support the work of the cooperative and the hamlet administrative system. Although the unwieldy and inefficient production brigades have been disbanded, the organization of labor retains a semi-collective character as the customary institutions of inter-household labor exchange are still employed to mobilize additional workers during land preparation, transplanting, and harvesting of the paddies.

Contemporary Social Organization of Resource Management in Tat Hamlet

In recent years, management of resources has become more complicated and now involves the cooperative, local government, and mass organizations, the state and its specialized agencies, as well as the individual households. The cooperative has retained control over some forestland but most land has been allocated to individual households. The state plays an increasingly visible role in forest management by imposing legal restrictions on resource exploitation, supervising allocation of forestlands to households, and collecting taxes on land and resources. The households make their decisions about resource management within this institutional context.

The roles of the cooperative, hamlet government, and mass organizations

Although it no longer plays an important direct role in the management of labor and distribution of agricultural production on the basis of labor points, the cooperative retains important functions in regulating land use, taxation, and maintenance of social order. All households in the hamlet must belong to the cooperative in order to live in Tat hamlet. If someone not native to Tat hamlet wants to come to live there he must be accepted as a member of the cooperative in order to stay. Every household member age 18 or older has the right to vote in cooperative elections. The cooperative in its present form has six officers, a chairman, a vice chairman, an accountant, a treasurer, and two control cadre (*cán bộ kiểm tra*). The chairman is elected for a two-year term by members of the cooperative. His election must be confirmed by the Village People's Committee. The chairman is supposed to receive an allowance of 50-60 kg of paddy rice per harvest, paid out of a fund collected from contributions made by all member households in the cooperative. Because of poor harvests in recent years only 25 households have made any contributions so that the chairman has not received his full allowance.

The cooperative is coterminous with the hamlet government and informants rarely distinguish between the two although they are formally separate organizations. The hamlet has a headman (*trưởng bản)* who is assisted by an administrative committee composed of the leaders of the local mass organizations and the security and militia units. The headman receives a salary of VND 40,000 per month, half paid by the Village People's Committee and half contributed by the district government. Administrative Committee members do not receive a regular salary but receive a share of the paddy rice contributed by cooperative members each year. In fact, many households in Tat hamlet fail to make any contribution so that only a few dozen kg of paddy rice are collected each year. Thus, administrative committee members are essentially unpaid volunteers.

The hamlet headman and the cooperative chairman share a number of responsibilities including:

- Taking part in important social functions such as funerals and weddings
- Dealing with fights and conflicts among neighbors. The hamlet headman accompanies the security unit to the scene, serves as a peacemaker, and decides on appropriate punishment.
- Collecting contributions to the cooperative fund
- Attending meetings of the Village People's Committee
- Allocating payments to households of funds received from Program 327 for protecting forestland. (In theory each household was to receive VND 47,000 per ha per year from Program 327, but in 1997 the cooperative received only enough funds for 16 households so the chairman had to divide this amount evenly among 79 households.)
- Overseeing management of lands remaining under the control of the cooperative.

The cooperative retains control of 55 ha of hill slope land divided into four tracts. Two plots of brush and grass covered slope land at Cọ Nôm (20 ha) and Suối Mương (10 ha) are reserved for households to clear swiddens. Two tracts of old secondary forest totaling 25 ha are also protected by the cooperative, although hamlet residents are allowed to collect bamboo shoots and cut bamboo stems there.

Until recently the two tracts reserved for swiddening were used in an alternating cycle of approximately four years with one tract being cultivated while the other was in fallow. After the cooperative announced that a tract was open for cultivation households were free to pick their own plots to clear for swiddens. Once they selected their plot the household would erect simple bamboo markers on its boundaries to let others know that the plot was already taken. After clearing the field they could farm it until the cooperative decided that it was time to return the tract to fallow at which time they would clear a new field in the newly reopened alternate tract. They retained no residual rights in the abandoned field. Since the beginning of the 1990s, however, the alternating system has broken down under the pressure of growing population and resulting shortage of land for swiddening. Households now maintain swiddens simultaneously in both tracts. A household may periodically fallow a plot but retains control over it and may clear it again for a swidden whenever it is ready to be planted again.

The cooperative also attempts to protect remaining mature forest in the two tracts under its direct control as well as in areas designated by the District Forest Protection Office as protected watershed forest. Households are permitted to collect bamboo shoots and cut bamboo stems in these areas but clearance of swiddens is forbidden. Households that clear swiddens inside protected areas are subject to fines and are required to replant trees on the illegally cleared plots, but are first allowed to cultivate the land for two or three years. The fines are quite low (800 VND/m^2) and many land-hungry households are willing to risk the penalty since there is a good chance they will not be caught if they clear their plot in an area not visible from the road. In the last several years, illegal cutting of swiddens and destruction of forest by fires escaping from burning swiddens has been on the increase but the cooperative and hamlet leaders feel powerless to control these activities. Although encroachment on protected forestland frequently occurs, villagers are quick to admit that there would be no forest left at all if it were not for the efforts of the cooperative and the State to protect it.

In addition to the cooperative and the local government, five mass organizations (the Women's Union, Farmer's Association, Elderly People's Association, Veterans Association, and Youth Association, all organized under the control of the Fatherland Front), are active in Tat hamlet. Each organization is tasked with encouraging members to improve their livelihoods and to protect the forest. For instance, the Women's Union is divided into nine neighborhood groups, with each group enrolling 9-11 persons. Members pay monthly dues of VND 500, which is similar to other associations. The group in the center of the hamlet has collected enough money to buy three cattle that are cared for by different members in rotation. The main objectives of the Women's Union are: (i) transferring knowledge and technology to its members, (ii) helping each other to do farming; and (iii) managing a micro-credit fund that makes small loans to members to improve their farming. Before each cropping season, each group of women has a meeting to discuss what type of crops they will select for planting.

The role of the state in resource management

The state exerts a strong influence on resource management in Tat hamlet through a variety of direct mechanisms including imposition of legal regulations, formal allocation of land rights, taxation, and provision of public services and construction of infrastructure, as well as the indirect impact of its activities elsewhere such as the in-migration of people into the hamlet who were displaced by the construction of the Hòa Bình reservoir.

Legal regulations and enforcement

In principle, use of resources in Tat hamlet, as everywhere in Vietnam, is regulated by uniform national laws enforced by state agencies such as the Forest Protection Department of the Ministry of Agriculture and Rural Development.

District forestry cadre prepared a land use map for Tan Minh village identifying areas of protected forests. Logging and swiddening are totally banned in these areas. In areas that are not classified as protected forest, trees may be cut by members of the hamlet for purposes of building or repair of their houses but not for sale. They are supposed to obtain permission to cut a certain number of trees from the Village People's Committee and pay the natural resources tax to the district on this timber. In 1998 one informant, who had just completed building a new house using timber from the forest, said that he had paid a tax of 100,000 VND.

A low-level forest protection cadre is assigned by the district to the village to oversee forest protection but enforcement of regulations on land use is the responsibility of the Village People's Committee and the cooperative and hamlet leaders. Given the lack of means and will on the part of the authorities to enforce forestry regulations, many young men engage in illegal logging to earn cash. Households also clear swiddens in protected areas with little fear of being punished. The hamlet authorities are aware that such infractions are occurring but say that they cannot stop them because people are hungry and have no choice but to break the laws.

Allocation of forestland to households

Most forestland has now been allocated to households. Some forested hill slope land was assigned to households to manage even during the cooperative period, with one informant reporting that her family had been given control of a plot of regenerating secondary forest near her house in 1975. Palm tree gardens planted by the cooperative were transferred to households at the same time as the paddy land was placed under their control. Allocation of remaining forestlands and issuance of 'red books' to households is still in progress.

Forestlands have been allocated to households by two different district offices, the Forestry Office and the Land Office. In the early 1990s the Forestry Office allocated approximately 100 ha of protected forestlands on hilltops to households. The households are expected to protect the forest on the plots assigned to them and not to cut timber or clear swiddens. The Forestry Office has issued certificates to these households but does not pay them for protecting the forest. Later, the Land Office allocated about 100 ha of hill slope land to households under the 327 Program and subsequently issued 'red books' to recipient households. These lands are supposed to be kept under forest but are in fact frequently cleared for swiddens. According to the hamlet headman each household was supposed to receive two plots, one for protection and one for cultivation.

Land allocation has been imperfectly implemented in Tat hamlet. At the western end of the hamlet plots allocated by the Forestry Office and the Land Office overlapped so that in some cases the same plot was assigned to two different families, each of which received a certificate or 'red book' for it. A more serious problem resulted from the failure of the Land Department cadre to make an actual map or ground survey of plots allocated under the 327 Program. Rather than climbing the steep slopes to survey the fields they stayed on the road and asked people to point out which plots they wanted. Households were subsequently issued red books, which showed the outline of their plot but not its actual location on the landscape. In May 1998 the Land Office cadre returned to the hamlet to show people the base map on which the location of household plots was marked. The new hamlet land map was beautifully drawn but, unfortunately, largely inaccurate. Few of the plots assigned to households were in the correct location. One informant told us that her family had controlled the land behind the waterfall next to her house for more than 35 years, but this plot had been assigned to a household located across the valley, while the plots shown on the map as being assigned to her in reality belonged to other households. The Land Office cadre asked to take back the red books to correct them, but the people refused to surrender their books, since they did not trust the cadre to correctly rectify the errors. One unfortunate consequence of this administrative mistake was that many people began to rapidly cut trees on their plots since they feared that their rights to this land would not be recognized officially. Given their historical experience of frequent changes in land tenure policies, this seemingly shortsighted response cannot be dismissed as being wholly irrational.

Taxation

The state actively seeks to regulate resource exploitation in order to protect the environment in Tat hamlet but at the same time it profits from resource exploitation by means of a variety of land and resource taxes. Although taxation is used in many countries as a policy lever to influence the way in which resources are used (e.g., the oil depreciation allowances in the American tax code that are designed to stimulate exploration for new petroleum reserves and the 'carbon tax' being used in the EU to reduce fossil fuel consumption to fight global warming), Vietnamese land and resource taxes appear to be intended only to generate revenue for the state.

The villagers are subject to a variety of state taxes collected by the Inter-village Tax Station, which is located next to the People's Committee office. The Tax Station has five staff. The head of the station is employed by Tân Minh village but the other staff are assigned there by the district tax office. All of the tax officials are paid by the district. Six kinds of tax are collected in Tat hamlet:

1. Agricultural Land Tax: Before 2000, when the state abolished all agricultural taxes, paddy land in Tat hamlet was classified into three categories in descending order of productivity. Category 4 was taxed at 28 kg of paddy rice/1,000 m^2 year. Category 5 was taxed at 18 kg of paddy rice and category 6 at 5 kg of paddy rice/1,000 m^2 year. The tax was calculated in paddy rice but paid in cash. The amount depends on the market price of paddy rice. According to the record book at the Tax Office, in the first 6 months of 1997 households in Tat hamlet paid 2,447 kg of paddy rice in agricultural land tax. At that time paddy rice was valued at VND 1,400/kg. About 10 households had failed to pay any tax that year. They are permitted a one-year grace period but can lose their land if they still fail to pay. Late payments are calculated at VND 1,700/kg of paddy rice so there is a significant penalty for late payment.
2. Mixed Garden Tax: Swiddens are now classified as 'mixed gardens' and are taxed at the rate of 8 kg of paddy rice/1,000 m^2. In the first 6 months of 1997 households in Tat hamlet had paid 301.8 kg of paddy rice in this tax. In view of the fact that the area cultivated in swiddens is vastly larger than the area of paddy land, there would appear to be widespread evasion of this tax.
3. Residential Land Tax: House sites and surrounding home gardens are taxed at the rate of 8 kg of paddy rice per 1,000 m^2 of land. A total of 428.6 kg of paddy rice had been collected from Tat hamlet households in the first 6 months of 1997.
4. Natural Resource Use Tax: This tax is collected on products extracted from the forest including timber, firewood, bamboo and bamboo shoots.

Only forest products that are sold are subject to taxation. Different rates are charged on the value of different forest resources. Timber is taxed at a 40 percent rate, pulpwood at 20 percent, bamboo and bamboo shoots at 10 percent, and firewood at 5 percent. No records were available for the taxes actually collected on these products in Tat hamlet.

5. Slaughter Fee for Livestock: A fee of VND 30,000 per head of cattle and VND 25,000 per pig is collected for any livestock that are slaughtered to sell the meat. Smaller livestock (e.g., goats, chickens) are not taxed. In the first 6 months of 1997 4 cattle and 16 pigs had been subject to this tax.
6. Business Tax: A tax of 4 percent is collected on the sales volume of shops. In the first six months of 1997 a total of VND 1,668,000 was collected from shops in Tat hamlet.

Provision of public services and construction of infrastructure

In recent years, state investment in mountain area development has increased dramatically, including in Tat hamlet. Public services and infrastructure are largely funded out of the state budget. These expenditures are heavily subsidized from the central budget since the taxes collected in Tat hamlet cover only a small fraction of government expenditures in the hamlet. The health clinic, post office, and the primary school, despite charging fees, are heavily subsidized by the state (Plates 11.1 and 11.2). Salaries of hamlet and village officials and leaders of mass organizations are also provided by the state budget. This represents a not inconsiderable inflow of cash into the economy of the hamlet.

One of the largest state investments was the construction of a power line that passes through the hamlet on its way from the Hòa Bình hydropower generator to the more remote villages in the western part of Đà Bắc district. Initially, only the People's Committee office in the village center was connected to it but gradually most households have also been connected and must pay monthly usage fees. Another major state investment has been the recently completed upgrading of the road that connects the hamlet to the district to all weather status. This has greatly increased the accessibility of the hamlet. Several trucks and buses pass through every day providing a relatively cheap and convenient way of moving people and goods to and from the lowlands. Improved access to markets has greatly increased the intensity of exploitation of forest resources as people seek to obtain cash to buy food and consumer goods. When the harvests are bad, as they have been in several recent years, this pressure is increased as people desperately search the forests for anything that can be sold to obtain cash needed to buy rice. This has had a very negative impact on biodiversity as timber and medicinal plants are exploited at well beyond sustainable levels. From the perspective of the villagers, however, improvement of access to the market is seen as wholly

Plate 11.1: New post office in the village center (Nguyễn Đình Tiến)

Plate 11.2: New primary school in village center (Nguyễn Thanh Lâm)

positive. As one informant remarked, ‘now when there is a bad harvest we can collect bamboo and sell it to buy rice whereas before all we could do was to dig wild roots from the forest in order to survive.’

The state has also encouraged people in Tat hamlet to participate in a variety of tree planting projects. In the early 1960s, President Hồ Chí Minh encouraged people in the mountains to plant Melia trees. The people in Tat hamlet did not understand why this was necessary since at that time they still had plenty of natural forest but complied anyway. Later, the district chairman actively promoted planting of *Aleurites montana* (candlenut) trees that yield tung oil. Many farmers planted groves of these trees but have difficulty in disposing of the crop because the market is limited. Shops in the hamlet purchase the seeds for VND 1,000/kg. In 1989-90, eucalyptus planting was promoted by the World Food Program (PAM) which provided seedlings and gave farmers 370 kg of rice for each hectare planted (out of which they had to pay the program 50 kg for transportation fees for the seedlings). The trees have grown well but the market for the wood is poorly developed. In any case, farmers say they no longer plant eucalyptus because it is bad for the soil but they did not know that at the time they planted it.

Indirect effects of state actions

Construction of the Hòa Bình dam, one of the largest state projects in northern Vietnam, displaced thousands of households from their homes in the valleys flooded by the reservoir (Vũ Quyết Thắng, 1991). Three households relocated to Tat hamlet thus increasing already heavy pressure on land and resources in the hamlet.

The role of the household in resource management

Although the roles played by the local government and the state cannot be ignored, it is individual households that make most decisions about resource management in Tat hamlet today. Increasingly, their decisions are driven by the imperatives of involvement in the market economy but they are still predominantly subsistence-oriented rather than seeking to accumulate capital. Having enough to eat is the first priority and their use of resources is largely determined by that very immediate goal.

Each household is a largely autonomous unit of production, consumption, and capital accumulation. Like peasant households everywhere in the developing world, the major objective of each household in the hamlet is to reproduce itself (the successful raising of children who in turn form autonomous households modeled after those of their parents), not accumulation of investment capital and striving for upward social mobility.

Although a limited amount of sharing of food and other goods occurs among different households, especially those linked by kinship relations, each household is essentially a self-contained economic unit. Its success or failure in the game of life is wholly dependent on the skill and willingness to work of its members. There is no social safety net to catch households that fall into poverty as a consequence of bad luck (illness, natural disaster) or personal shortcomings (laziness, gambling or drinking). The margin between success and failure is very narrow and the slightest perturbation can undo many years of hard work and careful accumulation of household wealth. Thus, one household with which we are familiar went from being one of the wealthiest to one of the poorest in the hamlet in less than five years after the male head of household died unexpectedly. As a retired cadre he had received a substantial pension but this terminated after his death. His widow was left to raise two teenagers, a son and a daughter, and to care for her 80 year-old mother. Unfortunately, the son, freed of paternal guidance, contributed little to the maintenance of the family and squandered its resources on drinking and gambling. He married and after a short period of bride service at his in-law's house brought his new wife to live with his mother. The new daughter-in-law was a willing worker but quickly became pregnant so added little to the labor power of the family. She gave birth to two children in two years and only recently has begun to be able to work full-time in the fields. The daughter did her best to assist her mother until she became engaged and had to devote much of her effort to accumulating her dowry. Then she married and left the house to live with her husband's family. The female head-of-household, a woman already in her 50s, became the sole supporter of a family with six mouths to feed. She worked incredibly hard but her crops were devastated for three years running by unusually severe outbreaks of pests and bad weather. Thus, bad luck combined with the personal failings of one of its members undermined the economic status of this household so that it declined into a state of extreme poverty.

Unlike Kinh communities in the Red River Delta that maintain institutional mechanisms (e.g., the cooperative social fund) for assisting households that encounter misfortune, particularly those headed by elderly widows, Tat hamlet has no community-wide social welfare institutions. Asked to whom they would turn for help in the event of disaster, 88 percent of the 42 households we interviewed in 1998-99 replied that they would seek help from close relatives. Only one household would seek help from unrelated friends or neighbors.

Not only are households in Tat hamlet wholly self-reliant but they must constantly be on guard to protect their assets from depredations by fellow villagers. Theft is a constant concern and loss of crops and livestock common. One poor widow was eagerly anticipating harvesting the ginger in her swidden, which would provide some much needed cash to buy food for her hungry household. A few days before she planned to harvest the ginger the entire crop was stolen from her field during the night. She was convinced that the thieves were fellow villagers, which is highly probable since no outsiders would be able to follow the poorly marked mountain trails to her swidden in the dark. As harvest time approaches, many families assign a member to sleep in a field hut in their swidden, in order to protect the harvest from predators, both four and two-legged.

Informants also express considerable concern about their ability to protect the forest plots that are assigned to their care. Despite the fact that land has been allocated to households, many villagers still view the forest as an open access resource where they feel free to appropriate anything of value that they happen to find. On one occasion an informant terminated an interview in mid-course to go and check his plot when he heard the sound of axes in the forest near his land. Lack of security is a major constraint on investing in tree planting on plots located far way from the settlement where protection is difficult.

Destruction of tree seedlings by livestock left free by neighbors to forage in the forest is another important concern. One poor farmer in neighboring Ênh hamlet complained bitterly of how his newly planted tea garden had been trampled by buffalo belonging to 'rich families.'

Some cooperation does occur between households but under rigidly defined conditions. Neighbors make contributions of food, alcohol, and cash for weddings and funerals in the expectation that their donations will be reciprocated when they in turn must host such events. Families that need to replace the palm leaf thatched roof of their house, a task that must be done every ten years or so, invite 10 or 15 households in their labor exchange group to help in this difficult task (Plate 11.3). The host family provides one half of the 300 leaves needed to complete the job while the helpers bring the rest. Each household sends one member to work for one day and also contributes rice and alcohol while the host family provides meat for a feast to reward the workers.

Much of the work of cultivation of paddy fields is done by means of labor exchange groups involving members of several different households. Harrowing of the paddy fields is done by teams of several men with each of the men guiding a buffalo that pulls the wooden harrow. Transplanting is done by groups of 10 to 12 women. Swiddens are sometimes planted by large groups of 50 or 60 people (Plate 11.4). In that case, the host household provides a meal to the workers but no meals are provided for helpers in the paddy field labor exchange work. Rules for reciprocity in labor exchange are strict and a household owes one labor day to every individual from whom it receives assistance. A household that fails to fulfill its obligations can easily be sanctioned by refusal of others to provide help

Plate 11.3: Neighbors assist family in re-roofing house (Nguyễn Đình Tiến)

Plate 11.4: A labor exchange group planting a swidden field (Nguyễn Thanh Lâm)

when it needs it. Thus, the existence of labor exchange practices should not be seen as providing evidence for a strong sense of solidarity in the hamlet.

Allocation of community resources is not always done in a fair and unbiased manner. Holders of positions of authority can and do take advantage of their power to divert resources to themselves. On one occasion an office holder kept all of the most valuable varieties of fruit tree seedlings that were donated to the hamlet and only distributed the inferior varieties to the other households. A poor individual in a neighboring hamlet complained that 327 Program funds were only given to a few wealthy households including the vice chairman of the cooperative. Although distribution of 327 Program funds in Tat hamlet has been done much more fairly, people nevertheless suspect that some of the money has been misappropriated. Regardless of the actual honesty of office holders it is assumed by their neighbors that they use their positions to benefit themselves and their families.

Conclusions

Tat hamlet can be characterized as being a loosely structured community. The sense of solidarity among its component households is poorly developed and it has few effective institutions to organize collective action to achieve community aims. The household is the maximal unit in which trust among members is strong and altruistic behavior expectable. Relations among unrelated households are rarely close and spontaneous cooperation among them infrequent. Social capital is scarce, people have little sense of trust in the honesty and integrity of their fellow villagers, and community-level institutions for organizing corporate actions are rudimentary or wholly missing. In many particulars, the situation of Tat hamlet is comparable to that of the southern Italian village of Montegrano where the sociologist Edward G. Banfield (1958) formulated his concept of 'amoral familism' to describe a society where the absence of trust in social institutions other than the family severely limits the ability of villagers to cooperate to achieve collective goals.

Of course, as in all peasant societies, some inter-household cooperation occurs in the form of labor exchange groups. Seeing groups of people joking together as they harrow the fields or transplant the rice seedlings can give the impression to the casual observer that people in Tat hamlet are highly cooperative. But such mutual aid is always given under closely controlled conditions that ensure that participants meet their mutual obligations or suffer the sanction of exclusion from future exchanges.

The complex relationship between social organization and resource management has changed repeatedly over the course of the history of Tat hamlet. Within a single lifetime the hamlet has experienced three radically different forms of social organization. These systems share only the common characteristic of having been essentially imposed by forces outside of the local community. The absence of locally evolved institutions for collective action — the essential mechanisms for implementing community-based resource management — has been the one constant across several decades of rapid social change. Management structures above the household level have always been imposed from outside and decision-making authority wielded by the local agents of external powers, whether the *đạo xóm* of the feudal regime, the production brigade leaders of the cooperative period, or the Forestry Office cadre of the present period. The villagers themselves have never developed any community institutions for effective collective management of resources. Instead, resource management is left almost entirely in the hands of individual households. A situation of virtually unchecked competition among households for scarce resources is the norm and continuing degradation of the environment its consequence.

Acknowledgments

This chapter is revised version of a paper that was originally published in *Southeast Asian Studies (Kyoto)*, 39 (3) 2001, pp. 299-324. It is republished with permission of the publisher.

This chapter is based on information collected over the years since 1992 by a collaborative project of the Center for Natural Resources and Environmental Management (CRES) of the Vietnam National University, Hanoi, the Center for Agricultural Research and Ecological Studies (CARES) of Hanoi University of Agriculture (HUA), and the East-West Center. We have relied heavily on information shared with us by other project researchers, especially Nghiêm Phương Tuyến and Vy Tôn.

Funding was provided by grants to the East-West Center from the Ford Foundation, the Keidanren Nature Conservation Fund, the John D. and Catherine T. MacArthur Foundation, the U.S. National Science Foundation, and the Swedish International Development Cooperation Agency (SIDA), by grants to CRES from the Ford Foundation and SIDA, and by grants to HUA from the Ford Foundation and the Toyota Foundation. We would especially like to acknowledge the major contributions made by Lê Trọng Cúc, formerly the Director of CRES, in organizing and managing this project.

12
Diversification of Household Livelihood Strategies in Tat Hamlet

Nguyễn Thị Hải Ninh, Đặng Việt Quang, Nguyễn Đình Tiến, and Ingrid Foik

Introduction

This chapter focuses on the diversification of livelihood strategies of households in Tat hamlet that has occurred in recent years. Livelihood strategies are defined as those activities undertaken by smallholder households to provide a means of living. A key goal of livelihood strategies is to ensure household economic security. In this analysis, we emphasize the diversification of income sources of households employing different livelihood strategies. We identify differences in income diversification among three groups of households, each employing a different farming system.

The efficiency of different productive activities engaged in by households in Tat hamlet is also analyzed to reveal which activities are the most effective and efficient from an economic perspective. Differences in economic efficiency of different activities can help to explain farmers' choices of livelihood strategies as they shift from a largely subsistence economy to one that is semi-market-oriented, even though food security remains a major concern and capital investment in production activities is very low.

Farm Household Income Diversification and the Economic Efficiency of Agricultural Production Activities

Small subsistence-oriented farm households of the sort found in Tat hamlet normally get income from many different activities. Diversifying income sources can help farmers to mitigate the risks caused by natural calamities or market failures and cope with problems arising from poorly functioning markets. Moreover, income diversification is a very useful way for farmers to use their off-season labor. Diversification of income sources can help farm households satisfy their food requirements when the prices of products needed are expensive due to the high transaction costs (International Food Policy Research Institute 2003).

From another perspective, income diversification can be considered to be simply the expansion of the farm household's activities into off-farm or non-farm areas, which have been called 'supplementary activities.' These activities can utilize surplus labor capacity to provide supplemental income for farm households. Engaging in such supplementary activities, therefore, is a suitable device for diversifying income sources and reducing risks that may occur in either production or marketing (Brons and Ruben 2005).

Economic efficiency is also an important attribute of farm productive activities. Generally, households seek to maximize return to their limited resources, particularly family labor, land and capital (Harwood 1979). They must try to achieve their goals as efficiently as possibly in the face of physical, environmental, legal and socio-cultural constraints. In agricultural production, economic efficiency usually refers to either the combination of inputs that maximize individual farm objectives or to the ratio of valuable outputs to valuable inputs (Poll and Orazem, 1984).

Although the economic efficiency of agricultural production in terms of productivity or monetary profit is not an important consideration for the subsistence-oriented farm households in Tat hamlet, since ensuring food security is their overriding concern, it is still a useful criterion for measuring the relative performance of the different productive activities. This criterion can help farmers make decisions about what activities should be performed to maximize their returns on scarce productive resources, which activities might be improved, and which might be abandoned. Identifying promising points of intervention where economic efficiency can be improved may also be useful to research and extension activities aimed at eliminating hunger, reducing poverty, and raising living standards.

Research Design and Methodology

Study design

The three main types of farm households in Tat hamlet, each of which employs a different farming system, are compared to see how they differ in terms of income diversity and economic efficiency. The three groups, which are defined using the criteria presented in Chapter 10, are:

- Group 1: Farm households with rice swiddens but no paddy fields.
- Group 2: Farm households with paddy fields and cash crop swiddens, but no rice swiddens (modified composite swiddening).
- Group 3: Farm households with paddy fields plus rice swiddens (traditional composite swiddening).

Data collection

Four data sets have been used in our analysis of household income diversification and livelihood strategies. Three of these sets are panel data on the same sample households collected at different times. The first survey, done in 1998-99 (referred to here as the 1999 household survey), included 42 randomly selected households. The second survey, done in 2004, included 41 of the 42 households in the 1999 survey. The third survey, conducted in early 2005, included 39 of the households from the 1999 sample. These panel data are used to compare economic patterns among the three groups of farm households.

In addition to these panel data sets, we carried out an additional survey of 48 households in January 2006 to collect data on their income generating activities in 2005. These households were selected using a stratified random sampling approach to ensure that the three types of livelihood strategies were represented. All eight Group 1 households in the hamlet that did not cultivate paddy rice were included in this sample. Groups 2 and 3 include 20 households each.

Survey research has been supplemented by group interviews and observation while in the hamlet. In order to depict the economic patterns of these three groups of farm households and the transformation of the household economy in Ban Tat, we use both quantitative and qualitative methods.

Analysis

Analysis of Variance (ANOVA) is used to identify the significant differences among three groups of farm households. The Simpson equation for calculation of diversity index is used to calculate the diversification of household income in different groups:

$$ID = 1 - \sum_{i=1}^{n} \left(\frac{n_i}{N} \right)^2$$

Where ID is income diversification index, n_i is the income generated by the ith activity and N is the total income (Dillon and McConnell 1997).

Additionally, econometric models (Logit and Tobit regression) are employed to analyze the participation of farmers in off-farm activities and the effects of this on the transformation of farm income. The Logit model employed in this research to analyze the participation of farmers in off-farm activities can be written as follows:

$$\text{Prob}\,(Y=1 \mid X) = \frac{e^{\sum_{i=1}^{n} \beta_i X_i + \alpha}}{1 + e^{\sum_{i=1}^{n} \beta_i X_i + \alpha}}$$

Where, the left hand side is the dependent variable of the probability of farm household participation in off-farm activities. On the right hand side, Xi is the ith independent variable with the corresponding coefficient of βi and n is the number of independent variables representing to agricultural production, household size, dependency ratio, age, education and gender.

The Tobit regression model used in this research can be written as follows:

$$Y = \sum_{i=1}^{n} \beta_i X_i + \alpha \, (0 \leq Y \leq 100)$$

Where, Y is the dependent variable that represents the percentage of farm income. Because dependent variables are censored data, which cannot be lower than zero or larger than 100 percent, a Tobit model is employed. On the right hand side, Xi is the i-th independent variable with the corresponding coefficient of βi and n is the number of independent variables representing participation in off-farm activities, agricultural production and household characteristics.

Comparison of the Productive Resources of the Groups of Households

The productive resources of the households can be described in terms of their land, labor force, and productive assets.

Land resources of households

Table 12.1 describes the average land holdings of each group. Households in Group 2 (modified composite swiddening) have the largest average areas of both paddy fields (1,220 m^2) and forestland (12,207 m^2). Group 3 households (traditional composite swiddening) also have a relatively large area of paddy fields, but a considerably smaller area of forestland. Households in Group 1 (rice swiddens but no paddy fields) have the smallest endowments of land, with no paddy fields and only one-third as large an area of forestland as the Group 2 households.

Households in different groups employ their land resources to grow different mixes of crops. Table 12.2 shows the average area of different crops planted by the groups of households.

Cassava and maize were the most popular crops in terms of area planted for all groups. Canna, which had earlier been a very popular swidden crop, was planted by only a few households after an outbreak of disease in 2003. Ginger was planted in small areas by 17 of 48 households. Vegetables were grown by 35

Table 12.1 Agricultural land holdings of different groups of households in 2005

	GROUP 1 (N = 8)	GROUP 2 (N = 20)	GROUP 3 (N = 20)	TOTAL (N = 48)
Average area per household (m^2)				
Paddy field	0	1,220	1,195	1,006
Homegarden	243	190	298	244
Forestland (includes swiddens and fallow areas)	4,000	12,207	9,905	9,880
Fishpond	31	180	135	136
Average area per capita (m^2)				
Paddy field	0	271	249	218
Homegarden	54	42	62	53
Forestland (includes swiddens and fallow areas)	889	2,713	2,064	2,136
Fishpond	7	40	28	30

Source: Household survey, January 2006

Table 12.2 Average area of crops grown by different groups of households (m^2)

	GROUP 1 (N = 8)	GROUP 2 (N = 20)	GROUP 3 (N = 20)	MEAN (N = 48)
Paddy rice	0 (0)	1,220 (609)	1,195 (694)	1,006 (743)
Swidden rice	3,688 (5,365)	0 (0)	3,385 (2,966)	2,025 (3,691)
Maize	1,063 (1,972)	1,740 (2,271)	4,037 (6,204)	2,584 (4,401)
Cassava	2,500 (2,976)	3,610 (3,663)	3,470 (3,043)	3,367 (3,263)
Ginger	113 (189)	68 (95)	34 (113)	61 (122)
Vegetables	243 (299)	190 (237)	298 (412)	244 (327)

Source: Household survey, January 2006

Note: Numbers in parentheses are standard deviations

households, mostly for family consumption. The rest either did not have enough land or lacked time to produce vegetables.

Labor force of households

Table 12.3 provides information on average household size, number of working age members, average age and years of education of workers, and dependency ratios for each of the three types of households.

Table 12.3 Labor force characteristics of groups of households, 2006

	GROUP 1 (N = 8)	GROUP 2 (N = 20)	GROUP 3 (N = 20)	MEAN (N = 48)
Household size	4.5	4.5	4.8	4.6
Number of working age members	3.3	3.1	3.7	3.3
Average age of laborers	28.9	35.2	31.3	32.4
Average no of years of education of laborers	8.2	5.9	7.1	6.8
Dependency ratio	0.38	0.48	0.32	0.39

Source: Household survey, January 2006

Households in Group 3 (traditional composite swiddening) had the highest average number of working age members (3.7) and the lowest dependency ratio (0.32), whereas households in Group 2 (modified composite swiddening) had the lowest number of working age members (3.1) and the highest dependency ratio (0.48).

Productive assets

The agricultural activities engaged in by different households are influenced by their productive assets. These include tools and other equipment and large livestock for traction and supplying manure. Households in Tat hamlet generally have relatively few productive assets. Production largely depends on human and animal labor. The number of households having expensive production equipment like rice mills or water pumps is very small. Ownership of livestock, poultry, and fishponds is more common. As is shown in Table 12.4, households in Group 3 (traditional composite swiddening) keep the largest number of most kinds of animals. In particular, they have more buffalo and cattle, perhaps because their fallow swiddens provide grazing areas.

Table 12.4 Average number of domestic animals per household, 2005

	GROUP 1 (N = 8)	GROUP 2 (N = 20)	GROUP 3 (N = 20)	MEAN (N = 48)
Pigs	3.0 (3.4)	2.2 (2.5)	4.3 (7.3)	3.2 (5.2)
Buffalo and cattle	3.0 (4.4)	3.7 (2.6)	4.8 (3.7)	4.0 (3.4)
Chickens	13.1 (17.7)	14.1 (12.8)	22.4 (25.0)	17.4 (19.6)
Ducks	1.9 (4.3)	4.0 (7.5)	3.8 (7.0)	3.3 (6.7)
Fishpond (m^2)	31.3 -	179.5 (386.4)	128.3 (157.0)	133.4 (272.3)

Source: Household survey, 2006

Note: The figures in parentheses are standard deviations

Sources of Income and Income Diversification among Groups of Farm Households

In this section we identify the differences in household income from various sources and the extent of income diversification among the three groups of farm households.

Income sources

Table 12.5 presents information on the amount of income that households in different groups earn from different sources. Income from logging activity, much of which is illegal, and is not reported reliably by interviewees, has not been included when calculating net income.

The income from crops of households in Group 1 is significantly lower than that of households in Groups 2 and 3. These households are either newly established families or households that have come to Tat hamlet from other places and have not been given paddy land (Nguyễn Thanh Lâm *et al.* 2004). Because they do not have paddy land, which generates a large amount of income, their income from crops is much less than that of Groups 2 and 3. Because they produce much less rice than the other two groups, the value of what they themselves both produce and consume (their subsistence value) is also lower and they must obtain income from other sources to meet their consumption needs, especially to purchase rice. Households in Group 2 do not grow swidden rice, yet their income from crops is somewhat higher than that of Group 3 and more than double that of Group 1, reinforcing the conclusion that swidden rice is less important than paddy rice in contributing income to farm households in Tat hamlet.

For Groups 2 and 3, animal husbandry is the largest source of income, providing 39 percent of income for Group 2, and 35 percent for Group 3, but only 18 percent for Group 1. Households in Group 1 have on average three buffalo or cattle, Group 2 households have almost four head, and Group 3 households have almost five buffalo or cattle. This may be because farmers with paddy fields raise large livestock not only for sale but also for draft power and for manure for their paddy fields. Therefore, growing paddy rice and raising buffalo and cattle are complementary activities.

Crops and livestock contribute about 60 percent of household income in Groups 2 and 3, but only about 29 percent for Group 1. Group 1 households, in contrast, rely heavily on income derived from collection of non-timber forest products (NTFPs) (37 percent) and off-farm activities (21 percent), which together provide 58 percent of their total cash income, compared to only 29 percent for Group 2 and 35 percent for Group 3. While their income from these two sources is higher than that of the other two groups, there is considerable

Table 12.5 Income of households from different sources by groups in Tat hamlet in 2004

	GROUP 1 (N = 6)		GROUP 2 (N = 10)		GROUP 3 (N = 10)		ANOVA
	MEAN	CV (%)	MEAN	CV (%)	MEAN	CV (%)	P-VALUE
Total net income per household (000 VND)							
Total income	10,666	85	12,560	36	9,176	52	0.534
Crops	947	139	2,538	32	2,166	47	0.061
Animal husbandry	1,344	75	4,922	49	3,373	102	0.045
NTFPs	3,361	120	2,311	64	2,358	72	0.647
Off-farm activities	2,510	141	1,472	149	826	125	0.365
Pension and salary	2,475	213	1,281	143	427	80	0.361
Garden	29	206	36	261	27	167	0.954
Share in total income (%)							
Crops	10.9	71	22.3	40	24.4	38	0.026
Animal husbandry	17.8	78	38.7	34	34.6	64	0.069
NTFPs	37.3	58	19.2	60	27.6	56	0.037
Off-farm activities	21.5	69	9.8	111	7.9	110	0.089
Pension and salary	12.1	162	9.7	152	5.2	91	0.608
Garden	0.5	187	0.2	187	0.3	163	0.692
Cash income per household (000 VND)							
Total cash income	10,399	107	7,934	99	6,073	75	0.559
Crops	437	161	818	70	721	64	0.430
Animal husbandry	2,121	200	2,655	111	2,209	124	0.932
NTFPs	2,660	129	902	96	1,763	92	0.239
Off-farm activities	5,152	169	3,523	149	1,353	86	0.371
Garden	29	206	36	261	27	167	0.954
Income per capita (000 VND)							
Net income	2,252	73	2,050	62	1,863	66	0.534
Cash income	2,155	86	1,760	109	1,394	82	0.670
Investment of households in both farm and non-farm activities (000 VND)							
Total investment	983	-	3,025	-	1,753	983	-
Agriculture	816	202	2,255	34	1,653	53	0.047
Non-agricultural	167	-	770	-	100	-	-
Subsistence value	1,251	123	7,651	90	4,857	83	0.078

Source: Survey data in 2005

Note: CV= Coefficient of Variance

variability of income from these sources within Group 1, much more variability than in groups 2 or 3.

Table 12.5 shows that households in these three different groups have significantly different income patterns. Groups 2 and 3 get about 23 percent of their income from crops, while Group 1 households obtain only about 11 percent of their total income from crops. Households in groups 2 and 3 get 39 percent and 35 percent respectively from animal husbandry, while Group 1 households obtain only 18 percent of their income from this source. Thus, Group 1 households are getting only 29 percent of their income from farming activities, compared to 61 percent for Group 2 households, and 59 percent for Group 3 households.

With no paddy fields and fewer buffalo and cattle it is reasonable that Group 1 households would put more emphasis on generating income from NTFPs and off-farm activities. And we see that Group 1 households get 37 percent of their income from NTFPs, compared to 28 percent for Group 3 households and 19 percent for Group 2 households. Group 1 households get 21 percent of their income from off-farm employment, while households in groups 2 and 3 both get less than 10 percent of their income from non-farm activities.

It is understandable that Group 1 households have much lower income from crops than groups 2 and 3 because they do not have paddy fields, which generate a lot of income. But Group 2 households, even though they do not cultivate swidden rice, have a somewhat higher net income from crops than Group 3 households, even though income from crops provides a slightly higher percentage of total income for Group 3. Group 2 households also have a higher total net income from animal husbandry than those in Group 3.

Obviously, farmers must make decisions about their livelihood strategies based upon their own specific circumstances. Thus, even within each of the three groups, there may be significant variability. For instance, while households in Group 1 in general get much of their income from NTFPs and off-farm activities, these households show great variability in the income they derive from NTFPs. The average income from NTFPs for Group 1 has a Coefficient of Variance (CV) of 120 percent, whereas the CV of income from NTFPs for groups 2 and 3 is only 64 and 72 percent respectively. Furthermore, statistically, the three groups do not differ significantly in regard to income derived from off-farm activities and the CV for income from off-farm activities is between 125 and 149 for households in all three groups.

Income diversification

Income diversification is a very important part of a viable livelihood strategy. Table 12.6 shows the income diversification index (IDI) of households in the three groups by net income source. Pension and salary income are included

Table 12.6 Income diversification index (IDI) of households in different groups

INCOME SOURCE	GROUP 1 (N = 6)		GROUP 2 (N = 10)		GROUP 3 (N = 10)		ANOVA
	IDI	CV (%)	IDI	CV (%)	IDI	CV (%)	(P VALUE)
Crops	0.29	119	0.71	7	0.68	16	0.000
Livestock	0.13	290	0.62	27	0.54	29	0.002
NTFPs	0.63	37	0.52	34	0.61	26	0.409
Off-farm work	0.23	96	0.28	98	0.25	90	0.940
Overall IDI	0.78	17	0.86	7	0.87	3	0.031

Source: Household survey 2005

Note: IDI = Income Diversification Index

in the off-farm category but income from the homegardens has been excluded because its contribution was so small.

Because the households in Group 1 are earning less income from crops and animal husbandry, their income diversification in cropping and animal husbandry is significantly lower than households in the two other groups. This leads to households in Group 1 having considerably lower overall income diversity compared to households in the two other groups. Moreover, households in Groups 2 and 3 have almost the same level of income diversification and their overall IDI exhibits less variation than that of households in Group 1.

Households with different farming systems have differences in income diversification. Composite swiddening households of both types (Groups 2 and 3) display more diversification than households practicing swiddening alone (Group 1). As depicted in Table 12.6, the households in Groups 2 and 3 have higher levels of diversification of income generating activities than households in Group 1, while their income from crops and livestock is significantly higher than that of households in Group 1. Crops and animal husbandry, therefore, are the dominant income sources of households in Groups 2 and 3. Off-farm activities and collection of NTFPs are supplementary sources of income for them. The income from crops and livestock of households in Group 2 is higher and less variable than that of households in Group 3. Additionally, the households in Group 2, because they do not grow swidden rice, can spend more time to diversify their income sources into off-farm activities. Thus, they have more opportunities to earn cash than households in Group 3.

Although households in Group 2 are on average better off compared to households in the two other groups in terms of average cash and non-cash income, and extent of income diversification, it should not be concluded on this basis that modified composite swiddening is the best strategy for all households in the hamlet to adopt. It is important to bear in mind that Group 2 consists of households who decided to abandon the cultivation of rice swiddens and farmers who made that decision tended to have more paddy land as well as more forestland. It is not unreasonable to assume that these were relatively strong, well-established households who were probably self-sufficient in rice, or nearly so. Farmers who had no paddy land or less paddy land or other resources seem to have been reluctant to follow their example.

Off-farm employment as a factor in diversification of livelihood strategies

Increased participation of households in Tat hamlet in off-farm income generation activities is an important factor in income diversification. Table 12.7 shows the changes between 1998-99 and 2004 in the participation in off-farm income earning of households in all groups.

The number of households engaging in off-farm activities increased by about 10 percentage points between 1999 and 2004, despite the fact that the number of households receiving salaries or pensions from the government remained unchanged and the number of households that earned money from handicraft production declined to zero. Thus, participation in off-farm activities generally grew during the period of five years because more farmers are selling their labor and providing services, including milling rice, shop keeping, operating motorcycle taxi services, and working as carpenters.

Economic Efficiency of Agricultural Production

In this section we examine the economic efficiency of land and labor in farming activities among three groups of households employing different farming systems in 2005. We look separately at the economic efficiency of crop production and raising pigs, poultry, and fish.

Economic efficiency of crop production

Economic efficiency of crop production will be examined in terms of the productivity of different crops in terms of yield per hectare, gross margin to land, and gross margin of labor.

Productivity of crops

Table 12.8 presents the productivity of six crops to see the efficiency of land use in Tat hamlet. Productivity of paddy rice was 3.8 tons/ha. Levels of

Table 12.7 Percentage of farm households participating in off-farm income generating activities in Tat hamlet in 1999 and 2004

OFF-FARM INCOME SOURCES	1999 (N = 41)	2004 (N = 39)	CHANGE
Government assistance	46.3	53.9	7.5
All sources of off-farm income	56.1	66.7	10.6
Pensions	7.3	7.7	0.4
Government salaries	14.6	15.4	0.8
Wage labor	17.1	25.6	8.6
Handicraft production	9.8	0.0	-9.8
Provision of services	14.6	33.3	18.7

Source: Household surveys in 1998–99 and 2004

Table 12.8 Productivity of crops by household group (kg/ha)

	GROUP 1 N = 8	GROUP 2 N = 20	GROUP 3 N = 20	MEAN N = 48
Paddy rice	-	3,984 (0.409)	3,629 (0.000)	3,804
Swidden rice	422 (0.172)	-	543 (0.383)	527
Maize	389 (0.008)	1,579 (0.832)	1,324 (0.018)	1,334
Cassava	3,429 (0.008)	5,351 (0.634)	5,633 (0.000)	5,225
Ginger	3,667 (0.168)	8,603 (0.391)	7,941 (0.339)	6,939
Vegetables	1,852 (0.384)	11,317 (0.954)	6,057 (0.499)	7,478

Source: Survey 2006

Note: Numbers in parentheses are P-value of t-test between groups 1 and 2, groups 2 and 3, groups 1 and 3

productivity were different, but not statistically so, between the two groups with paddy land: Group 2 harvested almost 4 tons/ha while Group 3 harvested 3.6 tons/ha. Productivity of swidden rice was 422 kg/ha for Group 1 compared to 543 kg/ha for Group 3.

There was a significant difference in productivity of maize between Group 1 and the other two groups. Productivity in Group 2 was highest at 1.6 tons/ha, then Group 3 at 1.3 tons/ha. Group 1 harvested only 389 kg/ha. Although groups 1 and 3 had higher costs, their productivity was lower than Group 2.

The average productivity of ginger was 6.9 tons/ha, but Group 1 had much lower productivity than the other groups. Not every household has land suitable for ginger, which grows well in newly burnt fields with soft soil. Moreover, it has recently become difficult for farmers to sell ginger. As a result, only few households still grow ginger.

Based on the t-test, we can see that, per hectare, there are not any significant productivity differences between Groups 2 and 3 in using land for paddy rice, maize, cassava, ginger, and vegetables. Groups 2 and 3 achieve a higher level of productivity than Group 1 for maize and cassava. Groups 2 and 3 also achieve higher productivity than Group 1 in growing ginger and vegetables, but the differences are not statistically significant.

Productivity of land

Gross margin to land is the most important indicator to assess the profitability of land use. For paddy rice, the gross margin to land of Group 2 was somewhat higher than that of Group 3, but the difference has no statistical significance (Table 12.9). For swidden rice, although the gross margin to land of Group 3 was higher than that of Group 1, the difference s not statistically significant (P-value of t-test is 0.383). Income per one hectare of swidden rice of Group 3 was 302,000 VND higher than that of Group 1; and Group 3 also gained 84,600 VND/ha more from cassava than Group 1.

Income from maize was higher per hectare than for cassava. On average, each hectare of maize yielded 1.9 million VND while cassava generated 1.6 million VND/ha. Groups 2 and 3 planted maize and as cassava more effectively than Group 1. Gross margin to land of maize for Group 1 was very low because of losses caused by cattle. For example, one farmer planted 10 kg of seeds (valued at 250,000 VND) but due to destruction by cattle he harvested only 50 kg of maize. The efficiency of land use for ginger was highest at around 27.4 million VND/ha, with no significant differences between groups.

Productivity of labor

The gross margin to labor is calculated in terms of net income from one working day. Among all crops, ginger brought in the highest average income at 56,300 VND/day followed by maize at 30,400 VND/day (Table 12.10). The lowest gross margin to labor came from vegetables (14,600 VND/day on average), which is less than the opportunity cost of one working day in Tat hamlet of 15,000 VND. Only Group 2 had a gross margin to labor for vegetables exceeding the opportunity cost.

Maize and cassava had higher average gross margins to labor than swidden rice. However, farmers in Groups 1 and 3 still cultivate swidden rice for many reasons. People in Group 1 need swidden rice to ensure their daily food needs, and people in Group 3 need it to supplement their paddy rice production.

The P-value of t-test in Table 12.11 presents the differences of gross margin to labor for each of the crops between the three groups. In general, the gross

Table 12.9 Gross margin to land of crop production by household group (1000 VND/ha)

	Group 1 N = 8	Group 2 N = 20	Group 3 N = 20	Mean N = 48
Paddy rice	-	6,129 (0.607)	5,899 (0.000)	6,013
Swidden rice	1,055 (0.173)	-	1,357 (0.383)	1,318
Maize	244 (0.013)	2,321 (0.927)	1,870 (0.031)	1,889
Cassava	1,027 (0.009)	1,605 (0.634)	1,690 (0.000)	1,568
Ginger	14,667 (0.174)	33,824 (0.399)	31,765 (0.339)	27,483
Vegetables	1,778 (0.379)	27,044 (0.948)	14,271 (0.499)	17,564

Source: Survey 2006

Note: Numbers in parentheses are P-value of t-test between groups 1 and 2, groups 2 and 3, groups 1 and 3

Table 12.10 Gross margin to labor (1000VND/working day)

	Group 1 N = 8	Group 2 N = 20	Group 3 N = 20	Total N = 48
Paddy rice	-	23.7 (0.83)	22.9 -	23.3
Swidden rice	17.3 -	-	17.2 (0.99)	17.2
Corn	6.5 (0.15)	37.9 (0.77)	27.2 (0.29)	30.5
Cassava	16.9 (0.30)	22.4 (0.53)	19.7 (0.61)	20.7
Ginger	81.3 (0.66)	45.0 -	66.0 (0.85)	56.4
Vegetables	5.3 (0.09)	26.9 (0.21)	9.1 (0.45)	14.6

Source: Survey, January 2006

Note: Numbers in parentheses are P-value of t-test between groups 1 and 2, groups 2 and 3, groups 1 and 3

margin to labor is rather similar for all three groups for each crop. For paddy and swidden rice, there is no significant difference of gross margin to labor between groups 2 and 3, or groups 1 and 3.

Economic efficiency of pigs, poultry and fish

Table 12.11 shows that gross margin for pigs, chickens, and fish were not statistically different between groups. The average gross margin from pigs was 497,800 VND per household, and provided 46 percent of total income from raising small animals and fish. Chickens had an average gross margin of 237,300 VND per household and provided 22 percent of total gross margin from small livestock and fish. Fish brought an average gross margin of 358,600 VND per household.

Table 12.11 Average gross margin for pigs, poultry and fish by household group (000VND)

	GROUP 1 N = 8		GROUP 2 N = 20		GROUP 3 N = 20		TOTAL N = 48	
	GM (1–2)	%	GM (2–3)	%	GM (1–3)	%	GM	%
Pigs	680.5 (0.805)	71.7	484. 8 (0.894)	51.2	437.9 (0.766)	33.7	497.8	45.6
Chickens	267.6 (0.420)	28.2	110.2 (0.323)	11.7	352.2 (0.762)	27.1	237.3	21.7
Fish	0 (0.006)	0.0	350.8 (0.614)	37.1	509.8 (0.089)	39.2	358.6	32.7
Total	948.0	100.0	945.8	100.0	1300.0	100.0	1093.7	100.0

Source: Survey 2006

Note: Numbers in parentheses are P-value of t-test

Fish generated more cash income than chickens, although investment in raising chickens was 1.5 times greater than for fish. However, only 29 percent of households had fishponds and many of these ponds had no water.

Farmer Choices of Livelihood Strategies

Comparison of data on household livelihood strategies collected in 1999 and 2004 reveals the dynamic changes in farming systems that have accompanied the decline in the importance of swiddening in the livelihood of villagers. In particular, different households have adopted different strategies of income diversification. Table 12.12 shows the relationships between different farming system strategies (changes in areas of paddy fields and swiddens and herd size) employed by households and the probability of their participation in off-farm activities.

In order to identify the income diversification strategies of households, the starting point should be the dynamic changes in shifting cultivation during the preceding five years, which categorized households into two groups: 1) decrease in swidden cropping area (71%) and 2) increase in swidden cropping area (29%). Households that reduced their swidden area expanded their paddy field area and diversified their livestock production, including pigs and cattle. In other cases they reduced or kept constant the area of paddy fields but increased livestock production. Both of these income diversification strategies generate a positive correlation with importance of farm income. However, with these strategies, farmers would be less likely to participate in off-farm wage employment. In the cases of decreasing swidden area without paddy fields or with a reduction of paddy field area, they would be more likely to do off-farm activities. Households

Table 12.12 Farm households employing different strategies of income diversification

STRATEGIES	INCREASE NUMBER OF CATTLE	CONSTANT NUMBER OF CATTLE	DECREASE NUMBER OF CATTLE	TOTAL
Percentage of sample farm households employing different strategies				
Decrease swidden—Increase paddy	26.3	2.6	10.5	39.5
Decrease swidden—Keep constant paddy	5.3	0.0	2.6	7.9
Decrease swidden—No paddy	2.6	2.6	0.0	5.3
Decrease swidden—Decrease paddy	13.2	2.6	2.6	18.4
Increase swidden—Increase paddy	10.5	2.6	2.6	15.8
Increase swidden—Keep constant paddy	2.6	0.0	0.0	2.6
Increase swidden—No paddy	0.0	0.0	2.6	2.6
Increase swidden—Decrease paddy	7.9	0.0	0.0	7.9
Total	68.4	10.5	21.1	100.0

STRATEGIES	INCREASE NUMBER OF CATTLE	CONSTANT NUMBER OF CATTLE	DECREASE NUMBER OF CATTLE	AVERAGE
Estimated probability of participation in off-farm wage employment(%)				
Decrease swidden—Increase paddy	31.1	58.3	32.0	33.2
Decrease swidden—Keep constant paddy	56.7	-	3.2	38.9
Decrease swidden—No paddy	89.5	36.3	-	62.9
Decrease swidden—Decrease paddy	45.2	1.9	0.0	32.6
Increase swidden—Increase paddy	22.9	32.7	-	20.8
Increase swidden—Keep constant paddy	58.6	-	-	58.6
Increase swidden—No paddy	-	-	40.0	40.0
Increase swidden—Decrease paddy	65.8	-	-	65.8
Estimated probability of participation in off-farm self employment (%)				
Decrease swidden—Increase paddy	16.9	23.6	18.5	17.8
Decrease swidden—Keep constant paddy	73.9	-	21.7	56.5
Decrease swidden—No paddy	15.3	13.3	-	14.3
Decrease swidden—Decrease paddy	45.5	70.1	96.3	56.2
Increase swidden—Increase paddy	35.5	58.9	-	33.5
Increase swidden—Keep constant paddy	58.6	-	-	58.6
Increase swidden—No paddy	-	-	74.7	74.7
Increase swidden—Decrease paddy	45.7	-	-	45.7

Sources: Estimated from survey data in 2004 and 1999 (N = 38)

Note: Blank cells have no sample or underestimated probability

who expanded their swidden area in 2004 either expanded or decreased their paddy field areas, and raised either more or fewer buffalo and cattle, depending on the characteristics of their households. A total of 24 possible strategies emerge from dynamic changes in three dimensions—swidden area, paddy field area and number of cattle. Participants in each strategy are identified and the probability of their participation in off-farm activities is estimated by remodeling the Logit prediction.

Table 12.12 shows that the highest proportion (26 percent) of farm households decreased their swidden area, expanded their paddy field area, and increased their number of buffalo and cattle. This strategy was the most popular among the 24 possible strategies. The second most popular strategy, one followed by 13 percent of households, was to decrease the areas of both paddy fields and swiddens while increasing the number of large livestock. Two other strategies were equally popular, with each used by 11 percent of households: One strategy was to decrease swidden area, increase paddy field area, and decrease the number of buffalo and cattle; the other was to increase the area of both swiddens and paddy fields and also increase the herd size. Interestingly, the dependency ratio of all four households adopting this strategy is quite low. The probability of participation in the off-farm activities by farmers using this strategy is also relatively low because the expansion of their farming activities limits the labor they can devote to off-farm activities.

The choice of farming system strategies by households is associated with the probability of their involvement in earning of off-farm income. Households that had no paddy fields and decreased the area of their swiddens while increasing their number of buffalo and cattle had the highest probability (90 percent) of participating in off-farm wage labor, followed (at 66 percent) by households that increased their swidden area, decreased their paddy field area, and increased their number of livestock. Households that had a decreased swidden area, a constant area of paddy fields, and an increased number of buffalo and cattle had the highest probability (74 percent) of engaging in off-farm self-employment, followed by households (at 70 percent) who had decreased areas of both swiddens and paddy fields and kept a constant number of livestock.

It is evident that there are three main strategies followed by households in Tat hamlet: 1) decrease swidden area and increase paddy field area; 2) decrease areas of both swiddens and paddy fields; 3) and increase both swidden and paddy field areas. A large proportion of households in all categories also are intensifying buffalo and cattle production, in part because of the ready availability of loans to purchase livestock. In 1999, 34 percent of farm households used credit while this number in 2004 was 56 percent, with these loans mostly used to develop livestock production.

Conclusions

The livelihood strategies of households in Tat hamlet are in the process of changing from largely subsistence-oriented production of crops and livestock towards a more diverse set of activities intended to generate cash income. The three groups of households employing different farming systems have different levels of income diversification. Food security remains their first priority but obtaining cash is of rapidly growing importance, especially for Group 1 households that lack adequate agricultural land resources to meet their needs for rice. For households in this group, crops and livestock provide only about 20 percent of their total income, which is much less than comes from collection of forest products, off-farm activities, and pensions and salaries. In contrast, households in Groups 2 and 3 obtain about 60 percent of their total income from crops and livestock.

Although most households engage in similar agricultural activities, economic efficiency among the three groups differs depending on their land and labor resources. For paddy rice, there is no significant difference in efficiency of land and labor use between Group 2 and Group 3. This shows that both of these two groups devoted similar attention to paddy rice. Households in Group 3 spend more working days on cultivating swidden rice than those in Group 1, but the difference in efficiency of labor is not statistically significant. The largest difference in economic efficiency among the three groups is found in the cultivation of maize, cassava and ginger. Groups 2 and 3 cultivated maize and cassava more effectively than Group 1. Group 2 has the most efficiency in using labor for maize and the most efficiency in using of land for ginger. In conclusion, households in Group 2 and 3 with composite swiddening have high efficiency in cultivating maize and cassava, which are the two main swidden crops. Group 2 households have the greatest economic efficiency of labor used in raising pigs.

Farm households adopt different diversification strategies according to the areas of their swiddens and paddy fields, ownership of large livestock, and participation in off-farm activities. Out of 24 possible strategies, the vast majority of farm households in Tat hamlet focus on only three strategies: decrease swidden – increase paddy; decrease swidden – decrease paddy and increase swidden – increase paddy. A large proportion of farm households in all of these categories also are intensifying buffalo and cattle production.

Given current trends, it seems likely that diversification of household livelihood strategies will continue to increase, especially as growing integration of the community into the national economic and social systems offers more opportunities for villagers to engage in off-farm income earning activities.

13
The Evolving Role of the Market

Phạm Thị Mai Hương, Nguyễn Thanh Lâm,
Vũ Thị Thao, and Trần Đức Viên

Introduction

Economic development in Tat hamlet, as in other communities in Vietnam's mountainous areas, faces severe difficulties. The upland regions lag behind economically and are not catching up with the overall growth rate of the country (Jamieson *et al.* 1998; Đào Thế Tuấn 1998). The lack of stable markets for their products is an especially important factor limiting upland farmers' interest in intensifying and diversifying production (Trần Đức Viên *et al.* 2005).

In the past, under the central planning mechanism, the production and marketing of agricultural products were controlled by cooperatives and state owned enterprises. During that time, the supply of products depended mainly on planning figures rather than actual market demand (Đào Thế Tuấn 1998). After the adoption of renovation policies by the Government, agricultural production has rapidly developed to meet increased market demand but, even though there has been some success in the shift towards a market economy, farmers still face many challenges and barriers in the marketing of their products (Ngô Thị Thuận *et al.* 2004).

Previous studies found that land use in Tat hamlet has diversified in response to integration into the market, government development programs, and increasing population pressures (Trần Đức Viên 2004; Nguyễn Thanh Lâm *et al.* 2004; see Chapter 10 in this volume). The economy has been gradually shifting from subsistence to market orientation in character. Improved marketing of forest and agricultural products is helping local people to improve their economic situation, but at the same time may be having negative impacts on natural resources, especially forest resources. In this chapter, we present an overview of the history of the development of marketing channels in the hamlet and describe the currently functioning marketing channels of major agricultural and forestry products.

Methodology

Data were gathered by structured and semi-structured interviews with farmers and direct observation. Questionnaire-based interviews were conducted with 39 households in 2004 and early 2005 to explore changes in market accessibility, production, and marketing of agricultural and forestry products. Data from a 2004 survey of 30 households on forest management issues were also used. In order to analyze farmers' responses to market fluctuations, secondary data were obtained from other studies (e.g., Rambo 1998; Trần Đức Viên 2004; Nguyễn Thanh Lâm *et al.* 2004; Ngô Thị Thuận *et al.* 2004; and Bùi Thị Gia *et al.* 2005).

Target group interviews were also organized to collect data on how agricultural and forestry products had changed over time. The target groups represented various marketing channel participants such as producers, collectors, and traders. The history of market development and commodity trading was recorded by interviewing elderly informants. These informants also helped to clarify the impact of market development on natural resources. Secondary data were obtained from the previous studies (Rambo and Trần Đức Viên 2001a and 2001b). Hamlet regulations regarding market development were also reviewed.

Historical Development of Marketing Channels in Tat Hamlet

The historical development of market channels associated with material flows in and out of the hamlet is summarized in Table 13.1. Each of the major phases in the development of marketing channels will be described in detail below:

The colonial era (before 1954)

During the colonial period, Tat hamlet was isolated, contained only a few households, and was served by very rudimentary market channels (Figure 13.1)

The road had not been constructed yet so villagers used the Tat River as a main transport route or walked on foot over mountain trails to reach markets (Rambo and Trần Đức Viên 2001). Only a few ambulatory peddlers (*lái buôn*) from the lowlands walked to the hamlet and looked for customers living in small settlements along the Tat River. They carried salt, cloth, dried fish and meat, and knives in baskets on their shoulder poles. Sometimes, they came to Tat hamlet at night because they were afraid of the French army unit based in the then district capital at Chợ Bồ. Payment was either by barter or cash. At that time, salt was very expensive. One farmer recalled that a 3-year old buffalo was worth

Table 13.1 Major events in the historical development of marketing channels in Tat hamlet

YEARS	EVENTS
Before 1954	The trade in commodities was very limited Local people walked 2 days to Yên Lãng market to sell products Lowland traders occasionally visited the hamlet (3 times/month)
1960–62	The commercial and credit cooperatives were organized in 1962 Rationing regime was established. One shop was established in the village center NTFPs were bought from the farmers by the commercial cooperative under a contract from the district
1969	Road No. 433 was opened NTFPs and timber extraction increased Center for trade of timber and bamboo was established in Cao Sơn
1976	Cao Sơn market established People migrated to Cao Sơn from Hòa Bình hydroelectric dam area
1976–1978	Fixed cultivation and sedentarization program was implemented *Aleurites Montana, Styrax Tonkinesis* and tea were planted
1982–83	Road 433 was improved
1987	Household management period, free market economy Swidden area started to expand Logging increased
1994	Forest land allocation program to households
1995–1997	Ginger export began to Japan
1997	Bus route to Hà Đông was established NTFPs were extracted daily from the hamlet Timber resources declined
1998	Ginger export ended due to very low price
2000	Road 433 was hard surfaced National electricity grid was constructed New school was constructed Credit programs and extension programs promoted raising large livestock 13 collectors for NTFPs
2004	16 collectors for NTFPs The 661 program subsidized farmers to plant bamboo plantations

Source: Semi-structured interviews, 2005

Figure 13.1 Goods flows via the marketing channels in Tat hamlet during the colonial era

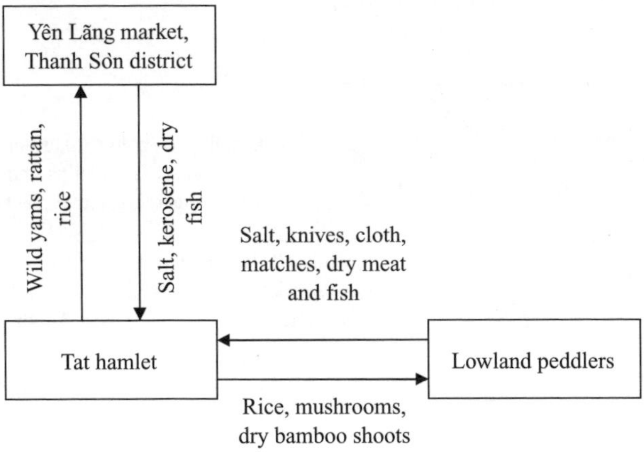

5 kg of salt. One cup of salt (10 ml) was exchanged for 3 kg of husked rice. The traders bought foodstuffs from the farmers such as dry bamboo shoots, rice, and mushrooms to carry back for sale in the lowlands. Villagers also sometimes walked for two days to Yên Lãng market in Phú Thọ province to sell rice, wild yams, rattan, and mushrooms and to buy salt, kerosene, and dry fish.

The early cooperative period (1959-1969)

A commercial cooperative was organized in Tân Minh village in 1962 with the objective of providing the villagers with access to basic goods that they had formerly had to buy at the distant market. The management board was made up of three people, including a chairman, a vice-chairman, and a bookkeeper. Although a cooperative shop was established in the center of the village, people could not freely buy what they needed. A rationing system limited each member household to buying only 2 kg of salt, 5 bags of tobacco, a small bottle of kerosene, and a small number of matches per month, and 4 m^2 of cloth per person per year. The ration was increased during the Tết lunar New Year holidays. Goods were supplied to the shop by the district state store. Because of insufficient supplies from the cooperative, local farmers still went to Yên Lãng market as they did in the previous period.

As specified by its contract with the District Commercial Department, the cooperative purchased a variety of NTFPs, including wild yams used for dye (*Artocarpus sp*), rattan, wild cardamon (*Amomum thyrsoideum*), lac resin (*Laccifer lacca*), *Cibotium barometz*, *Homalomena occulta*, palm leaves (*Caryota urens*), dry tea, and dry cassava chips (Figure 13.2).

Figure 13.2 Marketing channels in Tat hamlet from 1960 to 1969

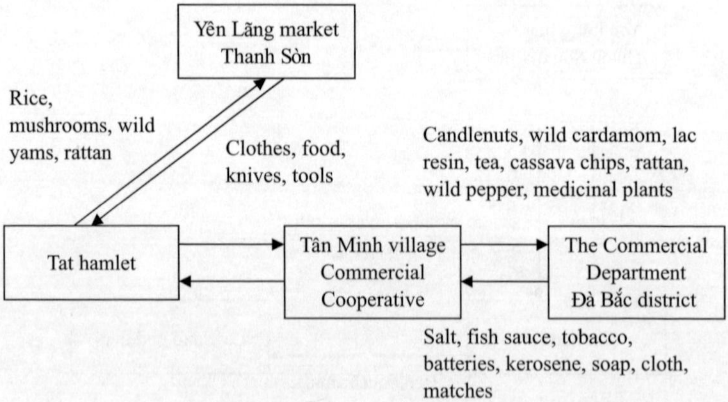

In this period, a special marketing channel for medicinal plants was established at the request of the state drug company through the cooperative management board. The cooperative received requests from the drug company to buy certain types of medicinal plants and this information was given to the villagers who then went to the forest to collect whatever species were requested. On set days, they had to bring their medicinal plants to the commune center and exchange their products for rice and miscellaneous goods. Due to the poor road, this channel was rarely implemented. Only a few medicinal plants were requested such as *Smilax glabra*, *Cibotium barometz*, and *Homalomena occulta*.

The cooperative organized the transfer of local products to the district and in return brought back consumer goods, including salt, kerosene and cloth, to the village shop once per month. To transport the products, the cooperative at first mobilized 10 people from each hamlet to walk all day carrying products on their backs to the district shop. This was extremely hard work so the cooperative asked the District People's Committee for help. In response, the district allocated six Mongolian horses for transporting goods between the district and Tân Minh village. Two to four tons of NTFPs per month were extracted from Tân Minh village. A credit cooperative also was established. Each member had to contribute 32 VND to build up the fund, which in turn gave credit to villagers. This was not an insignificant amount: At this time, 10 kg of husked rice had a price of 5 VND; a healthy buffalo sold for only 250 VND.

The late cooperative period (1969-1987)

The situation for marketing in the hamlet changed dramatically after Road no. 433 was constructed in 1969. It connected Tat hamlet to Cao Sơn and

Figure 13.3 Goods flows via the marketing channels in Tat hamlet from 1969 to 1987

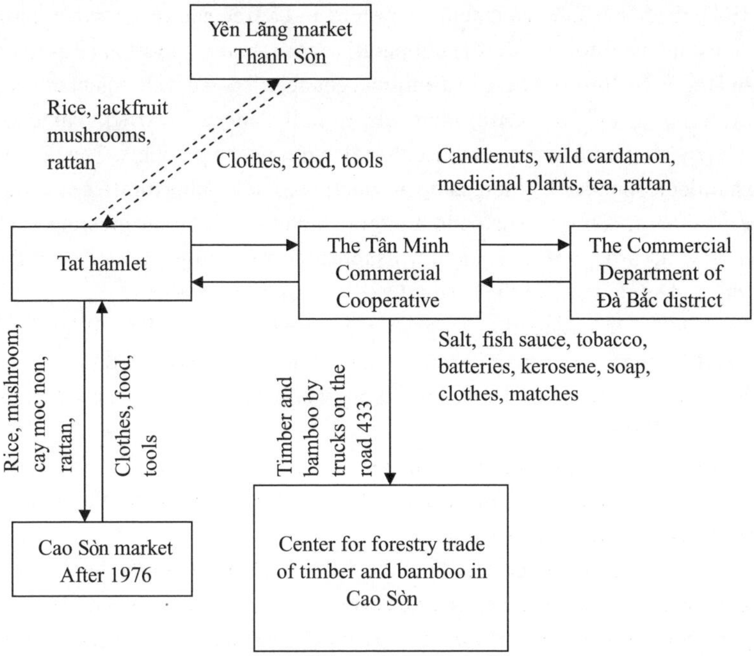

the district town. Trucks transported goods to the commercial cooperative and picked up forest products to carry back to the district in return. The quantity of NTFPs traded increased. Furthermore, a trade center for timber and bamboo was established in Cao Sơn. Timber was now being extracted from Tat hamlet. Construction of the Hòa Bình dam began in the late 1970s. Consequently, people migrated from the dam site to Cao Sơn, which increased the demand for forest products, particularly from Tat hamlet. The most important event was the establishment of the Cao Sơn market in 1976. People from Tat hamlet would walk the 12 kilometers to shop in the Cao Sơn market (Figure 13.3).

The fixed cultivation and sedentarization program (*chương trình định canh định cư)* was implemented by the government in Tat hamlet during this period. Farmers were assisted in planting *Aleurites montana*, *Styrax tonkinesis* and tea in their tree gardens. The government provided a market for these products for only a few years, after which farmers could no longer sell them. Road no. 433 was upgraded and expanded in 1982-83 so that people preferred to go to Cao Sơn market and stopped going to Yên Lãng market by 1985.

The household management period (1987 to present)

In 1987, the household management system and a free market economy regime were established to replace the cooperative management system (Rambo and Trần Đức Viên 2001a). These institutional changes led to development of a much more complex set of marketing channels. Initially, only a few trucks came to the hamlet every month to collect agricultural and forestry products. The people of the hamlet responded by increasing production as new opportunities opened up. The swidden area began to expand. Logging increased and some precious NTFPs such as wild cardamom and benzoin disappeared and the quantity of other forest products declined due to the gradual exhaustion of wild stocks.

In 1994, a land allocation program was implemented, followed by the 327 and PAM programs for planting trees on land that was considered to be 'barren hills' by the government. However, the market for planted forest products was not yet established.

In 1995-1996, ginger was planted over a large area because of a high demand for export. Many trucks from Ha Tay in the lowlands came to the hamlet to collect agroforestry products.

In 1997, a bus route connecting Tat hamlet to Hà Đông in the lowlands was established. At first there was one bus every two days but in 1998 daily service began. NTFPs were extracted daily from the hamlet and timber resources declined. Farmers could place orders with bus drivers to buy any products from the lowlands. In early 1998, the ginger market for export from the hamlet collapsed.

In 2000, even larger changes occurred in Tat hamlet because of infrastructure improvement and many state development projects were carried out in the hamlet. Road no. 433 was hard-surfaced, the national electricity grid reached the hamlet, and a new school was built in the village center. Farmers had opportunities to increase their crop production through participating in credit and extension programs. Extension programs for IPM and techniques for raising high-yielding crop varieties were implemented. There were 13 NTFP collecting shops located along the road and traders from the district town came to the hamlet and asked local shop owners to become their middlemen. During this period, the marketing flows both inside the hamlet and those linking Tat hamlet to external markets became much more complex (Figure 13.4). The key flows are described below:

Marketing flows within Tat hamlet: In 2004, Tat hamlet had 16 shops for miscellaneous goods and eight buses daily passed through the hamlet. Villagers could sell their products directly to the bus drivers or to collectors on the road. Two households specialized in selling miscellaneous goods and clothes on a daily basis.

Figure 13.4 Material flows in the marketing channels in Tat hamlet after 1987

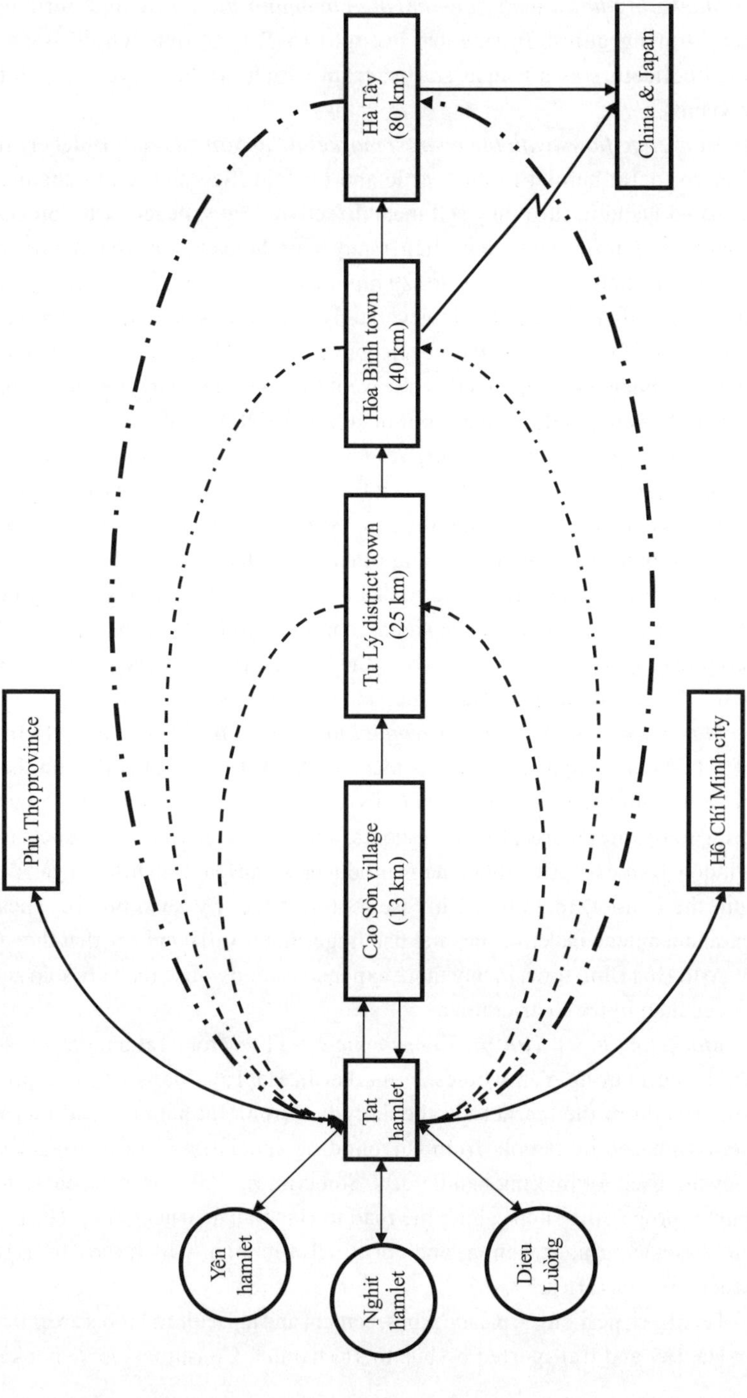

Marketing flows with neighboring communities: Cattle and buffalo from neighboring communities are brought to Tat hamlet, which is used by local collectors as a transit station from which products are shipped to the lowlands.

Marketing flows with the nearest market (Cao Sơn village): Butchers in Cao Sơn go to Tat hamlet to buy cattle and buffalo (See marketing channels for livestock), and in turn they sell meat directly to Tat villagers after processing. Thanks to improved transportation, many households now go to Cao Sơn market by bus (one hour) or motorcycle (20 minutes) to buy clothes, knives, seeds, and other miscellaneous goods. However, only households that own motorcycles go to the market often. Some have even gone as far as Thanh Sơn district to peddle clothes. One woman told us that she went to Yên Lãng market every 10 days to sell clothes and could earn a profit of 100,000 VND per day.

Marketing flows with the district town (Tu Lý): Many motorcycle hawkers come to Tat hamlet from Tu Lý to sell fresh meat and to buy pigs, cattle, buffalo, and maize. Commodities supplied to Tat hamlet include noodles, pork rind, slippers, batteries, flashlights, and other miscellaneous goods. Following the lowland system, the traders use trucks to deliver trading goods weekly to local shops, where the owners have signed contracts to become their agents. Some farmers have begun going directly to the Department of Agricultural Extension at the district to buy pesticides, medicine, and seeds.

Marketing flows with the provincial town (Hòa Bình): The next destination of NTFPs is Vo market, which is located 38 km from Tat hamlet, on the edge of Hòa Bình town. The main NTFPs are broom grass, medicinal plants for drug companies/stores, bamboo shoots, and processed timber products (doors, window frames). After 2000, many products from Hòa Bình town began flowing into the hamlet, including Chinese motorcycles, TV sets, pumps, speakers, parabola antennas, kerosene, and fish fingerlings. Villagers say that they prefer to go to Hòa Bình town to buy more expensive goods since they are sold at lower prices than in the district town.

Marketing flows with Hà Tây province: NTFPs from Tat hamlet are sold in the Đông Phương Yên wholesale market in Hà Tây. Many of these products are brought to the market via the daily bus from Tat hamlet. These products are purchased by people from surrounding specialized craft-villages where they are used for making handicrafts. Sometimes, timber is also sold to nearby timber processing shops along the road to Hanoi. The trucks from Hà Tây also buy cassava, ginger, canna, and corn in Tat hamlet, which they transport to processing factories.

Fertilizer, pesticides, plastic pipes, cement and agricultural tools are purchased in Hà Tây and transported by bus to Tat hamlet. Consumers and shopkeepers in Tat hamlet write down their requests on small pieces of paper and give them with money to the bus drivers. Then the drivers have the responsibility to buy goods according to the request and to deliver them to the front of the customers' houses. The drivers charge the customers service fees depending on the distance and the weight of the goods.

Sometimes, hawkers from Hà Tây come to the hamlet by bus bringing vegetables and fish that they sell to people at their houses. They have to stay overnight in Tat hamlet and return to Hà Tây early the next morning. The price of these goods is always about 20 to 30 percent higher in Tat hamlet than in Hà Tây. For instance, grass carp is sold at 25,000 VND/kg compared with 20,000 VND/kg in Hà Tây.

Characteristics of the Market for Agricultural and Forest Products

An active market for agricultural and forest products has begun to emerge in the hamlet. In this section we describe the specialized actors who have begun to emerge to ensure the functioning of the market system, the locations where farmers sell their products, and the ways in which they are paid in return for their products, and the prices they receive for them.

Actors in the market system

A distinctive characteristic of the development of this market is the appearance of local middlemen who act as collectors and traders. The increasing number of collectors shown in Figure 13.5 demonstrates that the market in Tat hamlet is developing and becoming more active year by year (Bùi Thị Gia *et al.* 2005).

Before 1987, under the central planning regime, the cooperative was responsible for collecting all agricultural and forest products from farmers. Thus, the market in that period contained fewer actors and was less active than in later periods. In order to remedy these shortcomings, many policies and laws aimed at improving trade and facilitating the marketing of agricultural and forest products were issued after the initiation of reforms in the late 1980s. These included the Land Law, Foreign Investment Law, and the Private Enterprise Law. The Land Law directly affected agriculture production by providing land use certificates to organizations and individual households. The Private Enterprise Law was promulgated to apply the development strategies of multiple sector economics, to stimulate private investment, and to protect the legal benefits of private enterprises. These laws encouraged trade development in the uplands and contributed to a steady increase in the numbers of private enterprises and

Figure 13.5 Changes in the numbers of collectors over the past 50 years in Tat hamlet

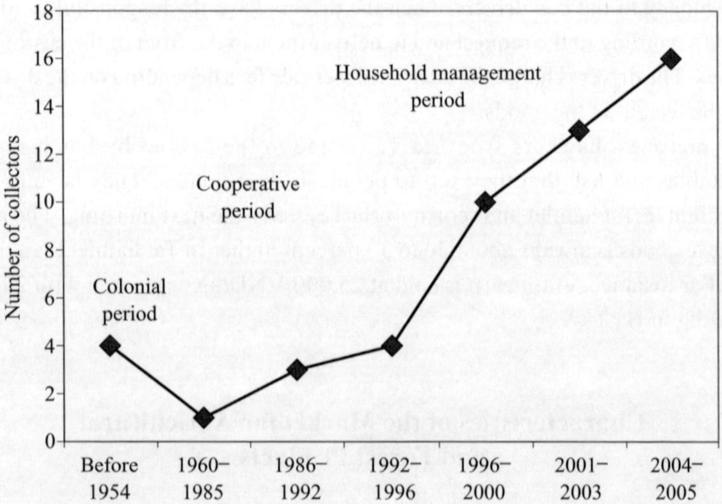

Sources: Bùi Thi Gia et al. 2005 (data from 1960 to 2003), Field survey in 2005 (data for before 1954 and after 2003).

individual households in the region participating in the market. In addition, marketing channels also became more diversified.

Government development programs, such as the 135 Program, which focused on improving upland infrastructure, have allowed farmers to more easily access markets and trade with outsiders. It was this program that upgraded Road no. 433, the hamlet's main connection with the outside world. This facilitated the movement of products between Tat hamlet and the lowlands. However, this increased market activity has been accompanied by greater exploitation of forest products and the disappearance of some valuable species and products. At the same time, the production of crops such as cassava and maize has increased because farmers have learned that these types of products can easily be marketed.

Traders living in the hamlet are people who have capital available to invest in purchasing agricultural and forest products from the farmers and then reselling them to wholesalers from the lowlands, especially Hanoi and Hà Tây. Their function is to gather the products from many collectors in large amounts, store them for two or three days, and then sell them to lowland traders or wholesalers. They are also responsible for transportation of products to other provinces. For example, they transport cassava and canna to Hà Tây province to sell. Two traders presently live in the hamlet.

Figure 13.6 Marketing channels for fresh cassava

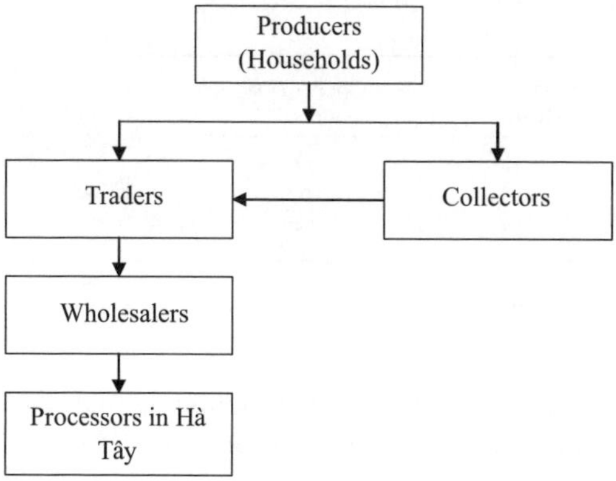

Collectors are residents of the hamlet who are hired by traders to collect products from farmers and bring it to the traders' shops. The collectors are paid on commission. Before they were collectors, they were farmers and mainly cultivated swidden fields, but the active market has attracted them to do business. In 2005, there were 16 households engaged in this activity. Their functions are to weigh and record the amount of products collected from each farmer. The majority of farmers sell products to collectors rather than directly to the shops of the traders.

Some products are sold to wholesalers from the lowland region by collectors and traders who reside in the hamlet. Other products, such as cassava and canna, are transported to processors in the lowlands by the traders. Figure 13.6 depicts the marketing channels of fresh cassava produced by farmers in Tat hamlet. Additionally, some vehicle owners also participate in marketing. They buy the products along the road at their convenience and transport them to market.

The rapid increase in the number of collectors in Tat hamlet is a good indicator of the growth of the market for commercial commodities. However, the scope of collectors' activities has been limited to the purchase of only those commodities that are available in large quantities.

Places where agricultural and forest products are sold

As is shown in Table 13.2, farmers in Tat hamlet sell their products at their homes, at collection points, at markets, in their fields and at other places along the inter-provincial road (Plate 13.1). Valuable agricultural products such as cattle and pigs are usually sold directly to the dealers at the farmers' houses. Bulky products, such as cassava and canna, however, are often sold by the farmers to collectors who live in houses next to the road; these collectors later resell them

Table 13.2 Percentage of households selling their products at different places (n = 38)

	HOME (%)	COLLECTION POINTS (%)	MARKET (%)	OTHER PLACES (%)
Crops	16	53	11	20
Livestock	55	0	34	11
Forest products	18	58	5	19

Source: Survey data, January 2004

to dealers. Villagers who sell their products directly to dealers at their houses receive higher prices than those selling to collectors (Plate 13.2).

There are five collection points, located at shops next to the road, that operate on a seasonal basis. Farmers usually bring their products to these collection places and sell them there without price negotiation because the market price has been fixed. If producers find the market price appropriate, they will sell their products. Prices among collecting places are usually the same.

Currently, there is no market in Tat hamlet so some households owning motorbikes bring their products (mainly livestock and crops) to the market in Cao Sơn to sell there at higher prices.

In addition to these three main selling locations, farmers sometimes sell their products in their fields and at other locations. They sell their products in their fields because their fields either are located near the road or they need to sell before the main harvest for cash. However, this type of selling has become unpopular because traders often buy crops at a low price and ask farmers to pay transportation fees. Sometimes producers sell their products by chance whenever they happen to meet a buyer (Ngô Thị Thuận *et al.* 2004).

Methods of payment

All crops, and three-quarters of livestock, are paid for in cash at the time of sale, but farmers who sell forest products often have to wait for three days after they deliver the goods to local traders and collectors to receive their money. This is because the dealers from the lowlands only come to the hamlet every three days, at which time they pay the collector for their previous orders. Then, the collectors in the hamlet distribute the payments to the farmers.

Prices of agricultural and forest products

Prices for some agricultural and forest products in Tat hamlet are shown in Table 13.3. The price paid for products differs among buyers. Previous studies found that there was a price difference among buyers that was influenced by trading

Plate 13.1: Bamboo stems stacked by the roadside waiting to be picked up by a deale (Nguyễn Thế Phương)

Plate 13.2: Woman selling bamboo shoots to buyer along the road (Nguyễn Thế Phương)

Table 13.3 Wholesale and retail selling prices of agricultural and forest products in Tat hamlet

PRODUCT	UNIT	WHOLESALE	RETAIL	DIFFERENTIAL BETWEEN WHOLESALE AND RETAIL
Rice	VND/kg	1,700	1,800	+100
Corn	VND/kg	1,600	1,800	+200
Cassava	VND/kg	300	400	+100
Canna	VND/kg	400	450	+50
Ginger	VND/kg	1,000	1,800	+300
Live pigs	VND/kg	10,000	11,000	+1,000
Chickens	VND/kg	24,000	25,000	+1,000
Bamboo shoots	VND/kg	1,000	1,500	+500
Bamboo stems	VND/piece	250	300	+50
Palm leaves	VND/leaf	250	300	+50
Broom grass (fresh)	VND/kg	700	800	+100

Source: Ngô Thị Thuận et al. 2004

relationships, seasonality, quality of products, as well as farmers' capability in communication (Ngô Thị Thuận *et al*., 2004; Bùi Thị Gia *et al*., 2005).

In the cooperative period, farmers always accepted any price set by the cooperative. In the new market-oriented economy, farmers are still usually price takers. In contrast, traders are usually price givers. After harvesting, villagers receive information about prices and they know clearly that prices of products in Cao Sơn market are higher than those in Tat hamlet. But because of lack of transportation facilities, they often choose to sell their products to collectors in the hamlet at the lower price (Ngô Thị Thuận *et al*. 2004). Additionally, farmers are not able to store many of their products and sell them later when the prices are better because they need cash immediately and, in any case, storage conditions are poor. As a result, farmers are inactive in setting prices and usually accept any prices set by lowland traders.

The power of farmers to negotiate prices is gradually improving, however. If the price offered is too low, farmers will now sometimes decide to wait until the price is increased. However this rarely happens. For instance, once, when the price of bamboo shoots decreased significantly, farmers jointly decided to wait to sell at a higher price. Their decision depended on the weather, however. If the weather was sunny, they could sell half of the shoots and dry the rest to sell at the higher price. If it was rainy, they would have to sell all the shoots immediately because they were not able to store them.

The market price differs according to the quality of products, which are graded by local collectors and wholesalers. For example, according to Ngô Thị Thuận *et al.* (2004), ginger has three grades: Grade 1 (Big and clean roots with length of over 5 cm could be sold for 2,000 VND/kg); Grade 2 (Clean roots with length between 2.5 to 3 cm or longer could be sold for 1,500 VND/kg); and Grade 3 (Small roots with length shorter than 2.5 cm could be sold for 1,000 VND/ kg). Similar situations can be found for bamboo and bamboo shoots. Price fluctuations also reflect seasonal characteristics with prices in the early harvest and late harvest season always being higher than in the main season.

Marketing Channels for Forest Products

Marketing channels for forest products have widely expanded in terms of the number of traders, collectors, and types of products. In 2004, Tat hamlet had 16 shops selling miscellaneous goods and acting as the collectors for agricultural products and NTFPs. All the shops are located along Road no. 433 so that they have easy access to passing trucks. The marketing channels for different types of plantation forest products and natural forest products are described below.

Plantation Forest Products

Farmers in Tat hamlet have a tradition of planting *Melia azedarach* in their swiddens and fallow fields to create small forest plantations (Trần Đức Viên 2007). Furthermore, planting of *Aleurites montana*, *Styrax tonkinesis*, and palm trees was officially encouraged from the 1960s to the 1980s by the fixed cultivation and sedentarization program; later, the planting of Acacia and Eucalyptus were supported by the PAM, 327, and 661 programs. More recently, farmers have been subsidized by the 661 Program to plant bamboo (*Dendrocalamus mebranaceus*), *Chukrasia tabularis*, and *Manglietia* to develop mixed plantation forests.

The marketing channels are complex for forest plantation products. Different products have different channels (Figure 13.7).

Only a few households sell Eucalyptus, Acacia, Styrax, and Aleurites as pulpwood because the market is not stable for these products. Only one household earns income by selling wood to the paper mill, earning only from 300,000 to 500,000 VND per year. Local producers face a serious disadvantage in marketing pulpwood since they have to patiently wait for specific collectors to buy their wood to transport the products to the paper mill. Also, although the price paid at the paper mill was 400,000 VND/ton (Võ Đại Hải, 2005), the farmers only received 200,000 to 300,000 VND/ton.

Figure 13.7 Marketing channels for plantation forest products in Tat hamlet

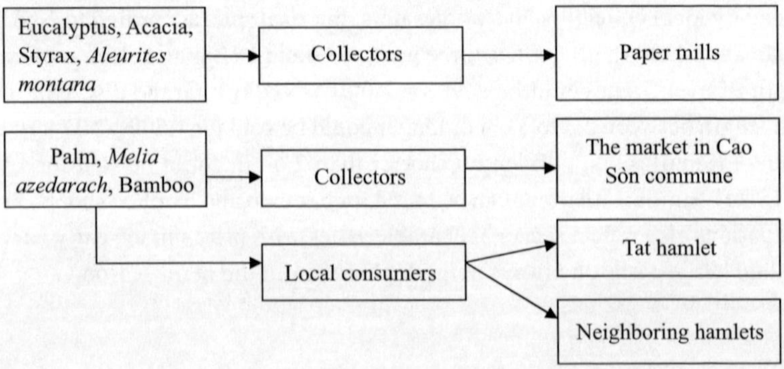

Palm trees (*Caryota urens*) were widely planted in tree gardens during the 1980s. Recently, palm leaves have been collected for sale to lowland consumers under market contracts. Two or three palm leaves packaged into one bundle are sold for 700 VND. Sometimes, palm leaves are sold to local people in Tat, Nghịt, and Yên hamlets. Recently, no one in the hamlet wants to plant more palm trees because they have enough leaves for their own consumption and they prefer other commercial products to palm leaves. In addition, the neighbors can ask producers for free palm leaves since many of them are relatives.

Most households have planted plantations of thick-wall bamboo (*Dendrocalamus mebranaceus*). The planted area has been increasing since 2000 because farmers prefer this type of bamboo because of its rapid growth and the high price of its shoots and stems. Other species of large bamboo (*Sinocalamus giganteus* and *Bambusa nutans*) are sold by growers to other villagers to use for fencing, pig pens, and roofing for 10,000 VND/stem.

Natural Forest Products

In recent decades, because of the pressure of population growth, high local demand, and expansion of market demand, natural forest resources in the hamlet have been severely depleted (Bùi Thị Gia *et al.* 2005). Forest products have become valuable trading goods, especially high quality timber and wild animals. Thus, forest products have become an important source of household income and play an important role in the community.

When transportation was poor and there was little lowland demand, farmers only collected forest products for use by their own families. Now this activity involves buyers coming to the hamlet from the lowlands and local intermediaries collecting and selling these products. According to 2004 survey data, the average income from sale of forest products was 3,465,000 VND/ per household. However, a household's participation in trading forest products depends on many factors, including market demand, seasonality, and the gender and age of its members.

Forest product collection depends on the seasonal availability of different species and the amount of time farmers have available, which in turn depends on agricultural activities during each season. The farmers can collect bamboo shoots from January to September (especially from April to September) and bamboo stems and palm leaves in every month of the year. However, in the crop cultivation period farmers devote their attention to farm work rather than forest product exploitation.

Timber

In 2004, about 90 percent of households reported that they had cut down trees for timber, both for their own permitted uses and to sell illegally. Of these, 60 percent cut trees in forest that had been allocated to them and 77 percent cut down trees on other forestland. In 2004, the number of households selling timber from their allocated forests was small, because they had been required to donate timber and bamboo to the primary school (2 planks, 10 small trees and 25 large bamboo stems per household member over 18 years of age).

After trees are cut down, their trunks are sawed into planks in the forest. Buffalo then drag the heavy planks back to the hamlet where they are stored under the floors of the loggers' houses. When lowlanders buy this wood, the loggers arrange to transport it to them using buses or trucks (Figure 13.8). People know that logging is illegal, so transport is carried out at night. In 2004, the average cash income of households from timber exploitation was 2,980,000 VND.

Figure 13.8 Marketing channels for timber products in Tat hamlet

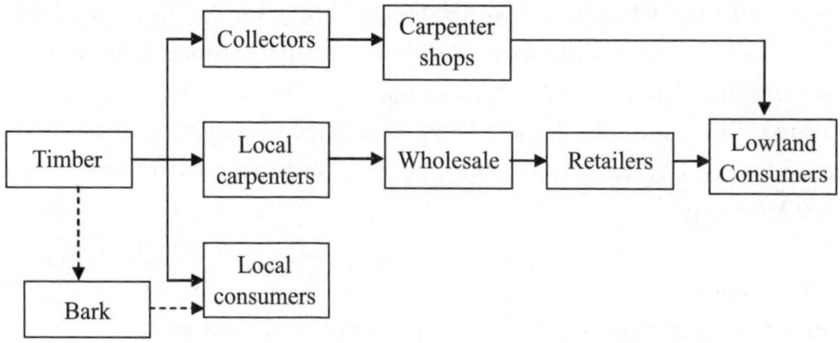

Firewood

Firewood is primarily used by the farm households as fuel for cooking and heating but is also sometimes sold, especially by poorer households, to earn cash. The demand for firewood has increased rapidly during the last five years. For instance, in 2004, 20 percent of the households engaged in selling firewood (100 VND/kg), whereas none had done this in 1998-99. In 2004, the average household income from selling firewood was about 920,000 VND. Trucks come every week to Tat hamlet to buy firewood for use in tea processing in the midlands and lowlands.

Bamboo stems

Exploitation of bamboo stems mainly depends on market demand. Normally, bus drivers and lowland traders ask collectors to buy bamboo stems from the farmers. Both men and women go to the forest to collect bamboo. Each person can collect 60-70 bamboo stems per day for which they receive 200 to 300 VND per stem from the collectors along Road no. 433. Local authorities have recognized the decline in bamboo resources and have established regulations to reduce bamboo exploitation. Farmers are allowed to cut down bamboo only when it is mature. However, the hamlet headman has said that no one obeys the regulation and bamboo continues to be depleted.

Bamboo shoots

Bamboo shoots are intensively exploited in Tat hamlet. The shoots have a seasonal character, with three major types of bamboo shoots: bitter bamboo shoots (February to March), big bamboo shoots (April to September), and small bamboo shoots (April to October). Women are the main collectors of bamboo shoots, while a few men and older children also sometimes participate in this activity. During the rainy season, people go to the forest in Thanh Sỏn district, Phú Thọ province or in Yên and Nghịt hamlets to collect big bamboo shoots. They usually go into the forest from 7:00 to 15:00, and then carry the shoots back to their houses in pack baskets (Plate 13.3). They then boil the bamboo shoots before selling them to collectors either at their houses or at the collecting point. The collectors sell the bamboo shoots to the lowland traders according to the contracts (Figure 13.9). The price of boiled bamboo shoots is about 2,500 VND/kg early in the season, but later the price decreases to 1,200 to 1,500 VND/kg.

Medicinal plants

In addition to general traders, specialized traders also came to the hamlet to contract with local collectors to purchase particular medicinal plants. They

Plate 13.3: Women carrying bamboo shoots in pack baskets

Figure 13.9 Marketing channels for bamboo stem and shoot in Tat hamlet

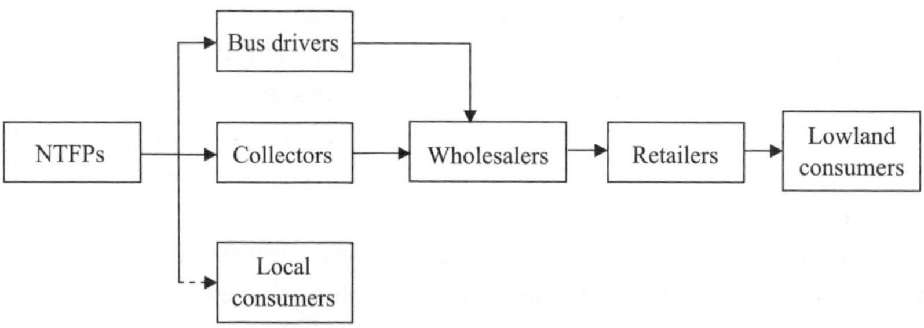

deposit some money with the collectors (100,000 to 200,000 VND) who, after they obtain a sufficient quantity from the village women who gather them in the forest, package the desired medicinal plants, and arrange with the bus drivers to take these products to the address of the traders along the road to Hanoi.

Collecting medicinal plants is frequently incorporated with other activities such as going to the swiddens to work, collecting pig fodder, or looking after livestock grazing in the forest and on fallow land. For example, we met a woman returning from her swidden with 5 kg of *Coscinium fenestratum* roots in her basket. She said that she would sell them to collectors for 1,000 VND/kg. Some local people worried about the decreasing availability of medicinal plants. They point out that now they hardly ever find some formerly common medicinal plants such as *Homalomena occulta* and *Smilax glabra*.

Agricultural Crops

Most swidden crops are transferred from the farmers to the local collectors and then to wholesalers, and finally to the food processing enterprises in Hà Tây (Figure 13.10). The recent development of a livestock feed processing industry in Hà Tây province has greatly increased the demand for fresh cassava and, due to better access to transportation, farmers can sell their fresh cassava right after harvesting it. The wholesalers receive orders from the food processing enterprises and then they make contracts with the collectors in the hamlet to buy specific crops. After making the contract, the wholesalers have to advance some money to the collectors. On the planned day, a truck will come to the hamlet to pick up the products. Sometimes, the wholesalers have to complete their contract with a high price, even though the actual price in the lowlands has fallen. The collectors say that farmers would stop selling them their products if the price went down suddenly. Prices for agricultural products also fluctuate according to seasons and market demand.

Livestock

Livestock numbers have recently increased in Tat hamlet. The farmers prefer to raise buffalo and cattle, mainly because of high productivity, but also because they are a liquid savings mechanism. Large livestock are considered to be like a silver ingot because cattle or buffalo can be readily sold to get cash needed to pay for house construction, motorcycles, TVs, weddings, loan payments, or medical expenses.

If a farmer wants to sell his large animals, he has to inform local shop owners and his neighbors, and then the information is spread to everyone in the hamlet and to outsiders. Normally, farmers sell directly to local collectors or consumers, who need a large animal for breeding at a price from 1,000,000 to 3,000,000 VND/head (Figure 13.11). There are two or three better-off households that serve as collectors; they will buy cattle from villagers at any time they want to sell them and then resell the cattle to lowland traders (wholesalers) when they have gathered a sufficient number of animals. If a farmer wants to buy cattle for either breeding or to fatten for resale, he can buy them either from collectors or from his neighbors who want to get cash.

Farmers face serious risks in raising cattle or buffalo because the animals may die from sickness or falling off cliffs while grazing in the mountains. Then, the owner asks a butcher from Tu Lý district town or Cao Sơn market to come to the hamlet to buy the sick or dead animal. After slaughter, the butcher sells

Figure 13.10 Marketing channels for agricultural crops in Tat hamlet

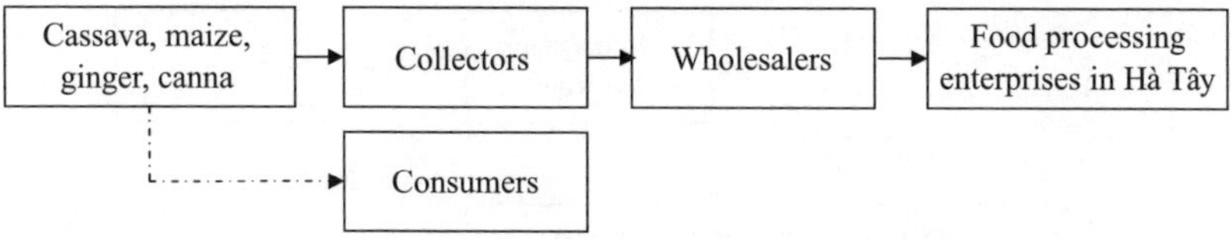

Figure 13.11 Marketing channels for large livestock in Tat hamlet

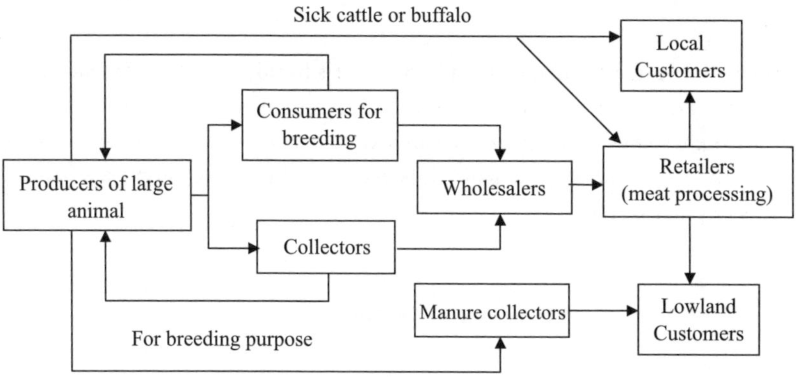

the meat at the farmer's house. Everyone in the hamlet knows what happened and comes to purchase fresh meat for about 30,000 VND/kg (only three-fourths of the price in the lowlands).

Large livestock also produce manure, which has recently become a cash crop. Several lowland collectors from Hà Tây province have come to Tat hamlet to buy manure at the price of 50,000 VND per m^3.

When a farmer wants to sell a pig he informs one of the motorcycle hawkers who operate along Road no. 433. A simple verbal contract is made with the hawker to buy the pig on a specific date. After slaughtering the pig, the motorcycle hawker sells the pork directly to consumers in Tat hamlet as well as other nearby hamlets (Figure 13.12). One household in the hamlet once tried to do its own butchering of pigs but abandoned this activity after one month due to the shortage of mature pigs in the hamlet.

Fish and poultry are normally consumed within the households raising them but some fish and poultry are sold to other villagers at the price of 25,000 VND/kg for carp, and 35,000 VND/kg for chicken. Hawkers from Hà Tây province in the lowlands come to Tat hamlet by bus bringing fish and meat to sell to the villagers. Early every morning, several motorcycle hawkers pass through the

Figure 13.12 Marketing channels for pigs, poultry, and fish in Tat hamlet

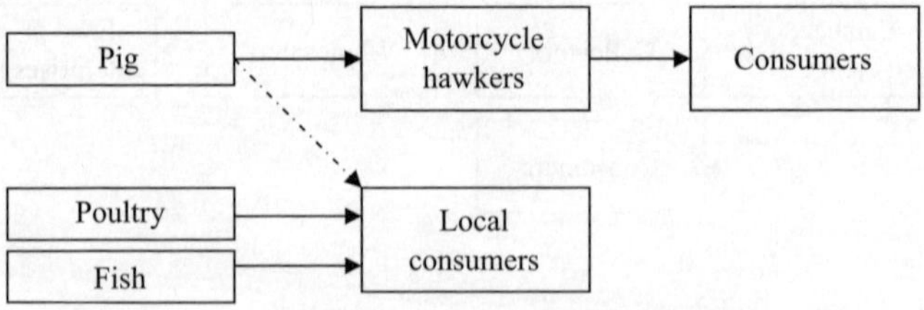

hamlet selling fresh meat to people who wait beside the road. However, the villagers can only pay a low price and demand for fresh meat is not high. Two shop owners sometimes buy two or three kg to resell to their customers. During the bamboo shoot season, demand falls even lower because people eat bamboo shoots in place of meat.

Conclusions

The expansion of markets in the lowlands and the improvement of transportation in Tat hamlet have enabled farmers to easily market their products and increase their household incomes. The marketing of agricultural and forest products has improved over time because more channels for goods have been created and the number of collectors has increased. Market integration in Tat hamlet has increased remarkably.

Farmers in Tat hamlet supply a considerable number of products to the lowland market, including (1) agricultural products such as cassava, maize, and canna, (2) livestock like buffalo, cows, and pigs, (3) forest products such as timber, palm leaves, bamboo stems and shoots, and medicinal plants. However, some agricultural products like rice, fish, chickens, and ducks are rarely marketed and are usually consumed by households in the hamlet.

Many kinds of agricultural and forest products are transferred to the lowlands using different marketing channels. Agricultural products are sold to traders via contracts between collectors and traders. Timber is sold to local carpenters or paper mills and NTFPs and medicinal plants are sold to collectors or bus drivers. Large livestock are sold to collectors or neighbors and pigs are sold to local hawkers who then sell the meat to villagers. With only a few exceptions (i.e., ginger), marketing channels for agricultural products are relatively stable. For forest products it is often supply problems that limit market activity; although in a few cases, especially pulpwood from plantation forest, the market is unstable.

Marketing channels for agricultural and forest products to final consumers in the lowlands are not complex and do not contain many intermediaries. The major intermediaries are local collectors, traders and processing factories. In these marketing channels, the farmers are usually price takers and their power to negotiate better prices is weak, whereas collectors and traders are usually price givers and price makers, respectively. As a result, the price gap between what the farmers receive for their products and what these products cost their ultimate consumers in the lowlands is sometimes in excess of 40 percent. At the same time as they receive low prices for the products they sell, people in Tat hamlet have always had to buy any goods transferred from the lowlands at very high prices compared with the price of the same goods sold in the lowlands.

Acknowledgments

The authors would especially like to acknowledge the contribution of Amanda Allbritton, Ingrid Foik, Neil Jamieson, and Stephen J. Leisz for their helpful comments on earlier drafts of this chapter. Many colleagues at CARES contributed to this research, including Đặng Việt Quang, Trần Nam Anh, Trần Trung Kiên, and Lăng Văn Trung.

14
Gender Roles in the Household Economy in Tat Hamlet

Nguyễn Thế Phương, Phạm Thị Dung, Nguyễn Thanh Lâm, and Nguyễn Thị Minh Hiền

Introduction

In this chapter we describe gender roles in the household economy and management of natural resources in Tat hamlet. We pay particular attention to gender differentiation in types of work, labor time allocated to different tasks, and decision-making.

Tat hamlet displays considerable differentiation in gender roles within households and the community as a whole. This is important because it influences the way in which resources are managed and the costs and benefits of agricultural activity distributed. Trần Công Hùng (1999) argued that one of the major reasons for the unsatisfactory results of many rural development projects in Vietnam in the 1990s was the neglect of social relations within the community, especially the roles of women. In recent years, however, gender roles in rural villages have received growing attention, especially the roles of women in the household economy and in natural resource management. Trần Thị Nhung *et al.* (1999) found that on average the women of Kinh, Muong, Dzao, and H'mong ethnic groups in Phú Thọ province worked harder than men did. Women were involved in many household production activities and played an important role in making decisions in their families, but their participation in community meetings was still limited (Trần Công Hùng, 1999; Trần Thị Nhung *et al.*, 1999; Hoàng Thị Sen and Lê Văn An, 2006). Hoàng Thị Sen and Lê Văn An (2006) have also explored ways to strengthen the social capital of women and the poor in upland communities in central Vietnam, near Hue.

Methodology

This chapter is mainly based on three primary data sets. A structured survey of 42 of the 112 households resident in the hamlet was conducted in early 2006. In this survey, respondents were asked about current gender issues and also asked to recall the situation in 1995. For purposes of this analysis, six households headed by widows were excluded because we assumed these women had to perform both male and female roles.

The second data set was collected from interviews with 30 households in late 2004 to examine gender roles in forest management. The third data set used was the 1998-99 CARES survey of 42 households. We also conducted field observations and in-depth interviews with key informants during study visits.

Gender-based Division of Labor in Livelihood Activities

Men and women play somewhat different roles in various livelihood activities, including crop production, rearing of livestock, and exploitation of forest resources.

Gender-based division of labor in crop production

Several types of agricultural labor were found in Tat hamlet. Household labor (self-employment) was by far the dominant type for agricultural activities. Exchange of labor between households was often observed in Tat hamlet, almost always to reduce labor shortages in paddy fields during transplanting and harvest periods. Some people did hire labor, but only occasionally, to help with paddy rice and vegetable production.

Table 14.1 shows the amount of labor time needed to produce crops in different components of the agroecosystem and the relative amounts of labor performed by men and women.

Overall, women contribute much more labor to crop production than men do. Labor in swidden fields was widely reported to be harder than work in paddy fields, and women contribute more than two-thirds of the labor in swiddens. Most men feel that cassava, maize, and vegetable production is women's work. Women perform almost all (87 percent) of the work in vegetable gardens and over seventy percent of the work in cassava and maize swiddens. But they also do around two-thirds of the work in rice swiddens and paddy fields. In only a few households do men contribute most of the labor in crop production. This is usually because the women are ill or busy caring for young children.

Paddy Fields

In general, men and women work together in the paddy fields and many tasks are done interchangeably by workers of both sexes. These include repairing paddy bunds, seeding the rice nursery, applying manure and fertilizer, and harvesting

Table 14.1 Gender division of labor in crop production

	PADDY FIELDS	RICE SWIDDENS	CASSAVA SWIDDENS	MAIZE SWIDDENS	VEGETABLE PLOTS
Mean area per household (m^2)	1,288	5,096	3,803	4,143	175
Labor days/mean area per hh	37	39	47	34	29
Female labor (%)	63.0	69.0	72.0	76.0	87.0
Male labor (%)	36.0	31.0	28.0	24.0	11.0
Hired labor (%)	1.0	0.0	0.0	0.0	2.0

Source: Tat Hamlet Household Survey, 2006

rice. Other tasks are gender specific: Land preparation (harrowing) and spraying pesticides are carried out primarily by men; while women do almost all of the transplanting, weeding, and water control activities in the paddy fields.

Swidden fields

The division of labor in the swidden fields is undergoing changes. Informants recalled that in 1995 men played the major role in selecting the sites for swidden fields and cutting the vegetation, but now women share these tasks with men more frequently. In fact, it is now more likely to be the wife than the husband who selects the site for a new swidden field. Burning the cut vegetation is performed by both men and women (Plate 14.1). Final clearing of the fields after burning has recently become a task performed mainly by women. In planting, it is usually men who make holes in the ground with dibble sticks, but by 2005 women had largely taken over the task of inserting seeds into the holes, work that had been shared about equally between men and women in 1995. Once the field is planted, women have primary responsibility for managing the swidden fields by controlling the weeds (plate 14.2). They go to tend their swidden fields at least once a week. On these walks the women also collect firewood (Plate 14.3) and sometimes teach their daughters to identify medicinal plants. Women do most of the weeding and harvesting, as well as carrying the harvest back to the house. It has been and still is almost entirely women who plant vegetables in the swidden fields.

Gender-based division of labor in rearing livestock

Raising livestock has become an increasingly important component of the composite swidden agriculture system. This trend adds significantly to the workload of many women, as can be seen in Table 14.2.

Plate 14.1: Man setting fire to slash in swidden field (Nguyễn Thanh Lâm)

Plate 14.2: Women weeding their rice swidden field

Plate 14.3: Woman cutting firewood in her swidden field

Pigs are mostly the responsibility of women whereas cattle and buffalo are cared for by both men and women as well as by grandparents and children. Free grazing is still the main method of raising cattle and buffalo and both men and women look after their livestock whenever they go into the forest for other reasons.

Gender-based division of labor in exploitation of forest resources

Women play an active part in forest management and the percentages of women participating in exploiting different forest resources are almost always higher than those of men (Table 14.3). However, men usually do more logging and hunting, while women usually spend more time collecting bamboo stems and shoots, broom grass, and medicinal plants, and gathering firewood.

Comparison of Workloads of Men and Women

Villagers generally view women as having heavier workloads than men (Table 14.4). Moreover, the workload of women is seen as increasing over time. Thus, in 1999, less than half of villagers believed that, on the whole, women worked harder than men, whereas in 2006 a clear majority shared this view.

Several key informants suggested that more crop diversification and raising new livestock breeds require more labor, which tends to be done by women,

Table 14.2 Division of labor among household members in caring for livestock

	Pigs (m = 28 HH)			Cattle (m = 21 HH)			Buffalo (m = 32 HH)		
Share of workload %	W	H	G&C	W	H	G&C	W	H	G&C
None (0%)	4 (14.3)	17 (60.7)	20 (71.5)	11 (52.4)	4 (19.1)	15 (71.4)	13 (40.6)	12 (37.5)	20 (62.4)
Some (1–49%)	0 (0.0)	9 (32.2)	2 (7.1)	2 (9.5)	5 (23.8)	2 (9.5)	5 (14.6)	7 (21.9)	2 (6.3)
Half (50%)	4 (14.3)	2 (7.1)	2 (7.1)	3 (14.3)	3 (14.3)	0 (0.0)	5 (14.6)	5 (14.6)	0 (0.0)
Most (51–99%)	8 (28.6)	0 (0.0)	2 (7.1)	5 (23.8)	2 (9.5)	1 (4.8)	7 (21.9)	3 (9.4)	3 (9.4)
All (100%)	12 (42.8)	0 (0.0)	2 (7.1)	0 (0.0)	7 (33.3)	3 (14.3)	2 (6.3)	5 (14.6)	7 (21.9)

Source: Tat hamlet household survey, 2006

Note: W = wife, H = husband, G&C = grandparents and children

Table 14.3 Gender division of labor in forest activities (n = 30 households)

ACTIVITIES	NO ONE IN HH (%)	MEN (%)	WOMEN (%)	BOTH SEXES (%)
Planting and protecting forests	3.3	0.0	6.7	90.0
Logging	10.0	60.0	3.3	26.7
Collecting firewood	3.3	6.7	56.7	33.3
Collecting bamboo stems	13.3	26.7	20.0	40.0
Collecting bamboo shoots	3.3	3.3	76.7	16.7
Hunting	70.0	26.7	0.0	3.3
Collecting mushrooms	63.3	0.0	10.0	26.7
Collecting broom grass	60.0	3.3	16.7	20.0
Collecting medicinal plants	56.7	0.0	36.7	6.7
Other activities	36.7	0.0	16.7	46.7

Source: Tat Hamlet Household Survey, 2004

raising their workloads. Also, many informants thought that although women work more hours than men, men tend to perform heavier work. On this basis, some informants think that a balance has been established in the gender-based division of labor in Tat hamlet.

Gender Roles in Household Decision-making

Table 14.5 shows the division of responsibilities among household members in decision-making about agricultural activities and sale of farm products.

Decisions about farming activities are shared by both spouses to a significant degree. Husbands and wives jointly make decisions about the selection of crops (60%) and animals (60%), pesticides (50%), taking out loans (67%), sale of crops (45%) and livestock (57%). Women play a major role in selecting swidden fields (62%) and selling farm products (45%).

Decision-making about forest management

As shown in Table 14.6, decisions about forest management are made jointly by spouses in 40 percent of the households, while men are the principal decision-makers in 37 percent of the households. In 13 percent of the households, all family members participate in making these decisions.

Table 14.4 Villager assessments of the comparative workloads of men and women (n=42)

	Number and percentage of households	
	1999[a]	2006[b]
Husband works harder	9 (21.4)	6 (14.3)
Wife work harder	19 (45.2)	24 (57.1)
Spouses work equally hard	14 (33.3)	12 (28.6)

Source. a: Rambo and Trần Đức Viên, 2001b, b: Tat Hamlet Survey Data, 2006

Table 14.5 Responsibilities of men and women in household decision-making about farming activities

	Farming activity					Sale of products	
	Selection of swidden fields	Selection of crops	Selection of animals	Pesticide	Credit	Crops	Livestock
Husband	8 (19.0)	12 (28.6)	14 (33.3)	11 (26.2)	13 (31.0)	4 (9.5)	9 (21.4)
Wife	26 (61.9)	5 (11.9)	3 (7.1)	10 (23.8)	1 (2.4)	19 (45.2)	9 (21.4)
Both	7 (16.7)	25 (59.5)	25 (59.5)	21 (50.0)	28 (66.7)	19 (45.2)	24 (57.1)
No answer	1 (2.4)	0 (0.0)	0 (0.0)	0 (0.0)	0 (0.0)	0 (0.0)	0 (0.0)

Source: Tat Hamlet Household Survey Data, 2006

These are mostly households that contain married children living with their parents.

Decision-making about household management and children

Table 14.7 shows that most decisions about household management, such as building or repairing houses, buying furniture and use of financial resources, are made jointly. Table 14.8 shows that decisions about children attending school or getting married are mostly made by both spouses, by the children themselves, or by the mother. The households where women are the primary decision-makers tend to have husbands who have little education or bad health. These results are quite different from those of a previous study in Tat hamlet, which found that men tended to monopolize household decision-making (Rambo and Trần Đức Viên, 2001b).

Conclusions

In Tat hamlet, women typically work longer hours and have less leisure time than men (Plate 14.4). There has also been a clear tendency for the workload of

Table 14.6 Gender roles in decision-making about forest management (n = 30 households)

PRINCIPAL DECISION-MAKER	NO. OF HOUSEHOLDS	PERCENTAGE OF HOUSEHOLDS
Husband	11	36.7
Wife	3	10.0
Both spouses	12	40.0
All members of household	4	13.3

Source: Tat Hamlet Household Survey, 2004

Table 14.7 Responsibilities of spouses in decision-making about household management (n = 42 households)

	HOUSE CONSTRUCTION	BUYING FURNITURE	MANAGING FINANCES
Husband	11 (26.2)	8 (19.0)	16 (38.1)
Wife	2 (4.8)	2 (4.8)	17 (40.5)
Both spouses	29 (69.0)	32 (76.2)	9 (21.4)

Source: Tat Hamlet Household Survey data, 2006

Table 14.8 Decision-making about children (n = 42 households)

PRINCIPAL DECISION-MAKER	NO.OF HOUSHOLDS	PERCENTAGE OF HOUSEHOLDS
Husband	6	14.3
Wife	9	21.4
Both spouses	14	33.3
Children	13	30.9

Source: Tat Hamlet Household Survey Data, 2006

women to increase over the past ten years. Several key informants have observed that more cropping diversification and raising new breeds of livestock require more labor, which tends to be provided by women.

During the same period of time however, the role of men as dominant decision-makers in Tat hamlet appears to have been reduced. Women are participating more in making decisions in the household about everything from farming activities to forest management to housing, household finances, and whether or not children will continue their schooling or get married. Comparable changes in decision-making are not so evident at the community level and this may have significant implications for natural resources management.

Because of the gender-based division of labor in Tat hamlet, men and women each have their own specialized knowledge of the local environment. Men know

Plate 14.4: Women sewing clothes

Plate 14.5: A community meeting with men as the dominant participants (Trần Đức Viên)

more about forest trees and wildlife, for example, while women know more about swidden fields and non-timber forest products. Yet, while women do a significant portion of the work in both the fields and the forests and thus often have a superior knowledge of the local environment, they typically have had little voice in making decisions about natural resources management at the community level. It is almost always men who dominate community meetings (Plate 14.5). Even when women are present, men usually do almost all of the talking.

Thus, in organizing meetings to elicit community opinions, or in planning development projects, it is important that a special effort be made to encourage the participation of women. This includes soliciting their presence, making them comfortable as active participants, and ensuring that they have opportunities to share their insights. This kind of gender imbalance in the flow of information to and from both men and women may have profound implications for the ways in which natural resources are managed.

PART IV
THE ECOLOGICAL SUSTAINABILITY OF COMPOSITE SWIDDEN AGRICULTURE

Introduction

The two chapters in this part present the findings from our intensive effort to empirically assess the ecological sustainability of composite swidden agriculture. Based on preliminary investigations in Tat hamlet from 1992-1995, it had been suggested that composite swiddening was an ecologically more sustainable farming system than the pure swiddening systems that were found in many parts of the NMR (Rambo 1998). It was argued that because the paddy fields in the valleys were highly productive, providing at least one-half of the grain requirements of the households, the farmers did not need to clear nearly as large a swidden area each year as they would have otherwise needed. Consequently, for a given population density, they could maintain a much longer fallow cycle than would have been possible if they were using a pure rotational swidden system. We expected that this would minimize the extent of deforestation and land degradation and permit the fields used for swiddening to regain nutrient balance over the course of the whole multi-year swidden-fallow cycle.

It was also suggested that positive functional linkages between the swiddens and the paddy fields contributed to the sustainability of the composite swidden system as a whole (Trần Đức Viên 1998). This was because nutrients that were eroded from the swiddens on the hillslopes around the valley were transported into the paddy fields by run-off water and because livestock that grazed in the fallow swiddens provided manure that the farmers used to fertilize the paddy fields. In our subsequent research we have attempted to assess the ecological sustainability of composite swidden agriculture in Tat hamlet. Two different methods, nutrient balance analysis and analysis of changes in vegetation cover, were employed in this research.

Chapter 15 measures nutrient balances of the paddy and swidden field components of the composite swidden agroecosystem in order to assess the extent to which composite swiddening is sustainable. The verdict is a mixed one: The nutrient balance in the experimental swidden plots was negative for nitrogen (N), phosphorous (P) and potassium (K) during the cropping period. Over the four years during which the fields were cultivated, about 124 kg of N, 11 kg of P, and 1,578 kg of K per ha were lost. These losses occur in the process of rice straw burning, the removal of crops, leaching, and most importantly, as a result of soil erosion from the steeply sloping field. Lost N and P can be replaced by means of aerial deposition (nutrients in dust and rain) and uptake from deep soil levels by the roots of fallow phase vegetation during a sufficiently long fallow period. In the case of Tat hamlet, it is estimated that a fallow lasting a minimum of 15 years is necessary to regenerate soil fertility back to the original level, and possibly as long as 30 to 40 years for N, if measurement of the swidden cycle is started from the initial burning of the secondary forest vegetation. The very high losses of potassium during the cropping phase cannot be replenished by even an extended fallow period but we suspect that K is unlikely to act as a limiting factor on crop growth because the reservoir of K in the soil is sufficiently large to sustain multiple cropping cycles. This is plausible because the parent bedrock materials are sandstone and schist, with some quartz and mica bearing granite, which contain average to high levels of potassium (Ziegler et al., 2004: 127).

In contrast to the swidden field plots, the paddy field plot maintained a positive balance over the course of the annual cropping cycle. This was the result both of inflow of nutrients in the irrigation water and heavy inputs of manure and chemical fertilizer by the farmers. Although the quantity of nutrients carried into the plot by the run-off water from the swidden plot was sufficiently large to meet the needs of the rice crop, additional inputs were still required because the loss of nutrients in water flowing out of the field was also very large. We suspect that some of these nutrients are captured by paddy fields lower down in the valley, so contribute to maintaining the overall balance in paddy fields within the hamlet watershed, but lack the data to confirm this. Recent research in a cascade of terraced paddy fields on a hill slope in neighboring Sơn La Province, however, has found that nutrients tend to concentrate in the mid-level terraces in the toposequence (Schmitter et al. 2008).

The nutrient balance analyses reveal that, under the current swidden cycle of 3 or 4 years cultivation followed by 3 to 5 years fallow, the swidden component is not sustainable. The quantities of nutrients lost from the field during the prolonged cropping period are much too large to be fully restored during the greatly abbreviated fallow period. It can be expected with some confidence that the already very low productivity of the swiddens will continue to decline in the future unless a way to improve fallow management so as to accelerate nutrient accumulation can be found.

Chapter 16 employs remote sensing data (aerial photographs and satellite images) combined with ground truthing to measure the extent of changes in vegetation cover in Tat hamlet from 1952 to 2003. The most important finding is that the total area under forest cover has not appreciably diminished over the past 50 years. Most of the land surface of Tat hamlet still remains under closed vegetation cover, either fallow swiddens where forest vegetation is in the process of regeneration or secondary forest. Indeed, since 1995, there has been a net increase in the area under more mature forest vegetation and a net decrease in the area of upland agricultural fields. That the extent of the area covered by forest has not diminished is all the more remarkable when it is remembered that the human population has increased considerably during this period.

Although the total area covered by forest vegetation has not appreciably changed, the quality of the forest has been seriously degraded compared to its original state. In 1952, large areas of the hamlet were covered by high quality mature forest that contained a high volume of valuable timber. Today, the few remaining patches of mature forest have been degraded by selective logging and grazing by livestock. The quality of vegetation on regenerating fallow plots is also generally poor with very low amounts of biomass even 10 or 12 years after being fallowed. Of course, this has major implications for the availability of nutrients in the next swidden cycle. But, despite the reduction in the quantity and quality of natural vegetation in the fallowed plots and secondary forest, it still provides excellent protection against erosion, hence preventing the occurrence of severe land degradation.

Taken together, these chapters provide a quite comprehensive basis for evaluating the ecological sustainability of composite swidden agriculture as it is practiced in Tat hamlet. Although composite swiddening is not now, under conditions of high population density and increased demand to produce cash crops, a fully sustainable agroecosystem, it still represents a more sustainable system than some commonly used alternative systems in the northern mountains. Certainly, deforestation and land degradation would be much more serious than they actually are in Tat hamlet if the farmers relied on pure swiddening. This reflects the fact that a relatively small area of paddy fields, because of high rice yields per hectare, can substitute for a much larger area of low-yielding swidden fields. At current yield levels, a one-hectare paddy field produces as much grain as 18 hectares of rice swiddens. The substitution effect of all the paddy fields in Tat hamlet reduces the area needed for swiddens by more than 300 hectares per year (Table IV.1).

Despite these very real advantages, it is clear that composite swiddening in Tat hamlet is becoming less sustainable as this type of farming system is pushed

Table IV.1 Substitution effect of paddy field area for rice swidden field area in Tat hamlet

	1993[A]	2003[B]
Paddy field rice yield (kg/ha/yr)	4,900	7,600
Swidden field rice yield (kg/ha/yr)	590	430
Ratio of paddy yield to swidden yield	8.3	17.7
Total paddy area (ha)	8.4	17.7
Total substitution effect of paddy on area of swidden fields (ha/yr)	-69.7	-313.9

Source: a = Rambo 1998; b = Nguyễn Thanh Lâm

beyond its adaptive limits. An agroecosystem that was highly sustainable when population density was lower and forestland more abundant is unable to cope with a higher level of population density and a reduction of the forest area where farmers can clear swiddens. In Tat hamlet, population density has increased from 52 persons/km^2 in 1993 to 69 in 2005. In order to maintain the same per capita production of rice, farmers would have had to clear a somewhat larger area of swiddens each year. That has not happened, however, because of the scarcity of suitable forestland to make new swiddens. The government has prohibited clearing new swiddens in large areas within the hamlet, forcing the farmers to concentrate swidden cultivation in certain limited areas. Because of this reduction in the land available for swiddening, the farmers have had to lengthen the cropping period and shorten the fallow period in the attempt to maintain total output of crops. But the shortened fallow period does not provide sufficient time for the field to regain nutrients lost during the extended cropping period. Consequently, swidden rice yields have declined to levels where many farmers no longer even try to grow hill rice. Instead, they plant cassava and other less nutrient demanding cash crops in their swiddens and use the income from their sale to purchase supplementary rice from the market. It can be expected that in future years more and more households will shift from traditional composite swiddening (paddy field + rice swidden) to modified composite swiddening (paddy field + cash crop swiddens but no rice swidden) (see Chapter 10); ultimately, composite swiddening of the type that first attracted our attention to Tat hamlet will no longer be practiced there.

15

A Nutrient Balance Analysis of the Composite Swiddening Agroecosystem

Nguyễn Văn Dung, Trần Đức Viên, Georg Cadisch, Nguyễn Thanh Lâm, Aran Patanothai, A. Terry Rambo, and Trần Mạnh Tường

Introduction

This chapter presents some of the findings of a long-term study of nutrient balances in swidden and paddy fields in Tat hamlet. Analysis of nutrient balances is an operationally feasible method for empirically measuring the sustainability of farming systems (Patanothai 1998), including the composite swidden system (Trần Đức Viên 1998; Trần Đức Viên *et al.* 2004).

The central hypothesis of this research is that the farmers of Tat hamlet manage the swidden and paddy field subsystems of their composite swidden agroecosystem in ways that maintain a balance of soil nutrients in each subsystem. In the case of the paddy fields, a positive nutrient balance is probably maintained over the course of each annual cropping cycle, with nutrient inputs equal to nutrient outputs for each crop. In the case of the swiddens, the question is much more complex and must be studied over the duration of the entire multi-year cycle of cultivation and fallowing.

The sustainability of the land use system as a whole very much depends on existence of strong positive functional linkages between the swiddens and wet rice field subsystems (Rambo 1998). Runoff water from the hill slope swiddens carries nutrients into the paddy fields and livestock grazed in the fallow swiddens provide manure that is used to fertilize the paddy fields. Developing a better understanding of these functional linkages among key sub-systems in the composite swiddening agroecosystem was one of the key objectives of this research project. Our study was designed to measure nutrient balances of each of the two key subsystems over the full rotational cycle of cultivation. This chapter presents the results of nutrient balance analyses of the swidden and paddy subsystems based on data collected over five years covering the cropping period in the rotational swidden cycle.

Research Design and Methodology

Conceptual framework for nutrient balance analysis

Analysis of soil nutrient balance has been widely used as a tool for assessing agroecosystem sustainability (Patanothai 1996; Patanothai 1998; Smaling *et al.* 1999). Nutrient balance is the difference between the sum of nutrient inputs and nutrient outputs in a given system. A positive nutrient balance (inputs > outputs), sometimes designated as nutrient 'surplus' or 'excess,' points to enrichment, and a negative balance, sometimes designated as 'deficit,' points to depletion of the stock of nutrients in the system. The basic assumption underlying the use of nutrient balances in the analysis of long-term sustainability of land productivity is that a negative balance would indicate a loss of nutrients from the system exceeding its self-regenerating capacity. If continued over a long period, the quality of the soil would be reduced and, consequently, productivity would be lowered. A positive balance may also affect land use sustainability if any kind of nutrient is accumulated to a level where it becomes toxic or creates a nutrient imbalance. Therefore, in assessing sustainability, we look for nutrients that are likely to be continuously depleted or continuously accumulated (Patanothai 1998).

To assess nutrient balances of a land use system, the system of concern needs to be clearly defined, all sources of nutrient inputs and outputs specific to the system need to be identified, and the quantities of nutrients in all inflows and outflows need to be measured. Figure 15.1 is a comprehensive model of possible nutrient inputs and outputs for a cropping field.

Sources of nutrient inputs to the cropping field include planting material, chemical fertilizers, manure applied by farmers or deposited by grazing animals, green manure, nutrients brought in by sediments and run-in water, irrigation water, flood water, rainfall, dust, and N_2-fixation by leguminous and non-leguminous plants and microorganisms. Sources of nutrient outputs are harvested products and straw or stubble removed from the field, loss from burning of vegetation in the field, grasses and weeds taken out of the field, loss by herbivores, erosion and runoff, leaching, lateral flow, and gaseous losses (e.g., ammoniac volatilization, de-nitrification). Of course, in any actual field, not all of these inputs and outputs will be found. Thus, the specific sources pertaining to the field or system of concern need to be identified. In practice, it will normally not be possible or desirable to measure or estimate the amounts of nutrients from all the sources pertaining to the system of concern, as some sources are difficult to quantify and some sources may contribute very little to the nutrient budget and can be omitted from the nutrient balance calculation without seriously compromising its accuracy. In rotational land use systems, such as the composite swiddening practiced in Tat hamlet, which are characterized by

Figure 15.1 Comprehensive model of nutrient inputs and outputs for a cropping field

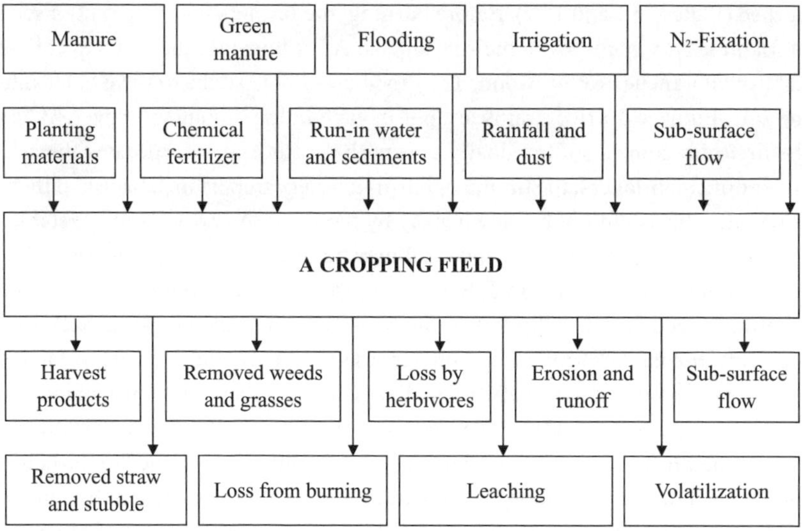

multi-year cropping and fallow cycles, nutrient balances need to be calculated both yearly and for the entire rotation period.

Because the composite swidden agroecosystem is very complex, our limited resources did not permit us to study the functioning of the total system. Instead, we focused on its two most important subsystems from the standpoint of human subsistence requirements: The swidden subsystem on the hill slopes and the paddy field subsystem in the valley. These two sub-systems are functionally linked to each other by the flow of water that transfers nutrients from the swiddens on the hill slopes to wet rice fields in the valley.

Sources of nutrient inputs and outputs for the swidden field subsystem

The farmers in Tat hamlet cultivate several different types of monocultural swiddens on sloping lands. Rice swiddens are located on the most fertile available land. Other swiddens are planted with cassava, maize, canna, or, more rarely, ginger. Most households cultivate several different swidden plots simultaneously. After the forest vegetation is cleared and burned, swiddens are normally planted with rice for one or two years, and then shifted to cassava or corn for one or two more years before being fallowed. Lately, the fallow period has been declining due to land scarcity and is now only three or four years on most sites.

After the *Tết* lunar New Year holidays in the early spring, the villagers go to the forest to select secondary forest plots (or fallow fields) to clear beginning in early May. The slashed vegetation is allowed to dry for 10 to 15 days and then burned (Plates 15.1 and 15.2). Before burning, the farmers may also scatter seed of *Melia azedarach* trees in their swiddens. After burning, the farmers collect unburned branches for firewood. They then use dibble sticks to poke holes into the soil. Eight to 10 rice grains are put in each hole without soil cover. When the first rain comes, soil gradually covers the seeds and incorporates the ash. According to villagers, the timing of burning is very important because, if there is no rain, the ash might be blown away by the wind. A few hills of vegetables (e.g., maize, squash, mustard, and taro) may be interplanted with the rice.

Weeding is done twice, in July and August, one and two months after rice planting. Weeds are usually left in the field, recycling the nutrients back into the field. During the cropping season, fences are constructed around most fields to help prevent damage by cattle. Fertilizers are never applied to the swiddens. Harvesting is done by cutting the rice panicles at about 20 to 30 cm below the base of their heads. The harvested panicles are bundled and carried to the house for drying and later threshing. The remaining rice stubble is left standing in the fields, which become open access pasture for cattle and buffalo from the hamlet. The swidden fields are then left fallow from November to May in the following year during which time short grass and Imperata grass regenerate.

Preparation for planting the second swidden rice crop starts the following June. The fields are weeded by hand pulling or cutting and all the weeds are left on the fields to dry. After several days, all the dried weeds are collected into small piles that are later burned. Management practices for planting, weeding and harvesting are the same as for the first-year rice crop. Usually, after two crops of rice, the swidden will be switched to growing cassava for one or two more years.

Depending on the amount of weeds, field cleaning and burning are required for cassava planting after swidden rice in the following year. The cassava-growing season extends from February until the following December or January. Planting is done with stem cuttings of 15 to 20 cm in length. Weeding is done twice, and no fertilizer is applied. Harvesting is done by digging out the roots and carrying the roots to the house. The remaining stalks are left in the field and are sometimes moved to the sides of the field during field clearing for planting the next cassava crop. Management practices for the second year cassava crop are essentially the same as the first year crop. No field clearing is done when the field enters the fallow phase of the rotation cycle of swiddening.

Key inputs for the swidden field include: (a) planting materials, (b) atmospheric deposition of nutrients contained in rainfall and dust, especially nitrogen fixed by lightening and ash from the surrounding burning fields, (c) manure deposited by grazing animals during the fallow period after rice harvesting, (d)

Plate 15.1: Burning swidden fields, Tat hamlet

Plate 15.2: Newly burned swidden field with ash on top of soil (Võ Hữu Công)

run-in water that carries sediments down from the fields higher on the slope, (e) sub-surface flow of water that brings in nutrients from the higher fields, and (f) nitrogen fixation by leguminous and non-leguminous plants (usually weeds) and microorganisms.

Key outputs from the swidden field are: (a) harvested products, (b) straw and stubble or stalks taken out of the field, (c) erosion and runoff, (d) sub-surface flow of water carrying nutrients out of the field, (e) removal by herbivores, e.g., cattle, rats, birds, insects, (f) leaching, (g) loss from burning, and (h) volatilization of nitrogen into the atmosphere in gaseous form.

Not all of these inputs and outputs were measured in this study. For the inputs, only planting materials and atmospheric deposition by rainfall and dust were measured. Normally there would be run-in water and sediments from the fields above. In this study, however, a concrete wall was built around the studied plot, preventing surface flow of water from entering the plot. Thus, this input source is not applicable to the situation in this study. The quantities of nutrients in manure deposited by grazing animals during the fallow period after swidden rice harvesting and in sub-surface inflow of water are expected to be small, as the nutrients in the deposited manure would be offset by the removal of plant materials from animal grazing, and the subsurface inflow would be counterbalanced by subsurface outflow; therefore, because these two sources are very difficult to measure, they are omitted from the nutrient balance calculation. Because only few leguminous plants were observed in the studied swidden fields, N_2-fixation was expected to be small and was also excluded from the nutrient balance calculation.

The outputs measured and included in the analysis were harvested products (which include plant parts taken out of the field with the products), erosion and runoff, leaching and loss from burning of remaining weeds and stubble in land clearing before planting the second-year swidden rice or first-year cassava. Cassava stalks taken to the edges of the field and loss from herbivores were not measured and are omitted from the analysis. Nutrient loss through volatilization was also expected to be very small in swidden fields and so it was not taken into account in the nutrient balance analysis.

Sources of nutrient inputs and outputs for the paddy field subsystem

The wet rice fields are built in a series of terraces rising like steps from the stream in the middle of the valley floor. Most paddy fields are planted with two rice crops per year. The spring crop season is from February to June, while the summer crop season is from July to October. Most farmers grow improved varieties of non-glutinous rice in both crops. Management practices for the two seasons are essentially the same. The land is prepared by harrowing the field three times

Plate 15.3: Woman carrying rice seedlings

before planting. During this time, manure from buffalo and cattle and green manure are applied. After the final harrowing, rice seedlings are immediately transplanted (Plate 15.3). Top dressing fertilizer is applied about one week after transplanting. Weeding is done twice at about 15 and 30 days after transplanting. Weeds are left in the field, thus, all nutrients from weeds are recycled back to the field. Wealthier households also sometimes apply additional fertilizer before rice flowering. Irrigation water is supplied from small streams coming down from the hills. Water is allowed to continuously flow into the upper paddy fields, and excess water then flows down into the lower paddy fields.

Harvesting is done by cutting the grain stalks at about 20 cm above the ground. Threshing is done by machine directly in the field. The straw after threshing and standing stubble are left in the fields, which then become open access feeding grounds for ducks and grazing area for buffalo and cattle. As the transition period between the spring rice crop and the summer rice crop is rather short (only about one month), rice straw of the spring rice crop is normally removed from the fields and left along the stream to avoid problems in land preparation for the summer crop. The transition period between the summer rice crop and the following spring rice crop, however, is much longer (extending from November to January). The straw and stubble of the summer rice crop are left to dry in the field, and whatever is left from grazing is burned before land preparation for the following spring rice crop.

Key nutrient inputs in the wet rice subsystem include: (a) planting material, (b) chemical fertilizers, (c) manure applied by farmers and deposited by cattle and ducks, (d) green manure, (e) irrigation water, (f) rainfall and dust, (g) N_2-fixation, and (h) sub-surface inflow.

Key nutrient outputs are: (a) harvest products, (b) straw removed in field clearing for summer rice, (c) loss from burning of remaining stubble and weeds in field clearing for spring rice, (d) loss to herbivores, (e) surface outflow of excess water, (f) leaching, (g) volatilization, and (h) sub-surface outflow.

Not all of these sources of nutrient inputs and outputs were measured and included in the nutrient balance analysis of the paddy field. Manure deposited by cattle and ducks during the inter-season fallow period, N_2-fixation, and subsurface inflow were not measured. For the outputs, straw removed in field clearing, loss from burning of remaining rice stubble and weeds, leaching, volatilization, and sub-surface outflow were not measured and therefore were not included in the analysis. These sources were not expected to make a substantial enough contribution to the nutrient budget to alter the interpretation of the nutrient balance analysis.

The study site

In order to collect nutrient flow data in a controlled fashion, a 3.54 ha experimental site was established in a small watershed at east longitude: 105 degrees 11' 92" and north latitude: 20 degrees 32'82''N (Figure 15.2). Of the total area, 3.23 ha is hill slopes and the remaining 0.31 ha is paddy fields. The site was intentionally selected to have both upland fields on the hillside with secondary forest at the stage that was ready to be cleared for swidden cropping and paddy fields directly below at the foot of the hill. The hillside has an average slope of between 29.3 and 36.4 degrees.

Soil samples were taken from different spots both in the hill slope and in the paddy fields. Both the upland soils and the paddy soils belong to the same soil group, being Ferralic Acrisols for the upland soils and Gleyic Acrisols for the paddy soils (Table 15.1). These soils are sandy and extremely acid. Total nitrogen, total phosphorous, available potassium, CEC, and organic carbon were all very low. Permeability of the upland soil was also determined to design the layout of the experimental plots.

Information on the vegetation communities found on the upland study site was obtained from a vegetation map with a scale of 1:500 (Lê Trần Chấn and Ngô Thế Ân 2000). That the vegetation of the study site had already been affected by human activity for a long time before the experiment is shown by the fact that among the 21 plant communities identified by Lê Trần Chấn and Ngô Thế Ân (2000), nine were types of vegetation that had been planted by the farmers (Table 15.2).

Table 15.1 Physical and chemical properties of the surface soil in the experimental site in Tat hamlet in 1999

	UPLAND SOILS	PADDY FIELD SOILS
Soil texture (%)		
Sand	47.5–58.2	51.2–62.7
Silt	24.5–33.0	18.0–27.9
Clay	12.1–24.4	9.4–30.8
Bulk density (g/cm^3)	1.0–1.3	1.0–1.6
Particle density (g/cm^3)	2.5–2.6	2.5–2.6
$pH_{(KCl)}$	3.6–4.0	3.7–4.4
Available P_2O_5 (mg/100 g soils)	0.3–4.0	1.2–5.4
Available or active ion iron (mg/100 g soils)	5.0–42.0	25.0–100.0
Available N (mg/100 g soils)	7.0–8.0	4.0–5.0
Available K_2O (mg/100 g soils)	6.4–28.0	7.3–17.7
SiO_2/R_2O_3	< 2.0	< 2.0
C/N ratio	< 13.0	< 11.0
K_2O (%)	4.8–5.4	4.7–4.8
N (%)	0.1–0.2	0.1–0.2
MgO and CaO	Poor	Poor
Organic carbon (%)	0.3–1.9	0.4–1.9
CEC (cmol kg^{-1} clay)	5.5–8.8	6.3–7.8
Aluminum saturation (%)	22.3–33.7	7.3–13.9

Layout of the experimental plots

Ten experimental plots were established in 1999 with three patterns of land use: (a) forest (bamboo and trees) on the top of the slope, and (b) swidden crops, i.e., upland rice, cassava, and fallow, on the middle slope, and (c) wet rice in the paddy field in the valley at the base of the hill slope (Figure 15.2). Forest covered 2.47 ha, swidden fields covered 0.76 ha (of which 0.21 ha was upland rice) and the paddy field covered 0.31 ha.

Of the ten study plots, nine were in the uplands and one was in the lowlands (Figure 15.2). Six upland plots were for swidden land use (Plots 1-6), and three for forest land use (Plots 7-9). The single lowland plot was for paddy land use (Plot 10). The forest and swidden plots were 5 m x 20 m, with the long dimension

Table 15.2: Vegetation communities in the upland study site

Natural vegetation communities
Woody trees mixing with bamboo (*Neohouzeanua dullooa*)
Shrubs
Shrubs mixing with scattered woody trees
Shrubs mixed with palms *(Livistona saribus)* and *Neohouzeanua dullooa*
Shrubs mixed with *Neohouzeanua dullooa*
Pure *Indosasa crassifolia*
Indosasa crassifolia mixed with palms and shrubs
Pure *Neohouzeanua dullooa*
Neohouzeanua dullooa mixed with scattered shrubs
Stachytarpheta jamaicensis mixed with shrubs
Microstegium vegans mixed with shrubs
Planted vegetation communities
Gigantochloa laevis mixed with *Melia azedarach*
Aleurites montana mixed with palms, *Indosasa crassifolia* and tea *(Thea chinensis)*
Aleurites montana mixed with shrubs
Aleurites montana mixed with palms, *Ormosia lalansae* and *Neohouzeanua dullooa*
Melia azedarach mixed with shrubs
Pure tea *(Thea chinensis)*
Styrax tonkinensis) mixed with scattered *Neohouzeanua dullooa*
Styrax tonkinensis mixed with *Ormosia lalansae*
Pure palms *(Livistona saribus)*

running up and down the slope. The size of the paddy plot was about 1,000 m^2, covering three paddy terraces.

Initially two cropping sequences were planned for the swidden plots, each with three plots to make three replicates. The planned cropping sequences were: (1) two years of rice, two years of cassava, and four years of fallow, and (2) two years of rice followed by six years of agroforestry (*Melia* trees intercropped with cassava which later became pure *Melia*). However, in actual practice, cropping sequences in the individual plots were changed according to the farmer's preferences as is shown in Table 15.3. In the first year (2000), all six plots were planted with upland rice. In the following year (2001), there were three different land use treatments: cassava, fallow, and upland rice interplanted with *Melia.* In the third year (2002), there were two types of land uses: Agroforestry *(Melia* intermixed with cassava) and fallow. In the fourth year (2003) some of the cassava + *Melia* plots were enriched with *Styrax*. In the fifth year, all of the plots were dominated by planted trees or grass and natural fallow vegetation. In the paddy field plot, two rice crops were grown per year in the spring and summer seasons in all years.

Figure 15.2 Map of the experimental site in Tat hamlet

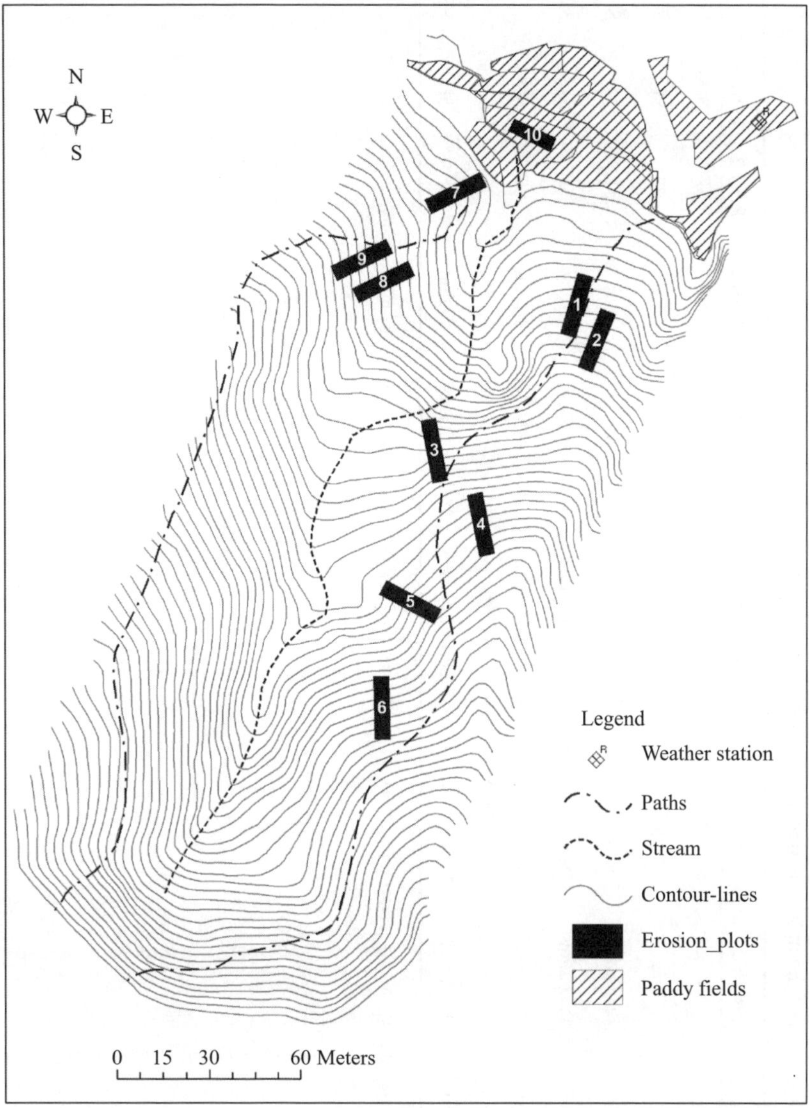

Measurement of nutrient inputs

During the experimental period nutrient inputs were measured as summarized below. A more detailed methodological description is given by Nguyễn Văn Dung *et al.* (2008).

Table 15.3 Cropping sequences of the upland experimental plots from 1999 to 2004

Year	Plot 1	Plot 2	Plot 3	Plot 4	Plot 5	Plot 6	Plot 7	Plot 8	Plot 9
1999	SF	SF	SF	SF	SF	SF	SF	SF	SF
2000	Rice	Rice	Rice	Rice	Rice	Rice	SF	SF	SF
2001	Rice + Melia	Tea Fallow	Rice + Melia	Rice + Melia	Cassava	Cassava	SF	SF	SF
2002	Cassava + Melia	Tea Fallow	Cassava + Melia	Cassava + Melia	Cassava + Melia	Fallow	SF	SF	SF
2003	Cassava + Melia	Tall grass	Cassava + Melia + Styrax	Cassava + Melia + Styrax	Cassava + Melia	Fallow	SF	SF	SF
2004	Melia + Palm	Tall grass	Melia + Styrax	Melia + Styrax	Melia + Styrax	Fallow	SF	SF	SF

Note: SF = Secondary Forest; Melia = *Melia azedarach*; Styrax = *Styrax tonkinensis*; Palm = *Livistona saribus*; Tea fallow = *Thea chinensis*

Nutrient inputs from rain and dust

These inputs were measured at the site in 2003-2004 using a 1 m x 1 m steel tray covered with a fine net to trap rainwater. Samples of rainwater collected in the tray were taken after each rainfall event and analyzed for nutrients. Amounts of nutrients deposited by rain and dust were calculated using rainfall data from the weather station at the site and measured nutrient concentrations. Two-year means of the individual nutrients were used as the inputs for all the plots in 2000-2002.

Nutrient inputs from planting material

For the swidden fields the seeding rate of swidden rice was obtained in 2001 in actual planting of three experimental swidden plots by weighing seeds in the planting bag before and after planting. Seed samples were sent to the laboratory for nutrient content analysis. For cassava, stem cuttings used for planting were counted and a sample of 10 cuttings was used to determine dry weights and nutrient contents. In intercropped fields percentages of the area planted to cassava were estimated, and were used to derive the amounts of nutrient inputs from this source for the respective fields.

For the paddy fields seedlings were grown in hard nursery beds and farmers used spades to cut off the top 1-1.5 cm of the soil that contained the roots with seedlings for transplanting (Plate 15.4). To measure nutrients brought in by planting materials, an area of 1 m^2 was marked on each of the 4 terraces of the field. The number of hills and plants in each quadrate was counted. Both the soil and seedling sub-samples were oven dried and analyzed for nutrient contents. In 2000, the seedling nursery was grown outside the studied paddy plot. Nutrient inputs included both nutrients from rice seedlings and from the soil attached to seedlings. However, from 2001 onward, the seedling nursery was located within the studied paddy plot and nutrient inputs were calculated only from rice seedlings.

Nutrient inputs from chemical fertilizers

The quantities of chemical fertilizers applied to the spring and summer crops were recorded.

Nutrient inputs from manure

These inputs were estimated by counting and weighing the baskets of manure applied to the paddy field in each season and a sub-sample taken to the laboratory for determining dry weight and nutrient content.

Nutrient inputs from surface inflow from the upland field to the paddy field

Inputs from surface inflow on rainy days were assessed by collecting in a tank

Plate 15.4: Nursery for rice seedlings with fence to protect it from livestock (Trần Mạnh Tường)

a part of the water from the upland field before it flowed into the paddy field. Three sub-samples were taken at three levels of the tank (top, middle, and bottom) after each rain event, mixed up, and sent for nutrient content analysis. On dry days, water samples were taken directly from the end of the stream before it flowed into the paddy field. The total amount of water flow from the upland field to the paddy field was estimated as the sum of the total amount of water losses from the paddy field through evapotranspiration (ET), percolation (P) and discharge (Q_{out}).

ET and P were measured with three pairs of lysimeters (one closed bottom and one open bottom in each pair) installed in the paddy field. The closed bottom lysimeter was used to measure ET, and the open bottom lysimeter was used to measure ET and P. The net evapotranspiration was determined daily. The ET was measured every day at 07:00 by a hook gauge on the closed and opened bottom lysimeters. P was calculated by the difference between water levels in the open bottom and the closed bottom lysimeters.

A weir system was set up at three positions on the paddy bunds (7 weirs at each position, totaling 21 weirs) to measure the discharge of surface runoff flow out of the paddy field. At each position, an automatic depth probe (Dataflow model 392, Dataflow Company, Australia) was installed to measure water levels. The discharge was determined as:

$$Q_{out} = 1838(L - 0.2H)H^{1.5}\ (l/s)$$

Where Q_{out} = Discharge (l/s), L = Crest width (m), and H= Difference between the crest and the water surface (m).

Measurement of nutrient outputs

Nutrient outputs in harvested products

Harvested outputs for swidden rice was determined by demarcating nine quadrates of 1 x 1 m, with three each on the upper, middle and lower sections of the plot. For paddy rice, six quadrates of 1 x 1 m were placed, two each in the three plots. All the plants in each quadrate were cut at the ground level and separated into grain and stover and then weighed to obtain fresh weight. Sub-samples were taken and weighed, and then sent to the laboratory and analyzed for nutrient contents. For cassava, three quadrates of 2 x 2 m were randomly placed in the upper, middle and lower parts of the slope, and fresh root weight determined before a sub-sample was analyzed for dry weight and nutrient contents.

Nutrient loss from burning crop residues and weeds

This loss of nutrients was measured only for swidden rice fields. In each upland swidden rice plot, 4 quadrates were marked before the time of field burning at the end of the 6-month fallow period. Just before burning, all biomass in each quadrate was cut and a composite sample of each plant part was taken for nutrient content analysis. Slashed biomass was laid on the ground to dry and then burned. After burning, ash in each quadrate was collected and weighed, and a composite ash sample (100 grams) was taken for nutrient content analysis. The amount of nutrients lost through burning was obtained as the differences between the amounts of nutrients in the biomass before burning and those in the ash remaining in the field.

Nutrient loss from leaching

Leaching losses were determined by installing four soil water samplers (Model 1900, Soil Moisture Equipment Corp, USA) in the experimental swidden fields at 4 depths (0-20, 20-40, 40-60 and 60-80 cm) to collect water samples in the soil profile. Water samples from the four profiles were combined and then analyzed for nutrient content (N, P, and K). Leaching was considered as downward movement of nutrients below 30 cm. The amount of leaching water was calculated for each rainfall event as the difference between the amount of rainfall and amount of water losses from runoff, evapotranspiration, and from the change in soil moisture content in the whole soil profile (0-80 cm). The amounts of leached nutrients were obtained by multiplying the amount of leaching water with its nutrient concentrations and summing over the period of estimation.

For the paddy field, amounts of percolation water (P) were measured by lysimeters installed in the paddy fields as described earlier. Water samples, however, were collected from three deep wells constructed in the paddy field (one for each plot) once a month for nutrient content analysis.

Nutrient loss from surface water outflow

Discharge water flowing out of the paddy field was measured by automatic depth probes (Model 392, Dataflow Company, Australia) that were installed at the three weir sites. A tank was also set up at each weir site to collect outflow water from the paddy field. To measure nutrient concentration of the water flow on rainy days, water samples were taken from the tank after each rain event. Three sub-samples were taken at three levels of the tank (top, middle, and bottom), mixed thoroughly, and sent for nutrient content analysis. On dry days, water samples were taken directly from the weir.

Nutrient losses from herbivores

Loss of nutrients from cattle and insects were measured using the cage method in the paddy fields in 2001. The cages were covered by metal mosquito screening that was small enough to prevent insects from entering. Four cages of 1 m x 1 m x 1 m were placed randomly in different terraces on the paddy field about one month after planting. At the time of the harvest, the rice plants, including straw, inside each cage, were harvested and weighed and a composite sample was taken to determine dry weight and nutrient contents. The rice plants from an equivalent area outside the cage were also harvested at the same time. Loss of nutrients was determined from the difference between amounts of nutrients in the harvested biomass inside and outside the cage.

Nutrient losses from erosion and runoff

These losses were measured using the nine experimental upland plots described earlier (6 swidden plots and 3 secondary forest plots) that were set up as runoff plots. Masonry walls were constructed around each of the 5 m x 20 m plots to block run-in water. Two collecting tanks were installed at the lower end of each runoff plot. A big tank (4 m × 1 m × 1 m) was used to collect soil, stones and sediments, while a small tank was used to collect runoff water. After each rain event, all sediments trapped in the big tank below each plot were collected, dried and weighed. Composite samples were sent for nutrient content analysis. In the small tank, where only water and silt were collected, water samples were collected from the top, middle and bottom of the tank by a small pump, and composite samples were analyzed for nutrient contents.

Nutrient losses from burning of forest biomass in the swidden field

The loss of nutrients resulting from burning of forest biomass after clearance of the swidden field in the first year of cultivation was determined by measuring the nutrient content of the biomass before burning and the amount of nutrients left on the field as ash after burning. This area had been under fallow for 12 years.

Based on the vegetation map of the study site, for each type of vegetation community, quadrates containing primarily that type of vegetation were selected. The sizes of quadrates and the number of replications varied depending on each type of vegetation. The forest was then cut and all of the slash from the plants growing inside each quadrate was placed inside the quadrate. For each quadrate, the wet weight of all the plant material inside was determined. The average weight of three quadrates with the same type of vegetation was used to calculate the total biomass of the whole swidden plot. A composite sample was taken for each vegetation type for measuring wet weight, dry weight and nutrient contents.

A sub-sample of the plant material in each quadrate was taken and weighed and a piece of aluminum foil was placed under this slashed vegetation. Immediately after the field was burned, all of the ash on the aluminum foil pieces was collected and weighed. Samples were taken from the ash of each vegetation type for nutrient analysis. All of the unburned plant materials on the aluminum foil pieces were collected and weighed. The ratio of burned to unburned plant material for each vegetation type was computed and used to determine the amount of nutrients remaining on the plot in the unburned vegetation.

The weight of unburned plant material removed from the field after burning by the farmer for firewood or fence building was also recorded. The amount of nutrients contained in the materials removed by the farmer was calculated and was subtracted from the estimated quantity of nutrients remaining in the field in unburned plant material. The total nutrient contents of the ash and unburned plant materials remaining on the field were calculated and compared with the estimated total nutrient contents of the biomass just before burning. The quantities of nutrients lost from burning were then computed.

Analysis of nutrients in soil, water and plant materials

The samples of soil, water and plants were sent for analysis at the JICA laboratory of Hanoi University of Agriculture.

Soil samples

To determine soil nutrient status of the paddy fields and swidden fields at different stages of cropping and fallow periods, soil samples were collected from each field every year after crop harvest and analyzed for major nutrients. In each field, 3 composite soil samples were taken, each from 10 soil cores (0-15 cm). Each sample was analyzed for organic carbon (Walkley & Black method), pH (1M KCl), total N (Kjeldahl), total P ($H_2SO_4 + HClO_4$ digest), total K (calcination at 1000°C with Na_2CO_3), available N (hydrolysis), available P (Oniani method), available K (NH_4CH_3CHOO 1M at pH=7), texture (pipette method), particle density, (picnomet method), and bulk density (metal ring).

Water samples

Water samples were filtered through a 0.45 μm cellulose acetate filter. The solid material accumulated on the filter, was dried at 105 °C until its weight was stable, and subsequently analyzed for organic carbon (OC) and nutrients using the above soil analysis method. Soluble nutrients in sediments were considered as available nutrients and analyzed as: Soluble N: steam distillation; Soluble K: flame photometer; Soluble P: colorimeter, using amonium-molybdate and $SnCl_2$.

Plant samples

Plant samples were dried at 80°C for 5 hours, ground and digested with a mixture of H_2SO_4 and $HClO_4$. N was determined by Kjeldhal method, P by colorimetric method and K by flame photometer as above.

Nutrient balance analysis

Yearly balances of N, P, and K were calculated for each experimental plot as the differences between nutrient inputs and outputs. Balances for swidden and paddy fields with successive years of cropping were combined to get the estimates of nutrient losses for the entire 5-year period of cropping.

Regeneration of soil fertility in swidden systems was also estimated based on balances of available nutrients and on soil organic matter regeneration. The contribution of biomass accumulated in the fallow vegetation to the replenishment of nutrients in the swidden field after it is cut and burned before the next cycle of crop cultivation was also evaluated.

Results

Rainfall characteristics at experimental site

Rainfall patterns during the experimental phase showed the expected distinct

Figure 15.3 Monthly rainfall distribution from 2000–2004

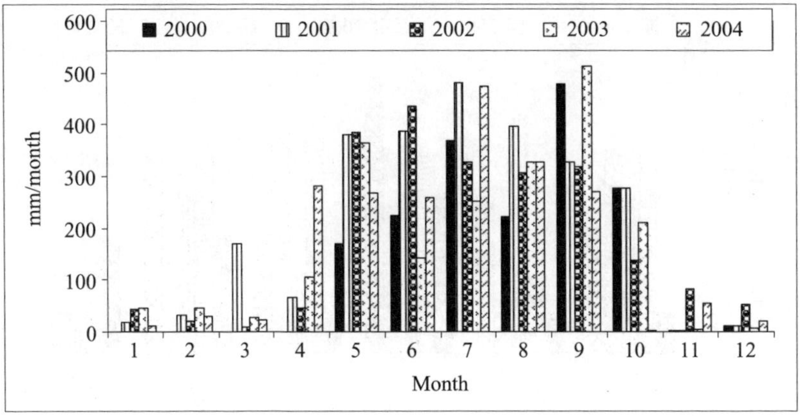

rainy season from about May to October followed by a dry season from November to February/March (Figure 15.3).

Cumulative annual rainfall at the experimental site amounted to 2547, 2163, 2048 and 2020 mm in 2001, 2002, 2003 and 2004 respectively. In 2000 rainfall measurements were only started in May with 1,756 mm recorded to the end of the year. The twenty-year rainfall average in the area (Hòa Bình provincial weather station) was 1,825 mm between 1978 and 1998, with a maximum rainfall recorded in 1996 of 2,380 mm and a minimum of 1,085 mm in 1991. Thus, rainfall during the experimental phase was well above long-term average rainfall and in 2001 it even exceeded the maximum annual rainfall recorded during the previous twenty years by 167 mm.

Analysis of rainfall during the rainy season confirms that rainfall during the experimental period was consistently higher than 20-year average. In particular, rainfall during the rainy season of 2001 was 665 mm higher than the long-term average suggesting that associated erosion/runoff events measured in 2001 represent upper values for this region. After 2001 rainy season rainfall declined continuously and reached a value similar to the one of the long-term average in 2004.

For assessing erosion and runoff risk, total rainfall is not necessarily relevant but rainfall intensity is. Thus rainfall data were clustered into different rainfall intensity groups. In 43-47% of the cases daily rainfall was greater than 10 mm per day and in 12-21% it was greater than 30 mm/day during the experimental years. The maximum daily rainfall recorded was 211 mm on September 11, 2000; 113 mm on July 22, 2001; 80 mm on October 23, 2002; 131 mm on September 10 in 2003 and 107 mm on July 23, 2004. Daily rainfall events greater than 30 mm/

Figure 15.4 Number of days in Tat hamlet with daily rainfall > 30 mm/day

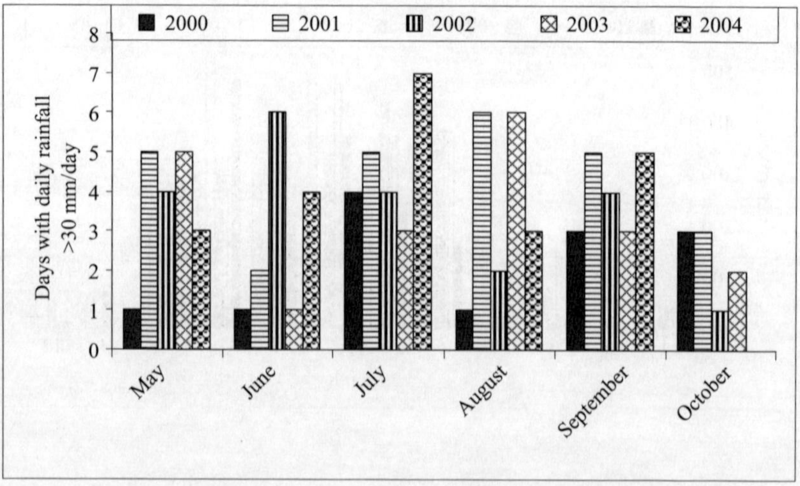

day were recorded on 13, 26, 21, 20 and 24 days in 2000, 2001, 2002, 2003 and 2004 respectively distributed between May and October (Figure 15.4).

Runoff and erosion from sloping swidden soil surfaces are more likely to be closely correlated with the critical rainfall (I30), which is defined as the amount of rainfall with duration of 30 minutes or more, than with total rainfall. Daily rainfall greater than 30 mm/day and lasting longer than 30 minutes was recorded on 6 days in 2001, 2002 and 2003 and on 7 days in 2004.

Runoff from swidden plots

Discharge of water from the erosion plots varied largely between years and treatments. A substantial amount of variation was accounted for by the variation in yearly wet season rainfall (y=0.84x – 825, R^2 0.9). However, there were even more pronounced effects of cropping patterns on runoff. For example, in 2002, average runoff from the secondary forest was 493 mm (plots 7-9) while average runoff from upland rice fields (plots 1-6) amounted to 667 mm (Figure 15.5).

Variation in discharge between the different swidden rice plots in 2000 was relatively modest with an average of 667 ± 51 mm. Also, no significant variation between different cropping systems in the same year was observed. For example in 2001, runoff from upland rice fields (plots 1, 3 and 4) averaged 1150 mm while that from cassava fields (plots 5+6) averaged 1080 mm. Thus, variations in runoff between land use systems were largely due to differences between fallow/secondary forest plots and cultivated plots. In 2002, yearly

Figure 15.5 Cumulative yearly amounts of rainfall lost in runoff from the different erosion plots

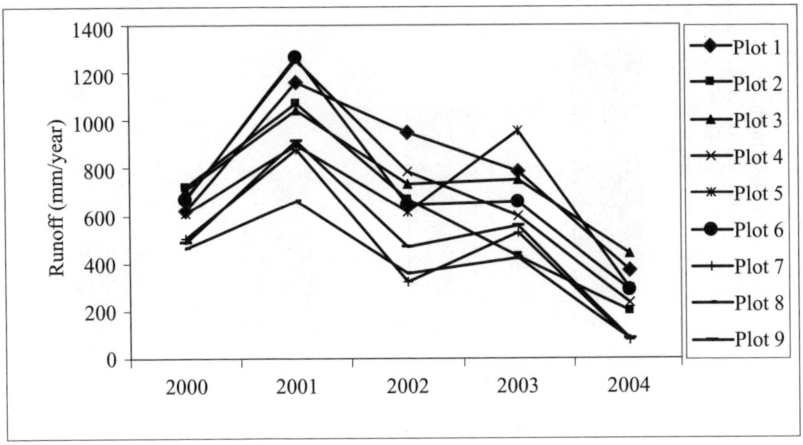

Key to plots: 2000: Upland rice (plot 1 to plot 6); 2001: Rice+Melia (plots 1, 3, 4), Tea fallow (plot 2), Cassava (plots 5, 6); 2002: Cassava+Melia (plots 1, 3, 4, 5), Tena fallow (plot 2), Fallow (plot 6); 2003: Cassava+Melia (plots 1, 5), Cassava+Melia+Styrax (ploys 3,. 4), Tall grass (plot 2), Fallow (plot 6); 2004: Melia+Palm (plot 1), Tall grass (plot 2), Melia+Styrax plots 3, 4, 5); Fallow (plot 6). All years: Secondary forest plots 7–9.

runoff from cassava, fallow and secondary forest, were 765 mm, 655 mm and 406 mm, respectively.

Putting land into fallow led to the slow establishment of a permanent vegetation cover that resulted in reduced runoff. As rainfall varied greatly from year to year a comparison of treatment effects is not possible across years. Thus effects of fallow duration on erosion and runoff were assessed in 2004 by comparing plots at different stages of fallow. Runoff in a one-year old fallow after cassava (Plot 3; Plot 5) was 437 and 369 mm/year (average: 403 mm/year) and decreased to 234 mm in Plot 2, which was under fallow for four years at that time (Figure 15.6). Runoff from the 4-year old fallow was still larger than that from the secondary forest (average 117 mm/year) indicating that nutrient losses only slowly equilibrate towards those of the older forest system.

Erosion from swidden fields

The amount of soil eroded from different plots has been continuously measured since the beginning of the 2000 rainy season. Figure 15.7 shows the amount of soil erosion on different days in 2003. Erosion events occurred scattered throughout May to September. Analyzing the monthly distribution of soil loss

Figure 15.6 Effect of fallow time on runoff estimated from data in 2004

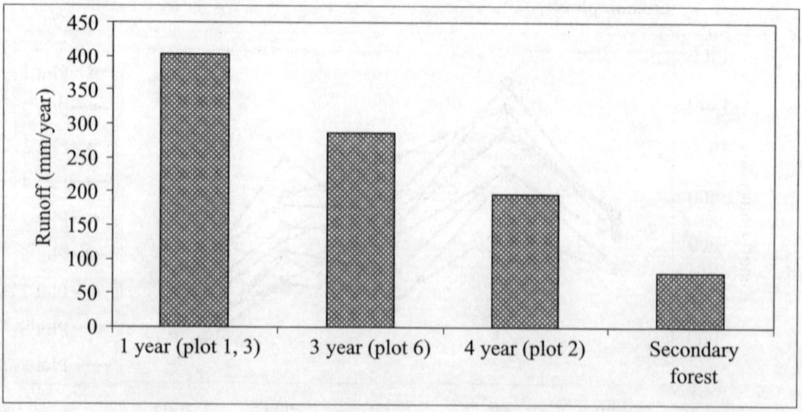

Figure 15.7 Daily erosion events during 2003 in the different erosion plots

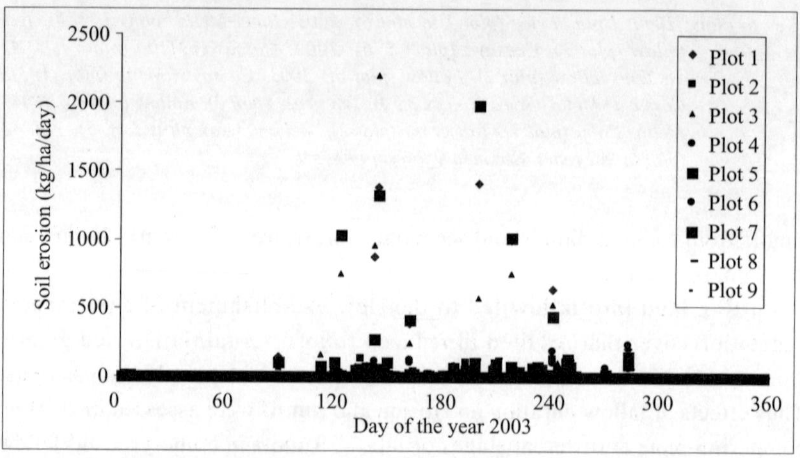

Key to plots: Cassava+Melia (plots 1, 5), Cassava+Melia+Styrax (plots 3, 4), Tall grass (plot 2), Fallow (plot 6), Secondary forest (plots 7–9)

showed that for upland rice plots highest soil losses occurred early in the rainy season after planting, i.e., in June and July with 11.5 and 9.9 tons/ha/month. The relationship between rainfall and soil erosion, although positive, was not as strong as for runoff. This was probably mainly due to the fact that erosion depended much more on larger storm events. Preliminary examination suggested that significant erosion occurred only with rainfall events >30-40 mm/day.

Soil loss through erosion was strongly related to land use management, probably relating to plant cover and tillage events. Forest land use had the lowest annual soil loss (Figure 15.8). For instance, in 2000, average soil loss

Figure 15.8 Yearly erosion from upland plots under different land uses

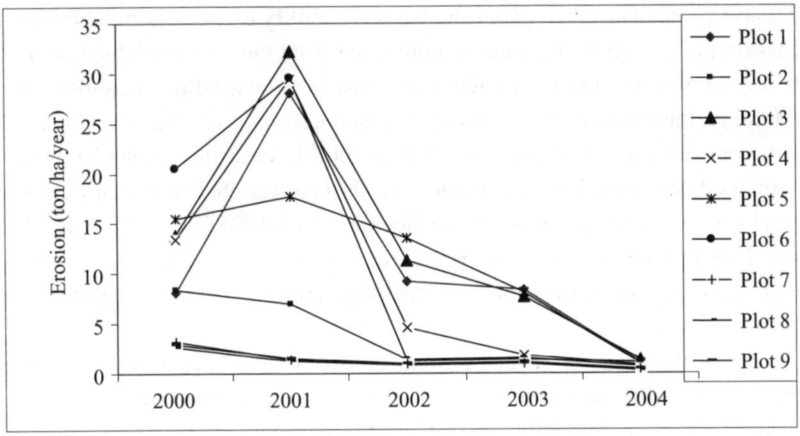

Key to plots: 2000: Upland rice (plot 1 to plot 6); 2001: Rice+Melia (plots 1, 3, 4), Tea fallow (plot 2), Cassava (plots 5, 6); 2002: Cassava+Melia (plots 1, 3, 4, 5), Tea fallow (plot 2), Fallow (plot 6); 2003: Cassava+Melia (plots 1, 5), Cassava+Melia+Styrax (ploys 3, 4), Tall grass (plot 2), Fallow (plot 6); 2004: Melia+Palm (plot 1), Tall grass (plot 2), Melia+Styrax plots 3, 4, 5); Fallow (plot 6). Secondary forest (plots 7–9).

from upland rice was 13 tons/ha, which was much higher than soil loss from secondary forest of only 2.9 tons/ha. Erosion in secondary forest plots declined further in subsequent years suggesting that the higher erosion in year 1 was probably induced by disturbance during the setting up of the erosion plots.

In contrast to runoff data, large variations in soil loss occurred in the same cropping systems. For example in 2000, erosion losses from swidden rice plots varied by as much as 12 tons/ha, i.e., from 8-20 tons/ha. The reason for this difference is not fully understood. The greater amount of erosion in plots 5 and 6 could be due to their location on steeper slopes (36°) higher in the experimental site compared to 29° for lower erosion plots.

In 2001 there were some changes in plot cultivation. Plots 1 and 3 remained under upland rice, while plots 4, 5, and 6 were changed from rice to cassava in February 2001 and plot 2 was abandoned and left to fallow. These modifications in land use led to significant changes in soil erosion, e.g., erosion on the plot left fallow (plot 2) dropped immediately to very low levels similar to secondary forest. Similarly, abandoning plot 6 in 2002 immediately reduced erosion from 29 tons/ha/year to 1 ton/ha/year (Figure 15.8).

There was also an apparent small decline in erosion when switching from rice to cassava, e.g., in 2001, average erosion in rice plots amounted to 30 tons/ha/year while the corresponding value for cassava plots was 26 tons/ha/

year. This could be attributed to the longer and earlier soil cover provided by cassava plants. However, given the large variability of erosion in the different cassava plots in 2001, it cannot be concluded from the data available that there is a general reduction in erosion with cassava compared to rice cultivation. Although there was a further general reduction in erosion events in 2002/3 in cassava fields compared to the rice data in 2000/1, these data cannot be directly compared due to the large variation in rainfall patterns between years. Erosion from cassava plots might also be reduced by interplanting trees (Nguyễn Huệ and Thái Phiên 2005), such as the planting of *Melia* and *Styrax* in some of the plots. However, again their impact cannot be quantified due to the variable rains between years.

During the first year after burning, the soil surface had a high density of plant roots and stumps and this may have helped to hold the soil together and reduce erosion. Additionally, water infiltration capacity may have been improved by the previous perennial vegetation, with old tree root channels inducing preferential flow, thus further reducing the potential for runoff (Van Noordwijk *et al.* 1991). However, by the second year, the tree plant roots and stumps had largely disappeared and were no longer able to protect the soil from the effects of large rain events. Erosion and runoff were smaller in the year when the plot was cut and burned in 2000 than in 2001, potentially supporting these hypotheses. However, due to the large increase in rainfall between years in 2001, a direct relationship between soil roots and surface status and erosion cannot be drawn in this case.

Loss of organic carbon from upland fields

Reduction of soil organic carbon (SOC) is a strong indicator that soil degradation is occurring. Our measurements indicated that the amount of SOC lost by erosion and runoff from the secondary forest plots was less than that from fallow, upland rice, and cassava plots (Figure 15.9). After four years, SOC loss was directly related to soil erosion and water runoff. In 2001, on average, upland rice plots lost 736 kg/ha of SOC while the cassava plots had lost 639 kg/ha of SOC. In 2002, the SOC lost from the cassava plots was 393 kg/ha. In contrast, soil loss by erosion and runoff in the fallow plots was only 5 kg/ha in 2002.

Loss of nutrients from upland fields

The amount of nutrients lost through erosion, surface runoff and leaching from the upland plots changed according to land use. Total nutrient losses, as well as plant available N, P and K losses, after each rainy day were calculated (Table 15.4). The amount of nutrients lost depended on the amount of rainfall, sediment yield, the amount of runoff, and the initial nutrient contents.

Figure 15.9 Soil organic carbon (SOC) losses by erosion under different land uses

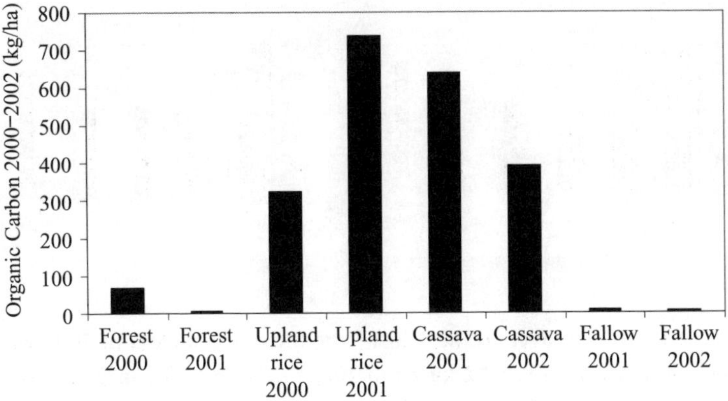

Table 15.4 Nutrient losses by erosion, runoff and leaching in the upland fields

	Total nutrient loss (kg/ha)			Available nutrient loss (kg/ha)		
	N	P	K	N	P	K
Upland rice 2000	31.9	3.7	298.4	10.8	1.2	111.6
Forest 2000	12.4	2.4	127.1	6.7	1.7	73.4
Upland rice 2001	88.4	9.2	718.0	18.8	1.8	64.0
Cassava 2001	72.7	9.2	613.1	17.9	1.7	58.7
Fallow 2001	40.0	4.8	119.3	16.6	1.6	56.0
Forest 2001	23.2	2.7	80.9	11.6	1.5	57.3
Cassava 2002	42.2	3.0	265.7	27.7	3.2	55.9
Fallow 2002	28.7	3.1	55.0	18.5	1.5	43.2
Forest 2002	31.5	3.1	31.4	23.7	1.5	27.2
Cassava 2003	43.8	12.4	267.1	18.4	4.9	53.6
Fallow 2003	27.2	6.3	66.6	14.7	4.9	39.5
Forest 2003	31.6	4.1	41.4	13.0	2.9	34.2
Fallow (2 yr.) 2004	16.2	1.3	38.0	7.7	0.7	35.6
Fallow (4 yr.) 2004	13.1	1.0	30.9	6.8	0.5	27.4
Forest 2004	9.6	1.1	13.1	6.4	0.8	11.9

Total nutrient losses were larger than losses of plant available nutrients. In 2000, in upland rice plots, total N and K losses were 31.9 and 298 kg/ha and in 2001, 88.4 and 718 kg/ha, respectively. Nutrient losses of the same land use type, but in different years were not the same because of different rainfall patterns

Figure 15.10 Nutrient ratio lost in different upland land uses by erosion, runoff and leaching in 2000

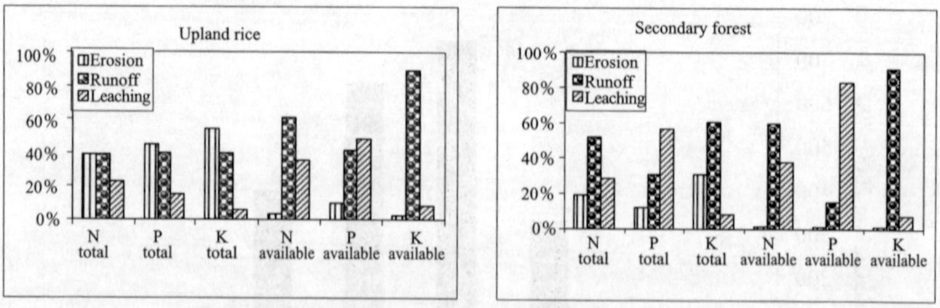

and extent of plant cover. Total N lost from the cassava plots in 2001 was 72.7 kg/ha, while it was only 42.2 kg/ha in 2002 and 43.8 kg/ha in 2003. Available K lost from the secondary forest plots in 2000 was 73 kg/ha, declined to 57 kg/ha in 2002, and was only 11.9 kg/ha in 2004.

The ratios of nutrient losses from erosion, runoff, and leaching differed strongly among different land uses. This was because total nutrient losses through erosion and runoff were higher than total nutrient loss through leaching for upland rice (Figure 15.10), which agrees with observations by Hashim *et al.* (1998). However, for available nutrients runoff and leaching are the dominant source of N and K losses. For forest or fallow plots, loss through erosion was small and leaching losses were often higher than or equal to nutrient loss through runoff. Reduced erosion in fallow and forest plots was due to the permanent vegetative cover of the soil and particularly due to the lack of tillage.

Nutrient losses from burning of forest biomass in the swidden fields

Table 15.5 shows the amount of nutrients in the forest biomass before the initial burning of the swidden field and the amounts lost through burning and removal of unburned materials from the field by the farmers.

After burning, quantities of ash and unburned vegetation remained in the field (see Plate 15.2). The sum of the amount of nutrients contained in these materials is the total amount of nutrients remaining on the field after burning. The amount of nutrients lost in the burning is calculated by subtracting the amount of nutrients remaining on the field after burning (unburned material 74%+ashes) from the amount of nutrients contained in the biomass before burning. The initial burning process causes almost complete N loss of 91%, P loss of about 82%, and K loss of about 70%. Thus our estimates of nutrient losses are higher than those of Hölscher *et al.* (1996), who found that 60% of N was lost to the atmosphere through slash and burning in Brazil; however, our

Table 15.5 Vegetation biomass and loss of nutrients from burning in the swidden field

	Dry matter (kg/ha)	Nutrient content (kg/ha)		
		N	P	K
Total plant material before burning	*24,299*	*134.9*	*15.3*	*100.3*
Removed before burning (Styrax and woody trees)	11,713	74.7	7.2	55.6
Left for burning	12,586	60.3	8.1	44.7
Total plant material after burning:	*2,718*	*16.3*	*3.1*	*32.5*
Unburned materials removed from field (26%)	583	3.8	0.4	2.7
Unburned materials remained on field (74%)	1,661	10.9	1.0	7.6
Ash remain on the field	474	1.5	1.8	22.2
Total loss	*22,749*	*126.3*	*12.9*	*73.2*
Loss by removal before burning	11,713	74.7	7.2	55.6
Loss by removal of unburned materials	583	3.8	0.4	2.7
Loss by burning	10,452	47.8	5.3	14.9

Figure 15.11 Changes in soil texture over time in the upland field experiments

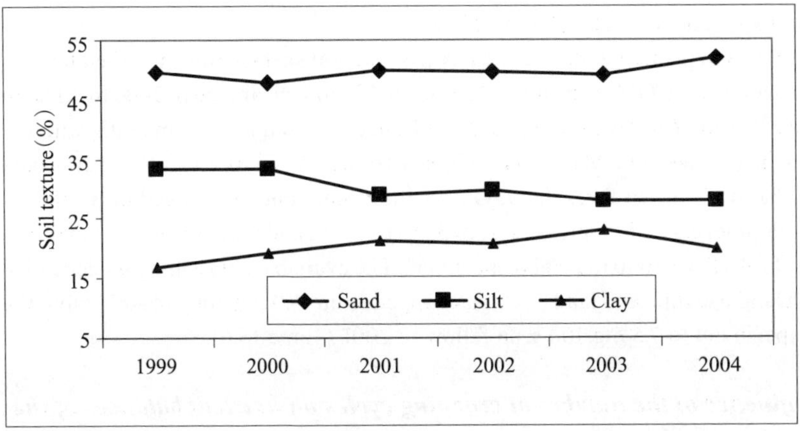

data include losses to the atmosphere as well as nutrient off-take in unburned plant material removed from the field.

Changes in texture and chemical properties of soil in the upland fields

The soil texture was found to have changed slightly at the end of the experiment. In particular, the amount of silt decreased from 33 percent to 28 percent, probably due to erosion (Figure 15.11).

Figure 15.12 Changes in soil pH from 1999 to 2004 in the upland experimental plots

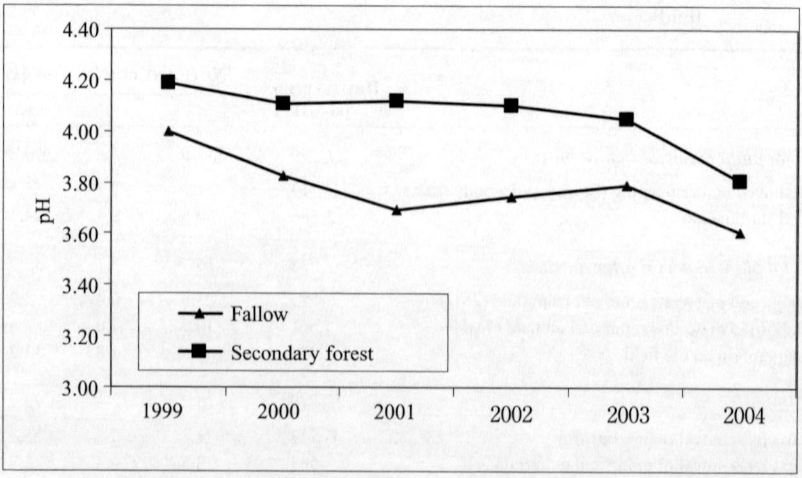

These changes in soil texture, along with leaching of cations, were probably the main causes for the observed decrease in pH from 4.2 in 1999 to 3.8 in 2004 in the secondary forest (Figure 15.12).

On land planted with rice and cassava, the soil surface was disturbed leading to decreases in OM in the topsoil from 5.0% before clearance in 2000 to 3.1% for rice in 2001. Cassava was 3.8% in 2001 but decreased to 3.6% in 2002, whereas on the fallow land OM increased slightly from 3.5% in 2001 to 3.6% in 2003 and 4.2% in 2004. On forest land, OM and nutrient content changed only slightly with time, e.g., OM increased from 5.0% to 5.4%, and total nitrogen from 0.16% to 0.19% from 2000 to 2004 respectively. The available potassium loss depended on land use time and land cover; it changed from 17.9 mg/100g of soil before the experiment to 7.5 mg/100 g on fallow in 2001 (Table 15.6).

Influence of the number of cropping cycles on nutrient balances of the swidden fields

Net nutrient balances in upland systems became more negative with an increasing number of cropping cycles (Appended Table 15.A1). This was particularly true for N (Figure 15.13) and even more so for K (Figure 15.14). However, the relationship was not linear but instead leveled off in the third and fourth cropping cycle. This can be attributed to several reasons: (i) nutrient rich topsoil and easily eroded soil fractions had already been eroded in the first cropping season, (ii) declining soil fertility after repeated cropping led to smaller yields and hence less nutrient export, (iii) in the third year there was a switch from rice to cassava which led to lower nutrient exports in harvested products

Table 15.6 Changes in the chemical characteristics of topsoil in the upland field after five years (2000–2004)

		TOTAL (%)			AVAILABLE (MG/100G)		
	OM (%)	N	P_2O_5	K_2O	N	P_2O_5	K_2O
Before 2000	5.01	0.16	0.05	2.77	7.96	1.50	17.92
Forest 2002	5.20	0.16	0.03	3.59	9.05	2.10	11.40
Forest 2003	5.31	0.18	0.05	3.44	10.0	2.80	17.6
Forest 2004	5.35	0.19	0.03	4.15	6.10	2.87	14.10
Upland rice 2001	3.10	0.17	0.04	3.57	10.50	1.50	9.10
Cassava 2001	3.80	0.13	0.04	3.15	10.80	1.80	8.60
Cassava 2002	3.56	0.13	0.04	3.99	6.23	2.17	15.13
Cassava 2003	3.63	0.15	0.04	3.17	9.60	2.18	11.90
Fallow 2001	3.50	0.16	0.03	2.05	6.50	1.50	7.50
Fallow 2003	3.60	0.17	0.05	2.96	15.10	2.70	13.70
Fallow 2004	4.24	0.17	0.04	2.81	8.10	2.37	10.30

and (iv) reduced rainfall in years 3 and 4 led to reduced runoff and erosion and hence lessened nutrient export.

Regeneration of soil fertility in swidden systems

Measured net nutrient balances of secondary forest sites were positive for N and P but apparently negative in several years for K (Table 15.7). Based on the average of these nutrient balances we can assume a minimum yearly recovery of about 7 kg N/ha and 4 kg P/ha. For K no estimate for recovery can be made since the long-term average net K balance was negative. We have to consider that, while the measured nutrient balances during the cropping phase are, apart from the gaseous losses, largely complete, the nutrient balances during the fallow phases do not include major inputs such as those of deep soil nutrient capture by trees and bushes and inputs through biological nitrogen fixation from leguminous species. Thus, these values represent minimum recovery estimates of secondary forests rather than their maximum potential rates. The recovery potential of younger fallow vegetation is likely to be lower, e.g., comparative yearly nutrient balances of fallows in plots 1-6 in 2004 (Appended Table 15.A1) were lower.

Taking into account this potential net yearly N accumulation rate we can estimate approximate recovery times for the swidden systems in plots 1-6 described above (Figure 15.15).

These simulations suggest that the time of the combined crop-swidden phase to complete N recovery (0 N balance) will be between 5 and 20 years. The shortest N recovery time was 4 years of fallow after only 1 year of cropping (plot 2). For longer cropping periods, the required fallow time to achieve a positive N balance increased to 11-16 years. However, if nutrients lost during

Figure 15.13 Effect of number of cropping cycles on aggregated N and P nutrient system balances

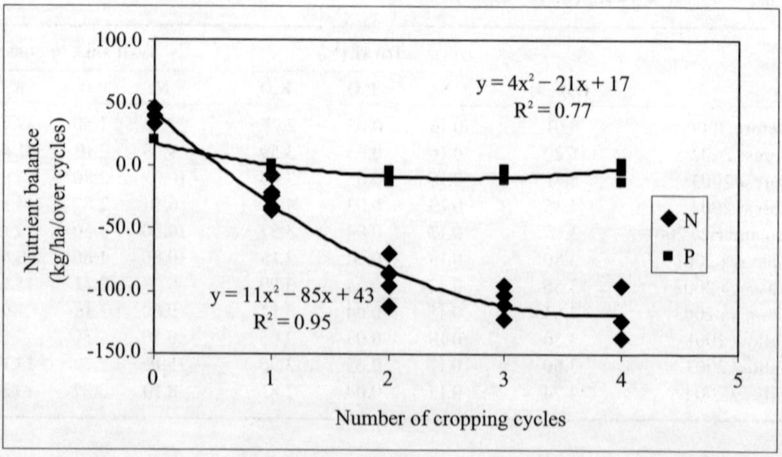

Figure 15.14 Effect of number of cropping cycles on aggregated K nutrient system balances

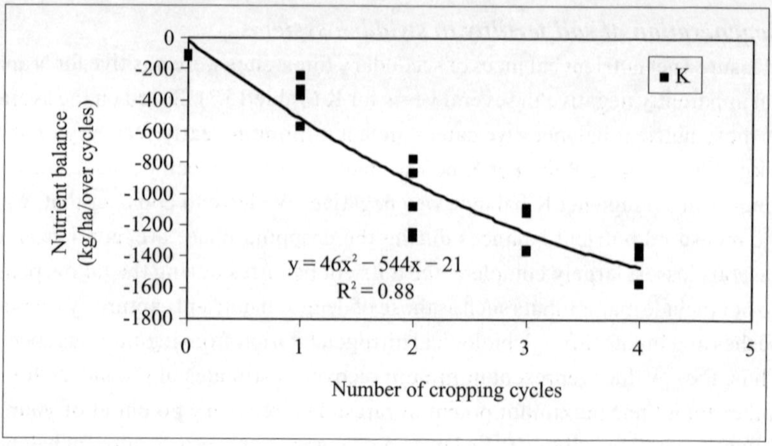

the initial burning of forest are taken into account, then full regeneration will take as long as 30-40 years.

Soil organic matter regeneration

The pattern and timescale of regeneration of soil organic matter content was assessed using a large soil data set of 120 samples taken from plots in the area that had been fallowed for different lengths of time and that had different vegetation communities. Although, individual samples were widely scattered (due to

Table 15.7 N, P and K nutrient balances of secondary forest (average of plots 7–9)

	SYSTEM NUTRIENT BALANCES (KG/HA/YEAR)		
	N	P	K
2000	16.9	4.3	–90.2
2001	3.2	3.8	–55.5
2002	–2.2	3.7	5.5
2003	–2.0	7.7	–25.5
2004	19.5	0.5	44.7
Cumulative 2000–2004	35.3	20.0	–121.0
Yearly average	7.1	4.0	–24.2

Figure 15.15 Regeneration patterns of systems N balance in different swidden systems

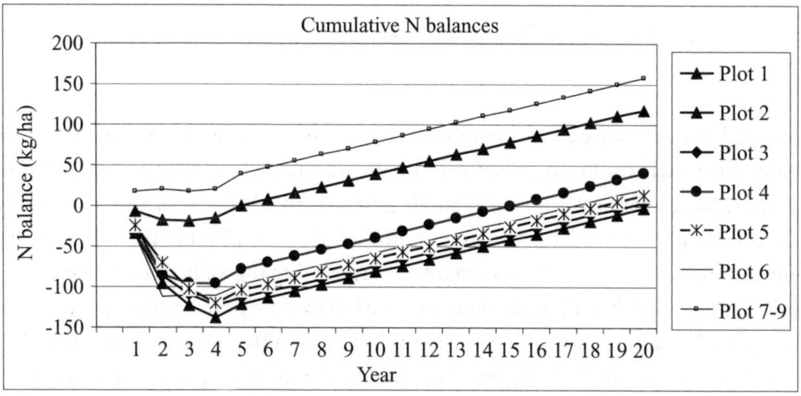

Key to history of plots: 2000: Upland rice (plot 1 to plot 6); 2001: Rice+Melia (plots 1, 3, 4), Tea fallow (plot 2), Cassava (plots 5, 6); 2002: Cassava+ Melia (plots 1, 3, 4, 5), Tea fallow (plot 2), Fallow (plot 6); 2003: Cassava+Melia (plots 1, 5), Cassava+Melia+Styrax (plots 3, 4), Tall grass (plot 2), Fallow (plot 6); 2004: Melia+Palm (plot 1), Tall grass (plot 2), Melia+Styrax plots 3, 4, 5); Fallow (plot 6). All years: Secondary forest plots 7–9.

different land use histories and vegetation communities) when the combined data set was clustered into age groups of 1-2, 3-4, 5-6, 10-12, 25, 30 and >50 years under fallow, it yielded a rough estimation of an average recovery time of about 12 years (Figure 15.16). This fits well with the estimated recovery time of 11-16 years to attain a positive N balance.

Build-up of soil organic matter requires N and P as essential molecules in organic matter structures. Based on the same soil data set as used for estimating the soil organic matter build-up pattern we found respective SOM:N:P ratios of 1:22:149. Thus, assuming a concurrent accumulation of these nutrients with

Figure 15.16 Regeneration of soil organic matter with ongoing fallow time

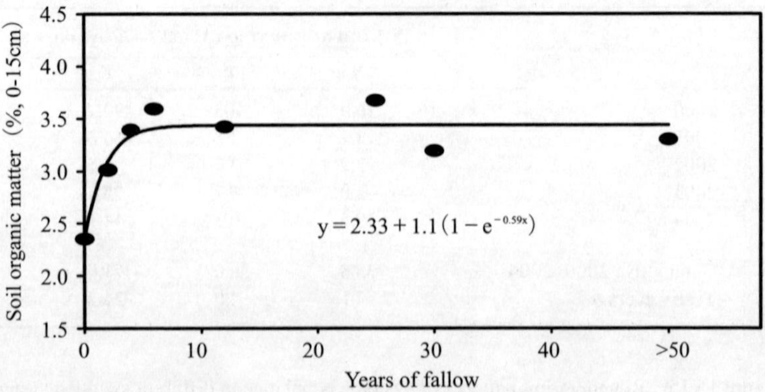

Source: Nguyền Thanh Lâm, unpublished data

matter build-up, we can estimate recovery of these nutrients as well, since we know the yearly build-up rate of soil organic matter. The respective estimations suggest large accumulations of N and P, i.e., around 800 kg N/ha and 100 kg P/ha after a 12-year fallow. Those estimates are substantially higher than those of the nutrient balances. It can be reasonably assumed that the additional N represents new inputs from N_2-fixation, deposition of manure by free-ranging livestock, and deep N capture, since little mineral N is available in the soil solution. The case for P is less clear. Part of the accumulated P in soil organic matter is likely to be derived from internal recycling, i.e., from the absorbed/complexed phase. Nevertheless, the results suggest that with the regeneration of soil organic matter enough N and P nutrients would be bound in organic form to offset losses during the previous cropping phase in the swidden system.

Flows of water, sediments, and nutrients from the upland fields to the paddy fields

Water discharge from the watershed

Water discharge from the experimental watershed was highest in the first two years, i.e., 104,000 m^3 in 2000 and 119,000 m^3 in 2001. In 2001 the largest discharge coincided with the highest rainfall during the experimental period. Subsequent declines in water discharge (32,000, 19,000, and 18,000 m^3 in 2002, 2003 and 2004 respectively) can only partly be explained by decreasing rainfall. For example the amounts of rainfall in 2000 and 2003 were similar but discharge was substantially reduced in 2003. Thus the ongoing decline in water discharge

Figure 15.17 Yearly total fine sediment flow from uplands to lowlands measured at the outlet of the watershed

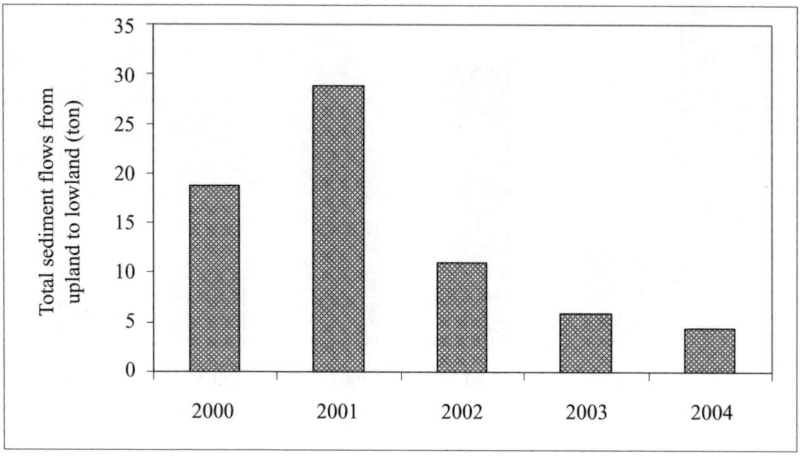

with time is likely also to be strongly associated with the change in land use. Over the course of the experiment, an increasing proportion of the total area of the watershed was becoming fallow, which has been shown above to substantially reduce runoff and hence water discharge.

Sediment flows from the upland fields to the paddy field

Each year, substantial amounts of fine sediment flowed from the uplands to the paddy fields (Figure 15.17). The amount of fine sediment flow increased from 19 tons in 2000 to 29 tons in 2001 but thereafter greatly decreased to 11 tons in 2002, 6 tons in 2003 and 4 tons in 2004. This pattern of sediment flow was closely correlated to rainfall patterns in the different years, e.g., rainfall was also highest in 2001. It has to be noted that these measured sediment flows represent only the fine sediment transport since larger sediment particles were not able to pass the fine mesh screen at the sampling point. Thus, actual total sediment flows were likely to have been somewhat higher.

Approximately 27-63 percent of the sediment discharged from the watershed settled in the paddy field. The amount of sediment retained in the paddy field increased from 5.2 tons to 15.4 tons between 2000 and 2001 and decreased to 6.7 tons in 2002 and to 2.1 tons in 2004 (Figure 15.18).

As a consequence of sedimentation, soil texture in the paddy fields near the discharge inlet was altered. The three plots in the paddy field were located at different distances from the inlet. Plot 1 was 10 m from the inlet at the base of the hill, Plot 2 was 20 m, and Plot 3 was 30 m. The sand ratio increased from

Figure 15.18 Yearly amount of sediment retained in the paddy fields

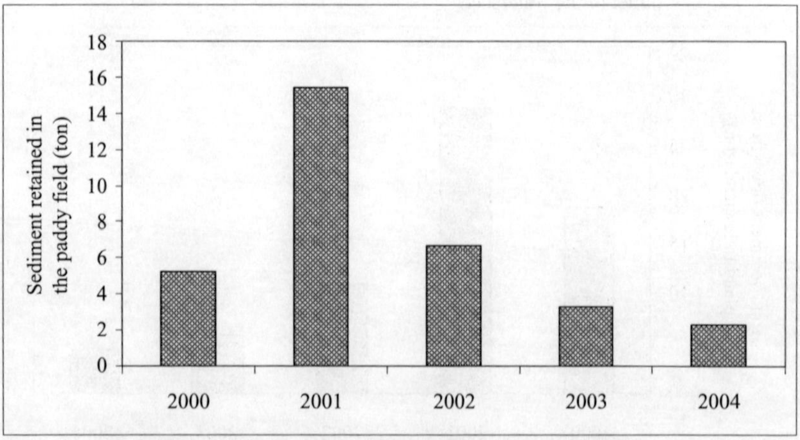

46% to 72% for Plot 1 and from 46% to 64% for Plot 2 from 1999 to 2000, mainly because the sediment that remained in the paddy field was predominantly sand. Silt content only changed in Plot 1 but clay content was not changed for the entire experimental period (Figure 15.19). The reason why Plot 1 was most affected was due to its position nearest the water inlet.

The amount of sediment deposited in the paddy field depended on the distance from the base of the hill and the velocity of the water flow. At a distance of 10 meters from the hill, sand accumulated, probably because silt and clay particles were carried out of the plot in out-flowing water (Figure 15.20). The proportion of sand near the inlet was 62% and decreased to 41% at 30 meters from the base of the hill.

Nutrient flows from upland fields to the paddy field

Nutrient flows from upland fields to the paddy field associated with sediment transport and in water dissolved nutrients showed a similar pattern (Figure 15.21). Hence, the nutrient flow was shown to be directly proportional to the amount of rainfall and prevailing soil surface/cropping conditions. In the year 2001, associated with large rainfalls, N, P and K losses in the stream from the watershed were considerably larger than in other years. In this year total N loss from the watershed amounted to 227 kg and available N loss was 79 kg. Even larger were total and available K outflows from the watershed, i.e., 380 kg and 253 kg respectively in 2001. These high K losses from the watershed are in accordance with the high K losses in the upland swidden systems. However, the relationship between K and N losses measured in the upland nutrient balances

Figure 15.19 Changes in soil texture over time in paddy field plots located at different distances from the inlet

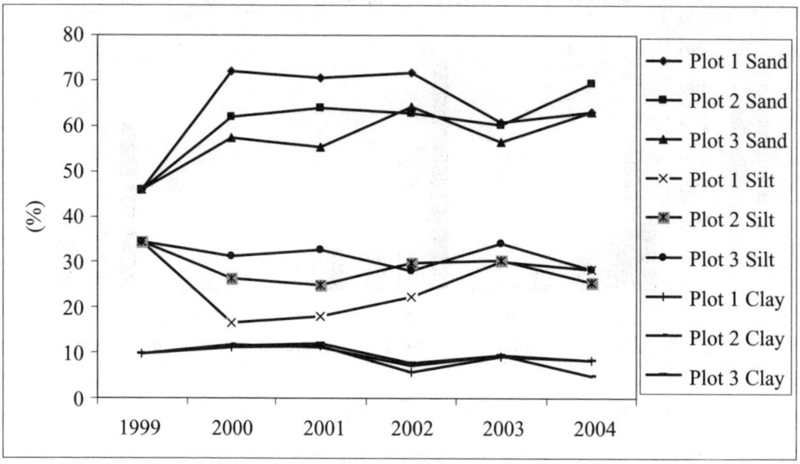

Figure 15.20 Effect of sedimentation on distribution of soil texture in the paddy field at different distances from inlet at the base of the hill

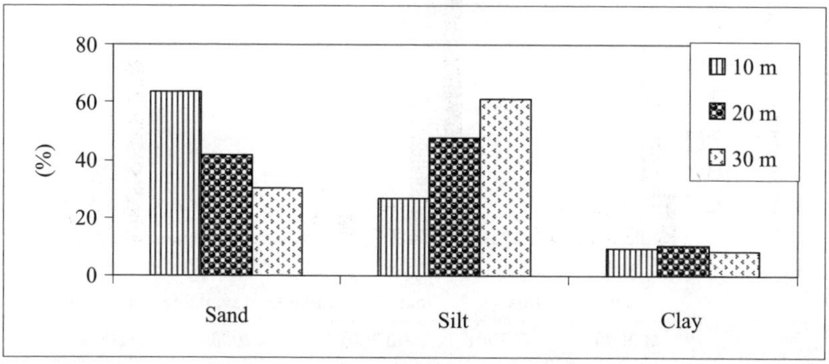

showed a wider K : N ratio suggesting that substantial amounts of K are deposited on lower areas during landscape erosion processes on their way to the stream. In contrast to K, discharges of N and P were largely in accordance with nutrient losses measured in runoff and erosion. Nutrients flows from the upland fields to the paddy fields in this experiment increased from 97 to 227 kg total N between 2000 and 2001 and decreased thereafter sharply to 78 kg N in 2002 and to 50 kg in 2004. P and K outflows from the watershed showed similar trends.

Nutrient losses through overflow in the paddy fields

In the small watershed under study, the ratio of upland area to paddy field area was high, i.e., the upland area was 3.23 ha, and occupied 91 percent of the study

Figure 15.21 Yearly nutrient flows from upland field to the paddy field

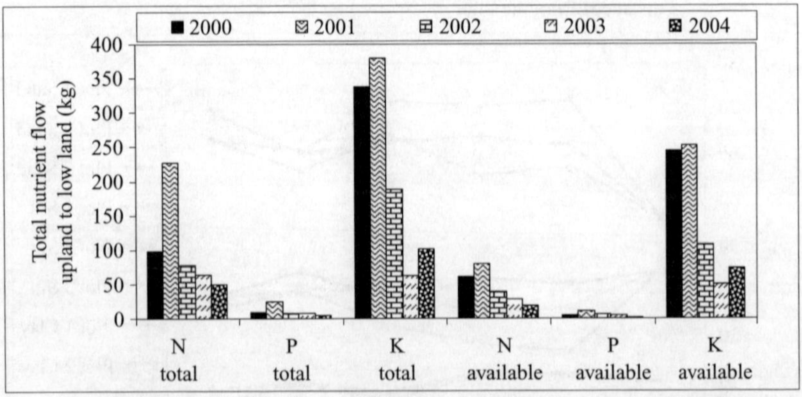

Figure 15.22 Yearly nutrient outflows from the paddy fields

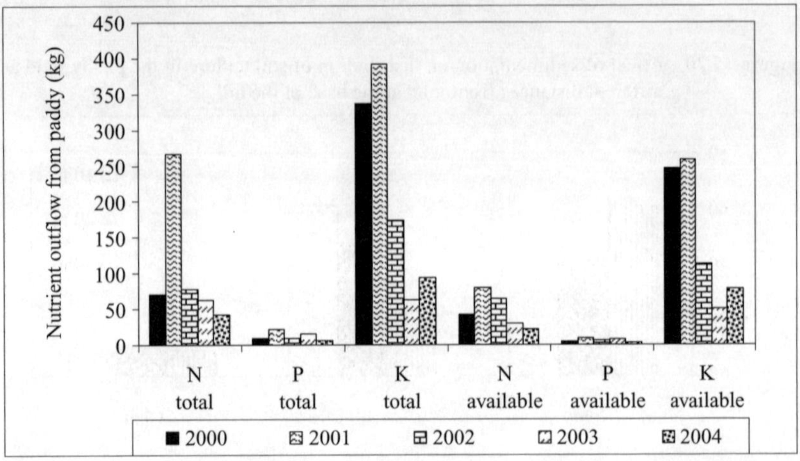

area. In contrast, the area of the paddy field where the outflow measurements were taken was only 1,020 m^2, which represented 3.4 percent of the total paddy field area associated with the watershed of 0.3 ha. The distance from the inlet at the edge of the paddy field next to the hill slope to the discharge outlet was on average 35 m. Thus, in cases of heavy rains the water velocity flowing in and out of the paddy fields was very high and consequently could carry large quantities of sediments and nutrients with it. The sediments flowing out of the paddy field amounted to 13.5; 13.4; 4.3; 2.6 and 2.1 tons in 2000, 2001, 2002, 2003 and 2004 respectively.

The highest nutrient losses associated with the overflow from the paddy fields were of potassium. The amounts of K flowing out of the three paddy fields were

Table 15.8 Average yearly contribution of available nutrients from the upland field to the paddy field (2000-2004)

Indicators	Units	N_a	P_a	K_a
Total inflow	*kg/year*	*45*	*6*	*146*
Inflow paddy (0.1ha)	kg/ha equivalent	453	58	1455
Inflow paddy (0.3ha)	kg/ha equivalent	151	19	485
Total outflow	*kg/year*	*47*	*6*	*143*
Outflow paddy (0.1ha)	kg/ha equivalent	466	61	1432
Outflow paddy (0.3ha)	kg/ha equivalent	155	20	477
Balance total 2000–2004	*kg/year*	*–1*	*0*	*2*
Balance 2000–2004 (0.1ha)	kg/ha equivalent	–13	–2	23
Balance 2000–2004 (0.3ha)	kg/ha equivalent	–4	–1	8

Note: 0.1 ha denotes measurement area and 0.3 ha total paddy area at bottom of watershed

392 kg total K and 258 kg available K in 2001 (Figure 15.22). Potassium loss from the paddy field was particularly high in the first two years but became relatively small in 2003 and 2004. Associated with highest rainfall, nitrogen loss was also largest in 2001, i.e., 260/80 kg total/available N respectively. These outputs decreased in 2003 and 2004 to only 62/42 kg total nitrogen and 30/22 kg available nitrogen.

Nutrient balances in paddy fields

On average, the inflow of available nutrients to the upper three paddy fields were equivalent to available nutrient additions of as much as 453, 58 and 1455 kg/ha for N, P and K respectively (Table 15.8). Even when considering the whole paddy area, potential inputs could on average account for up 151, 19 and 485 kg of available N, P and K respectively. However, in reality farmers would not divert all of the discharge into the paddy fields and potential output on average accounted for up to 155, 20 and 477 kg available N, P and K respectively. The average net balance between inflow and outflow of nutrients in the paddy field system was slightly negative, with -4, -1 and 8 kg/ha available N, P and K respectively.

Outflows from the paddy fields often offset nutrient inflows from the uplands so that the net balance between inflow and outflow of nutrients in the paddy system was often slightly negative (Figure 15.23). The net nutrient flow balance varied considerably between years. In 2000, N flow balances were positive for both total and available N. However, available N was reduced to -0.7 kg in 2001 and even to -21 kg in 2002. Net flow P balances varied least. Potassium net

Figure 15.23 Yearly net balance of inflow-outflow of nutrients in the paddy field

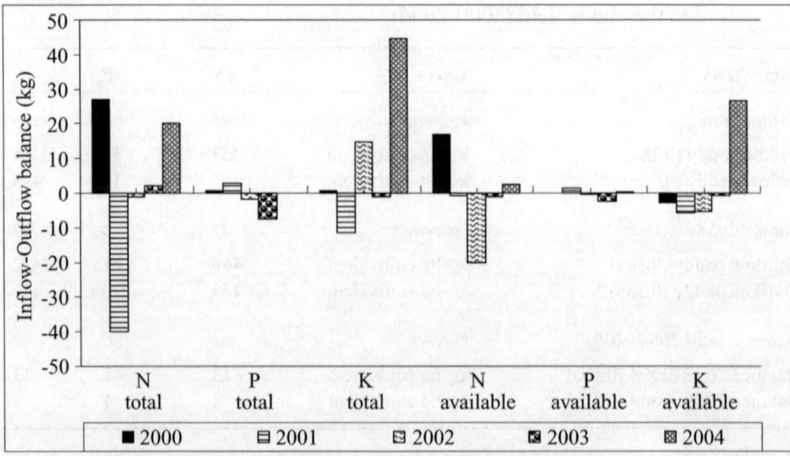

inflow-outflow balances showed similar large yearly variations as N. Potassium net flow balances, although mostly negative in the first years, became positive in the last year. There was no consistent relationship in the net inflow-outflow balances between the three nutrients tested.

These net inflow-outflow balances cannot fully account for the impact of the upland area on paddy soil fertility and plant nutrition because of the additional inputs occurring in paddy fields. However, it can be concluded that, due to the often fast water flow through the paddy system, the paddy field and its plants seemed to be able to catch only a small fraction of the nutrients provided by runoff, erosion and leaching events of upstream swidden fields. Figure 15.24 shows the inflow and outflow of total nutrients in water and sediments for the paddy field subsystem and its linkage to the upland swidden fields. It appears that for a small watershed like this one it might be necessary to supplement nutrients in order to increase yields of paddy rice. Figure 15.24 also shows that the likely nutrient contribution from swidden rice fields is, at the watershed level, smaller than that from cassava swidden fields, despite the higher losses per hectare from rice swiddens in run-off and erosion. This is because the average area under upland rice was smaller than that under cassava.

The complete nutrient balances for the paddy field revealed that the N and P balances were generally positive. In fact they were strongly positive for P (up to +11.58 kg P/0.102 ha in 2001) (Appended Table 15.A2). This was due to the high P fertilizer inputs. As excess P was not lost in the out-flowing water it is likely to build up in the soil system and would probably lead to reduced input requirements over time. N was also positive in most years by up to about 80 kg

Figure 15.24 Average yearly total nutrient flows from the uplands (via run-off and erosion) into and out from the paddy field

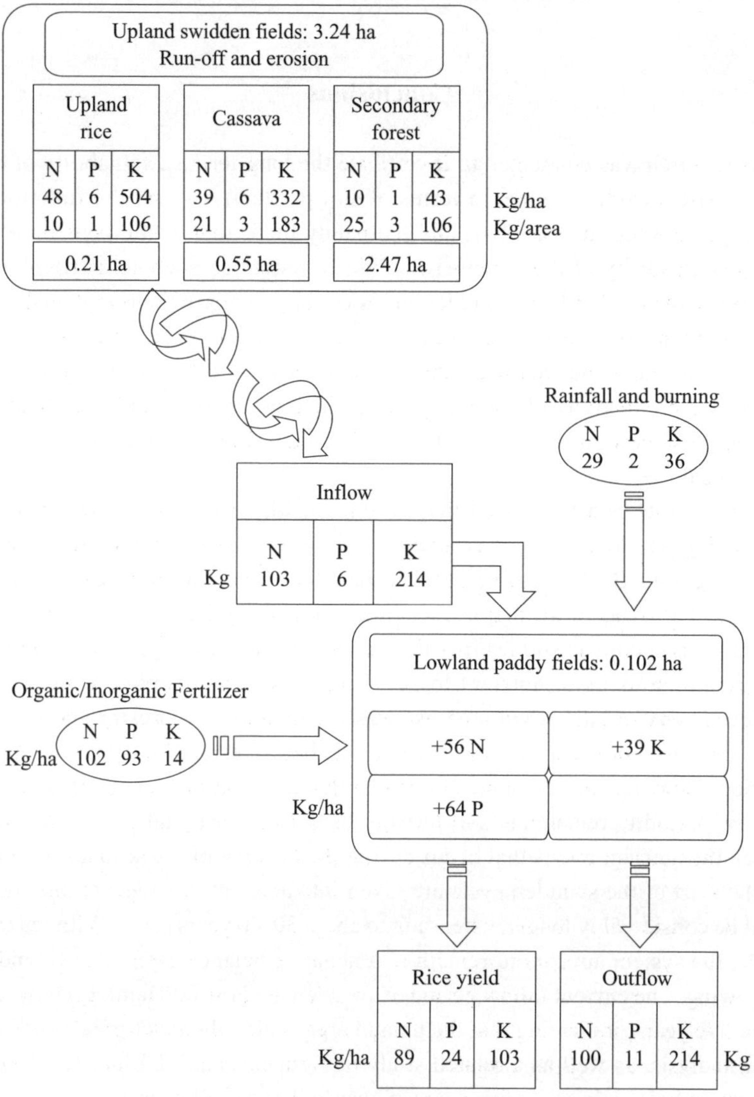

N/ha/year. However, unless this was accompanied by a build-up of soil organic matter, it is likely that a large proportion of this excess was lost by denitrification, which was not accounted for in the partial balance approach applied. K balances were negative, however, suggesting that despite the large upstream flow-through of K, not enough was retained in the paddy field to offset losses. It thus appears that in the current upland and lowland systems the nutrient replenishment patterns of K are least understood and need further investigation.

Conclusions

This research was conducted to investigate the long-term sustainability of the composite swidden system in terms of soil productivity and the functional linkage between the swiddens and the paddy fields that would contribute to the sustainability of the composite swidden system as a whole. Particularly, the nutrients eroded from swidden fields could potentially help replenish soil fertility in the paddy fields, leading to nutrient inputs more or less in balance with the outputs and hence maintain the soil productivity of the paddy fields. With increased productivity of the paddy fields, the pressure on swidden cultivation would be somewhat relieved, and the fallow period of the swiddens might be prolonged.

Our results clearly showed that swidden fields under crop cultivation are degrading. The longer the field is under cropping, the more nutrients are removed from the soil. Both negative nutrient balances and soil analyses indicated substantial losses of all major nutrients, particularly K, and consequently a continuous decline in soil fertility throughout the successive years of cropping. The major sources of nutrient losses were erosion and runoff, which were reduced substantially when land use was changed to agroforestry and fallow. Our data also showed that soil fertility was being restored during the fallow period. However, our estimates indicated that it would take about 15 years of fallow period to regenerate soil fertility back to the original level. However, when the nutrient losses that occur during the burning of the secondary forest at the start of the swidden cycle are taken into account, the regeneration time will be considerably longer, extending to about 30-40 years for N. With regards to K, the system appears to remain in a negative balance even after extended fallowing. The current fallow period of swidden fields at Tat Hamlet is now less than five years, indicating that the upland area is already under great pressure. Our analysis, as well as a related study of Nguyễn Thanh Lâm *et al.* (2005), clearly indicated that only five years of bush-tall grass fallow is not sufficient to restore soil fertility to its level at the beginning of the first cropping year. If the current practice is continued, the land will continue to suffer further degradation, making the swiddening component unsustainable.

The paddy fields showed positive balances for N and P. The strongly positive P balance was because of high P fertilizer inputs. As in the upland nutrient balances, paddy system K balances were negative. Our data showed that there were substantial amounts of nutrient inflows to the paddy field from eroded sediments and runoff from upland fields as had been hypothesized, but nutrient losses through water outflows were also quite large, and appeared to offset the inflows of nutrients. As water outflows also carried nutrients from applied manure and fertilizers, it was not possible to separate them from inflow nutrients from upland fields. Thus, the question of the beneficial effects of upland fields providing nutrients to paddy fields remains unanswered and requires further study. As K fertilizer was not applied, the negative K balance suggested that despite the large upstream flow-through of K, not enough was retained in the paddy fields to offset losses. It thus appears that in the current upland and lowland systems the nutrient replenishment patterns of K are least understood and need further investigation.

16
Does Composite Swiddening Cause Deforestation? Evidence from Analysis of Land Cover Change in Tat Hamlet from 1952 to 2003

Stephen J. Leisz, Nguyễn Thị Thu Hà, Võ Hữu Công, Trần Nguyên Bằng, and Nông Hữu Dương

Introduction

For more than one century, the Đà Bắc Tày farmers in Tat hamlet have relied on a composite swidden agricultural system (Rambo 1998). In the past, this system was found to be diverse, flexible, dynamic and sustainable. However, the ability of the community to continue to practice composite swiddening is being challenged by the pressure of population growth and the implementation of the Vietnamese government's policies on forest protection. The main question is whether the system remains sustainable or whether the inhabitants should be encouraged to replace it with an alternative farming system. The second option is the one currently favored by government officials who view swidden agriculture as the main cause of deforestation in the mountain regions. This negative view of swiddening has been questioned in the literature on farming systems and biodiversity over the past twenty years. The overall aim of this chapter is to provide input into the debate over whether swiddening, and composite swiddening in particular, causes deforestation or not.

The main research question investigated is:

> Does the land cover change history of Tat hamlet support the contention that composite swidden agriculture causes deforestation?

Sub-questions that guided the research are:

> How has land cover changed in Tat hamlet since 1995? How has it changed since 1952?

For purposes of this study the definition of deforestation is similar to that employed by Fox *et al.* (2000). We take as a starting point the FAO definition that deforestation is:

> The transfer of forest land to non-forest uses and includes all land where the forest cover has been stripped and the land converted to such uses as permanent cultivation, shifting cultivation, human settlements, mining, and building of dams (FAO 2000; Rao 1989).

Fox *et al.* (2000) point out that in this definition, the FAO regards deforestation as both a change in land cover and a permanent change in land use. This is different than forest degradation, which the FAO defines as 'a reduction in the extent and quality of forest cover due to such factors as indiscriminate logging, inappropriate road-making methods and forest fires' (Rao 1989). If land is cleared only once, then reverts to forest cover, and is never cleared again, there has been a one-time change in land cover but the land use has not permanently changed so the area is still classified as 'forest.' The key point is that in order for deforestation to take place there must be a permanent change in land use as well as a change in land cover. Thus, if a forest area that has been cleared regenerates after clearing, the process should at worst be called forest degradation, and not deforestation. If, however, there is evidence that periodic clearing for agriculture purposes has taken place in the past, then it should be categorized as 'agricultural' land and not 'forest' land according to the FAO definition, since the use of the land has been for agriculture, albeit in the form of periodic cultivation followed by fallow regrowth, for a period of time previous to the study. In this case the area cannot be considered deforested since the area was not devoted to forest prior to the study.

Materials and Methods

Materials

Landsat images of the study area from 1995, 1998, 2000, and 2003 were collected. For each year, at least one set of images was acquired. These images are all co-registered to an accuracy of +/- 15 meters root mean square error to an orthorectified Landsat image from September 2000. More than 200 ground truth points were collected during the field surveys in 1997, 1998, and 2003 and used to classify the images and create the land cover maps used in this analysis.

Ground surveys

Three ground surveys were carried out. The first survey was carried out in June-July of 1997, the second in November of 1998, and the third in June of 2003. During each survey upwards of 200 locations were visited in Tat hamlet and the surrounding Tân Minh village. At each point the location was recorded using a Global Positioning Receiver (accuracy +/- 10 meters), the land cover in the area was described, and, at a minimum, pictures were taken in the four compass directions. The first survey provided the ground truth that was used to classify the satellite image from 1995 (Fox *et al.* 2000). The first and second survey provided the ground truth for the 1998 land cover classification in this paper. Points that had been visited in 1997/98 were revisited for the last survey in 2003. The land cover in the vicinity of each point was described and pictures were taken in the same direction as the pictures taken in 1997/98.

Satellite image classification and analysis

For the land cover maps of 1998 and 2003, initial maps were made using an unsupervised classification technique. The resulting clusters of image pixels were then grouped into the appropriate land cover classes based on the ground truth points from the corresponding field visits. These land cover classes were aggregated into five general land cover classes. The result is a land cover classification for the whole hamlet. An accuracy assessment of the results was done using 100+ points that had not been used to do the land cover classification.

Since no ground truth points were available for use with the 2000 image, the 2000 land cover map was made by using a set of rules in a classification tree. The classification tree's starting point is the land cover map made for 1998 and end point is the land cover map made for 2003.

In addition to the 1998, 2000, and 2003 land cover maps, a land cover map for Tat hamlet for 1995 was also used in carrying out the change analysis. This map was made by one of the authors and the results were previously published (Fox *et al.* 2000). Before the 1995 map was used, it was verified that the map was co-registered to within 15 meters of the other maps. This was done in order to allow for a change analysis to be done between the 1995 and 1998 images.

The final step of the analysis was to perform a land cover change analysis for the 1995, 1998, 2000 and 2003 land cover maps in order to identify the patterns of land cover change over time.

Results

The land cover classification for Tat hamlet over the period 1995-2003 is shown in Figure 16.1 (a to d). The geographical distribution of the agriculture land (paddy fields and upland crops) changes over time. In 1995, upland agricultural land is found in almost equal amounts to the north and the south of the main area of paddy fields, which runs east to west through the approximate center of the hamlet. At that time there was more upland agriculture land in the western part of the hamlet than in the eastern part. In 1998, upland agriculture land is seen to have shifted to new areas and the total amount of upland fields decreases.

In 2000, agriculture land is shown to have shifted again, with the most obvious locations being a concentration of fields in two areas in the north, and one area in the south. Also more upland agriculture fields are opened up in the east, while the size of those in the western part of the hamlet decreases. In 2003, upland fields again are shifted to new areas and are less concentrated and more scattered, being almost equally distributed in all four quadrants of the hamlet.

A change is also found in both the location and amount of the paddy field areas. In 1995, as in Figure 16.1(a), the main area of paddy fields is located in the approximate center of the hamlet, running east to west, roughly parallel to the Road no. 433. There is also a small paddy field area in the southern valley of the hamlet. From 1998 the paddy field area is expanded, both in the eastern part of the southern valley, and also along the main road. Further, there is a small area of paddy fields in the northern part of the hamlet, in an upper valley approximately 0.5 kilometer from the border of Phú Thọ province. This area is present in 1998 and has been expanded since then. Overall, the paddy field area has increased from 15 ha in 1995 to 18 ha in 2003 (Table 16.1).

The tall trees and/or old mature bamboo land cover in 1995 is mainly found in the northern part of the hamlet on the steepest slopes, with one fragment of this type found in the southern part of the hamlet. This is the land cover type closest to what could be considered 'forest.' 'Old mature bamboo,' locally called 'giang' (*Dendrocalamus paterlaris*) and 'nứa' (*Neohouseaua dullooa*), is bamboo that has reached its full growth. These bamboos, when fully mature, indicate an area has not been cut for upwards of 10 years. Particularly, 'giang' may be found on hillsides and mountainsides that have never been cleared for swiddens, but may have been cleared for other purposes (logging, etc.). In 1998, the overall amount of this land cover class increases, but certain areas decrease in size (to the northeast and west), while other areas of regrowth 'bush and bamboo' land cover develop into tall tree/old mature bamboo (on the hills between the main area of paddy fields and the south valley). This pattern continues to 2000 and by 2003 the pattern changes and this category of regrowth increases in overall abundance and is observed in all quadrants of the hamlet. Generally, from 1995 to 2003, the tall tree/old mature bamboo land cover type is found to be fairly stable, increasing from 13% to 17% of the hamlet area. The upland agriculture class decreased in area by 60 hectares from 1995 to 1998, but from 1998 to 2003 has remained fairly stable, varying between 74 and 80 hectares.

Figure 16.1(a) Land cover map of Tat hamlet in 1995

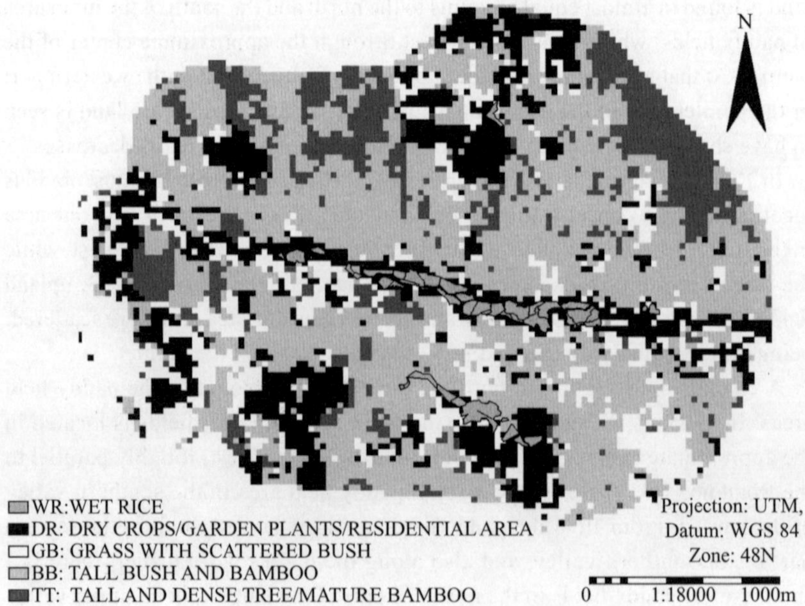

Figure 16.1(b) Land cover map of Tat hamlet in 1998

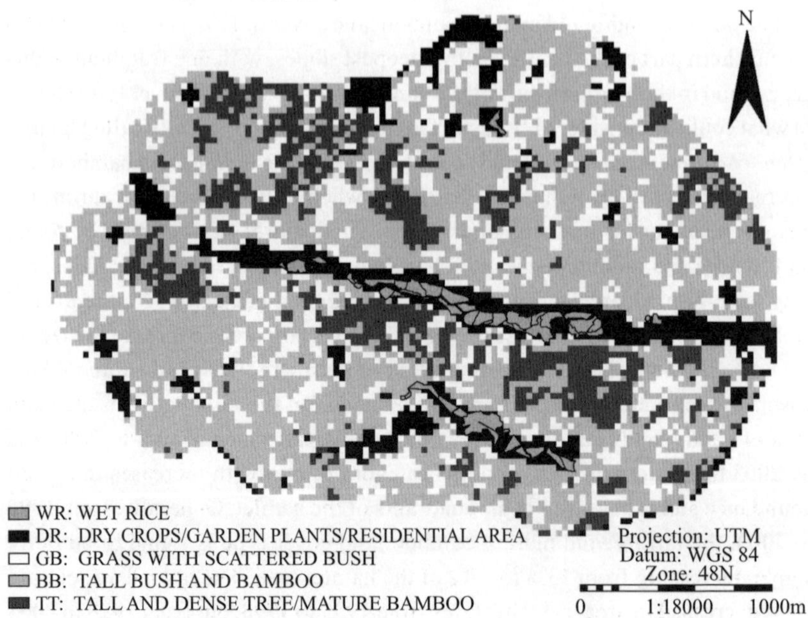

Figure 16.1(c) Land cover map of Tat hamlet in 2000

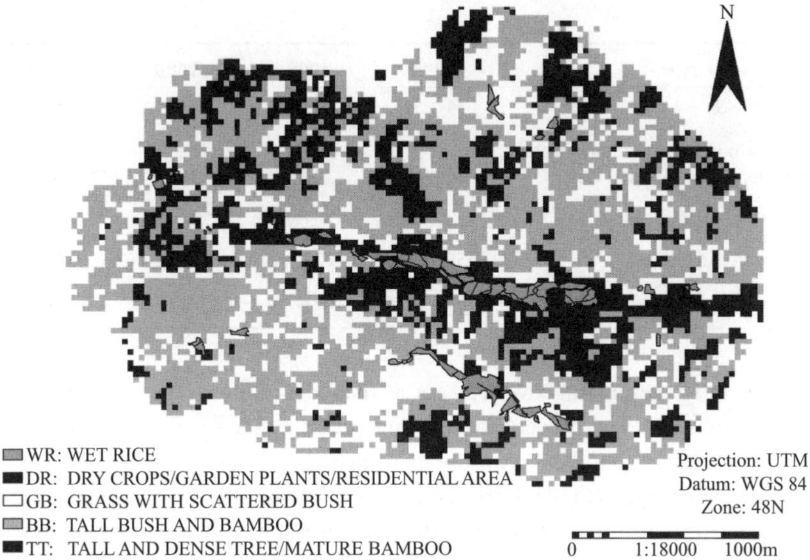

Figure 16.1(d) Land cover map of Tat hamlet in 2003

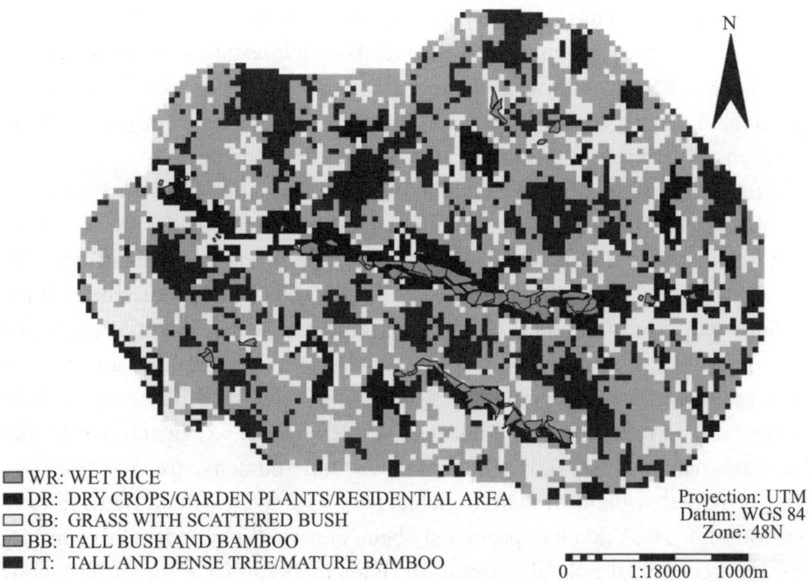

Table 16.1 Changes in area of different land cover types (ha) in Tat Hamlet from 1995 to 2003

LAND COVER TYPE	1995	1998	2000	2003
Paddy fields	15 (2.0)	15 (2.1)	18 (2.4)	18 (2.4)
Upland agriculture: (annual crops/ gardens) and residential area	136 (18.7)	74 (10.2)	76 (10.5)	80 (11.0)
Grass and scattered bush	128 (17.6)	150 (20.7)	203 (27.9)	151 (20.7)
Tall bush and bamboo	355 (48.7)	382 (52.5)	331 (45.4)	356 (48.9)
Tall and dense trees/old bamboo	95 (13.0)	105 (14.5)	101 (13.8)	124 (17.0)

The two general land cover categories that cover fallow regrowth, or could also be considered early secondary forest regrowth, 'grass with scattered bush' and 'tall bush and bamboo' are those that consequently lose area as they mature into the older regrowth of tall trees/old mature bamboo, but they also gain area as upland agriculture fields are abandoned. Grass and scattered bush increases from 17.6% of the area in 1995, to 27.9% in 2000, and then decreases to 20.7% of the area in 2003. Tall bush and bamboo increases from 48.7% in 1995 to a high of 52.5% of the area in 1998, then decreases to cover 48.9% of the hamlet in 2003. Overall, these two 'fallow regrowth' categories increase from covering 66.3% of the total hamlet area in 1995 to 69.6% of the area in 2003.

Figure 16.1 (a through d) and Table 16.1 do not provide the whole picture of how land cover has changed over time. In order to understand whether the composite swidden agricultural system leads to deforestation, another analysis was done to illustrate how land cover changes in specific areas. Tables 16.2 through 16.4 allow for an analysis of the movement, in terms of geographical location, of different types of land cover over the study period.[1]

Table 16.2 reveals that the small addition to the paddy field area occurred in what had earlier been grass and scattered bush area. 31 hectares of land that are included as 'upland agriculture land and residential land' in 1995 are found in the same places in 1998. This land is residential land and the various gardens and annual crops that are grown near residential areas and near paddy field areas. The remaining 43 hectares of upland agriculture land in 1998 is where swiddening was actually practiced that year. In 1995 16 hectares of this area was 'grass and scattered bush,' 21 hectares was 'tall bush and bamboo' and only 6 hectares was 'tall and dense tree/old bamboo.' This provides evidence that the vast majority of active swidden fields in 1998 are found in areas that had previously been cleared for swiddens. Examining the 105 hectares of upland agriculture fields in 1995 that were left fallow by

Table 16.2 Detected change in land cover from 1995 to 1998 in Tat hamlet

1998	1995					
LAND COVER TYPE (HECTARES %)	PADDY FIELDS	UPLAND AGRICULTURE LAND AND RESIDENTIAL AREA	GRASS AND SCATTERED BUSH	TALL BUSH AND BAMBOO	TALL AND DENSE TREE /OLD BAMBOO	TOTAL
Paddy fields	15 (2.0)	0	1 (0.1)	0	0	16
Upland agriculture land and residential area	0	31 (4.3)	16 (2.2)	21 (2.9)	6 (0.9)	74
Grass and scattered bush	0	27 (3.7)	28 (3.9)	77 (10.5)	19 (2.6)	150
Tall bush and bamboo	0	71 (9.7)	74 (10.2)	194 (26.8)	42 (5.8)	381
Tall and dense tree/old bamboo	0	7 (0.9)	10 (1.4)	61 (8.4)	26 (3.6)	104
Total	15	136	129	353	93	725

1998 finds that 71 ha of the area has regrown to 'tall bush and bamboo' and 27 ha has regrown to 'grass and scattered bush'. The patterns of clearing and of regrowth both exemplify the land cover changes one expects to find in a swidden cultivation system.

Table 16.3 shows that from 1998 to 2000 land cover changes for specific fields in a similar fashion to how it changed from 1995 and 1998. Similar to the expansion of paddy fields from 1995 to 1998, the small addition to the paddy field area was in what had previously been grass and scattered bush land cover. 32 hectares of land that is included as 'upland agriculture land and residential land' in 2000 are found in the same places as in 1998. This land is residential land and the various gardens and annual dry land crops that are grown near residential areas and near paddy field areas in 2000. The remaining 47 hectares of upland agriculture land in 2000 is the land where swiddening was actually practiced that year. In 1998, 15 hectares of this area was 'grass and scattered bush,' 31 hectares was 'tall bush and bamboo' and only 1 hectare was 'tall and dense tree/old bamboo.' This indicates that the vast majority of active swidden fields in 2000 are found in areas that had previously been cleared for swidden. Of the 42 hectares of upland agriculture fields in 1998 that were left fallow by 2000, 35 hectares have regrown to 'grass and scattered bush' and 7 ha have regrown to 'tall bush and bamboo' by 2000. Similar to the pattern seen in the clearing and regrowth between 1995 and1998, this pattern of clearing and regrowth is typical of the land cover changes that one expects to be found in a swidden cultivation system.

Table 16.4 shows land cover changes from 2000 to 2003. During this time paddy field area did not expand. 20 hectares of land that is included as 'upland agriculture land and residential land' in 2003 are found in the same place as in 2000. This land is residential land and the various gardens and annual dryland crops that are grown near residential areas and near paddy field areas in 2003. The remaining 58 hectares of upland agriculture land in 2003 are where swiddening was actually practiced that year. In 2000, 28 hectares of this area were 'grass and scattered bush,' 24 hectares were 'tall bush and bamboo' and 6 hectares were 'tall and dense tree/old bamboo.' This again indicates that the vast majority of active swidden fields in 2003 are found in areas that had previously been cleared for swiddens. Of the 56 hectares of upland agriculture fields in 2000 that were left fallow by 2003, 25 have regrown to 'grass and scattered bush' and 28 have regrown to 'tall bush and bamboo'. Similar to the pattern seen in the clearing and regrowth from 1995 - 1998 - 2000, this land cover change pattern of clearing and regrowth is typical of a swidden cultivation system. Figure 16.2 shows the land use dynamics over the study period.

Table 16.3 Detected change in land cover from 1998 to 2000

2000	1998					
Land cover type (hectares %)	Paddy fields	Upland agriculture land and residential area	Grass and scattered bush	Tall bush and bamboo	Tall and dense tree /old bamboo	Total
Paddy fields	15 (2.1)	0	2 (0.3)	0	0	17
Upland agriculture land and residential area	0	32 (4.4)	15 (2.0)	31 (4.3)	1 (0.1)	79
Grass and scattered bush	0	35 (4.8)	58 (8.1)	93 (12.9)	13 (1.8)	199
Tall bush and bamboo	0	7 (0.9)	64 (8.8)	219 (30.5)	34 (4.8)	324
Tall and dense tree/old bamboo	0	1 (0.1)	12 (1.6)	33 (4.5)	54 (7.6)	100
Total	15	75	151	376	102	719

Table 16.4 Detected change in land cover in Tat hamlet from 2000 to 2003

2003	2000					
Land cover type (hectares %)	Paddy field	Upland agriculture land and residential area	Grass and scattered bush	Tall bush and bamboo	Tall and dense tree /old bamboo	Total
Paddy fields	18 (2.5)	0	0	0	0	18
Upland agriculture land and residential area	0	20 (2.8)	28 (3.9)	24 (3.3)	6 (0.8)	78
Grass and scattered bush	0	25 (3.5)	44 (6.1)	63 (8.8)	17 (2.4)	149
Tall bush and bamboo	0	28 (3.9)	100 (14.0)	174 (24.4)	46 (6.5)	348
Tall and dense tree/old bamboo	0	3 (0.4)	27 (3.7)	63 (8.8)	30 (4.2)	123
Total	18	76	199	324	99	716

Figure 16.2 Land cover dynamics 1995–2003

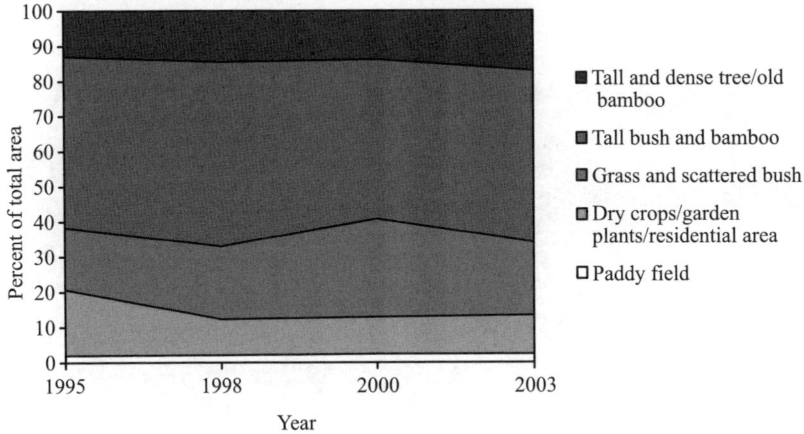

Looking at several specific points located along two south to north transects of the hamlet helps to demonstrate land use change dynamics. Land cover at point 1 in Figure 16.3 changed from upland crops in 1995 to grass and scattered bush in 1998 and in 2003 the area is cleared for swidden again. Point 2 in Figure 16.3 is in the paddy field area in the middle of the hamlet and no change in the land cover is seen. The land cover at point 3 in Figure 16.3 is more dynamic, in 1995 it is classified as grass with scattered bush, in 1998 the point is located on the edge of a swidden field, and in 2003 it is classified as 'tall bush and bamboo,' land cover that appears 3 to 5 years after cultivation of an area ends and the land is fallowed.

Figure 16.4 illustrates another transect walked during fieldwork in 1997 and 2003. Each of the seven points were first visited during 1997, their location recorded using a GPS and a picture taken at the point. Each point was revisited in 2003 and a picture of the same area was taken. The map shows the land cover for the two periods. The land cover in this area is fragmented and has exhibited change both in the type of land cover found along the transect during this time period and in the location of different land cover types.

Plate 16.1 shows the actual land cover as captured by pictures at each of the points. Of the seven photo comparisons, three have not changed land cover type and four have. Point 1, paddy field, is the same in both time periods; the upland agriculture field at point 2 has expanded; the land cover at point 3 (small trees and bamboo) and point 5 (grass/scattered bush) have stayed the same; the grass at point 4 has developed into denser bush and small trees; and agriculture fields on the mountainsides of points 6 and 7 have been covered with grass and bushes.

Figure 16.3 Land cover transect: 1995 to 2003

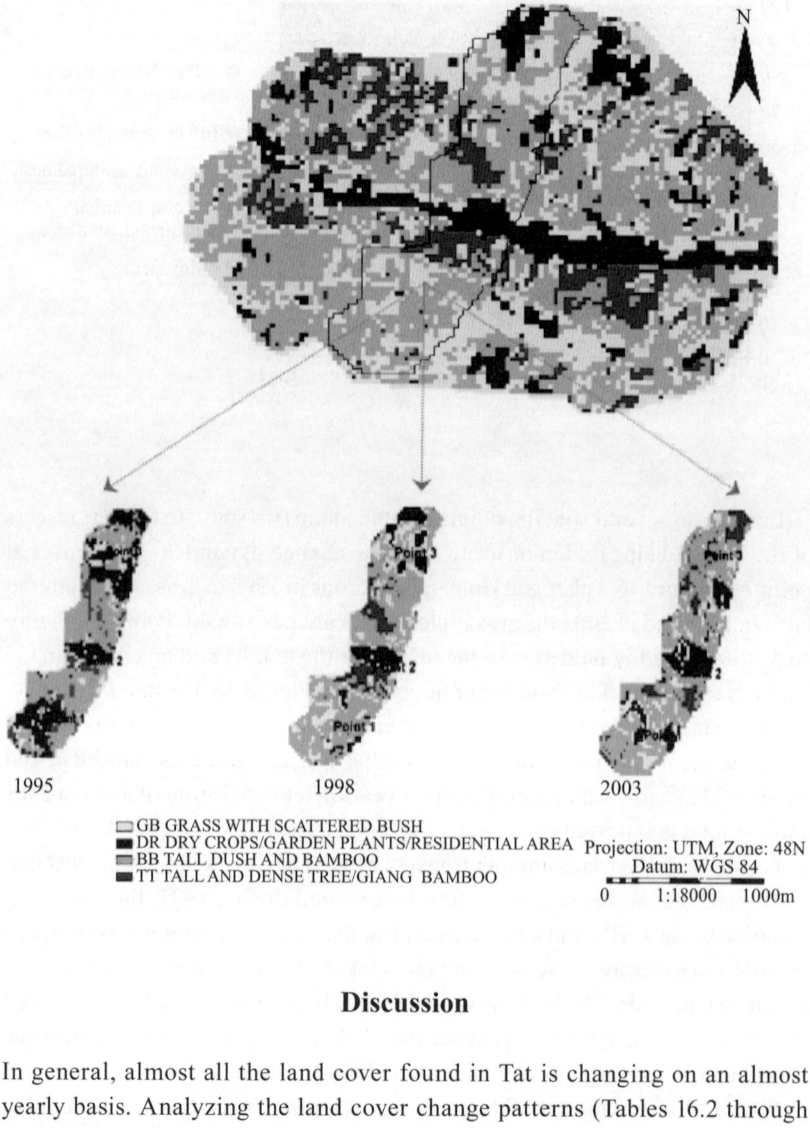

Discussion

In general, almost all the land cover found in Tat is changing on an almost yearly basis. Analyzing the land cover change patterns (Tables 16.2 through 16.4) and looking at the land cover statistics for each year show that there are three types of land cover that do not change (1) paddy fields, which only increase by a few hectares from 1995 to 2003; (2) approximately 30 hectares[2] of land that is residential and 'near residential annual dry land crop' area; and (3) approximately 30 ha of 'tall and dense tree/old bamboo' area. The 'tall tree/old bamboo' land cover that remains fixed and unchanging from 1995 to 2003

Figure 16.4 Transect map of land cover: 1998–2003

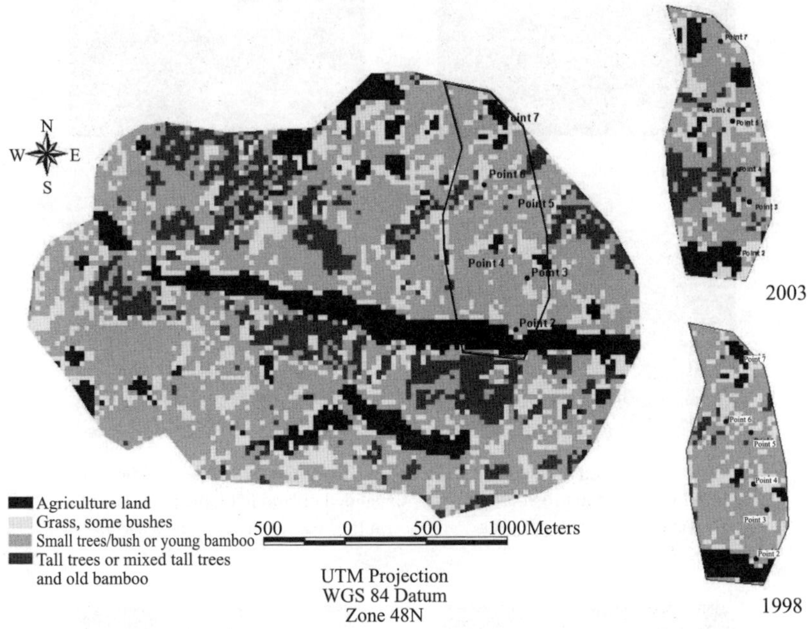

is found in two patches: One in the western part of the hamlet and the other located on a ridge in the eastern part of the hamlet. The rest of the land cover is constantly evolving from upland agriculture fields (swidden fields) to 'grass and scattered bushes', to 'tall bush and bamboo', and some 'tall bush and bamboo' matures into 'tall and dense tree/old bamboo' each year, before it is again cleared for upland agricultural fields. As noted previously, this pattern is expected in a community that relies on composite swidden agriculture. The change tables also clearly demonstrate that the landcover categories that are favored for clearing for upland agriculture fields are either 'grass and scattered bush' or 'tall bush and bamboo' categories.

The analysis shows that from 1995 to 1998, 60 hectares of land that had been included in the swidden/fallow rotation was removed permanently from cultivation. In 1995 104 hectares were used for active swidden fields, by 1998 this area had decreased to 43 hectares, in 2000 the area was 46.5 ha and in 2003 the area of active swidden cultivation was 57.5 ha. This decrease of areas cleared on a yearly basis for swidden fields coincides with changes in land tenure rules that went into affect during this period.

Point 1—1997—rice paddy

Point 1—2003—rice paddy

Point 2—1997—edge of upland field looking down the hillside—field is near the residence area

Point 2—2003—in the middle of the expanded upland field, the point is no longer on the edge of the field as it was in 1998

Point 3—1997—path in small trees and bamboo

Point 3—2003—path in small trees and bamboo; small trees are more developed

Point 4—1997—grass and small scattered bushes (pasture)

Point 4—2003—bushes and trees; grass that is left is taller, bushes are denser, small trees, and some larger ones, are now present

Plate 16.1: Photo comparisons of points on transect in 1997 and 2003 (Stephen Leisz)

Point 5—1997—edge of grass/bush looking south and down mountain to bush/small trees

Point 5—2003—edge of grass/bush looking south and down mountain to bush/small trees

Point 6—1997—upland field (foreground) then bushes and tall trees on top of mountain

Point 6—2003—upland fields are now grass/bush then dense bushes and then tall trees on top of mountain

Point 7—1997—upland fields on side of mountain; tall trees and old bamboo on the top

Point 7—2003—grass low on the mountainside, then bush and small trees midway up the side of the mountain, and tall trees and old bamboo on the top

Plate 16.1: continued

Given that the overall pattern of land cover appears stable over this nine-year period, the next question is to ask how the population has changed over time. It is usually suggested that swidden systems are stable when population does not change, but that as population pressure increases, more pressure is put on the resource base and the land cover in the area is degraded, i.e., the swidden

Figure 16.5 Changes in population (in number of people) and land cover (in hectares) from 1995–2003

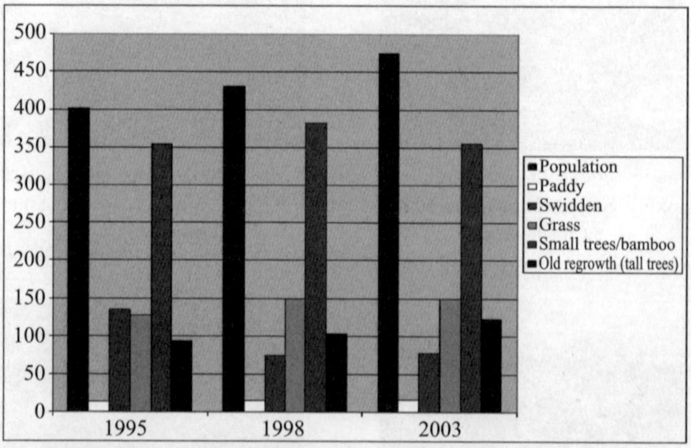

system evolves into a system that has a shorter and shorter fallow period until the land cover is dominated by grasses and the agriculture fields have become permanent; and/or more intensive agriculture is practiced.

Population in Tat hamlet has grown from 402 in 1996[3] to 476 people in 2003, representing an increase of 18.4 percent over the period, or a yearly rate of 2.1 percent. However, during this time, the amount of cultivated upland area has actually decreased, while the irrigated paddy field area has increased only slightly (3 ha). Figure 16.5 illustrates the changes.

The next question is whether Tat hamlet relies on its own land area for agriculture land, or are members of the hamlet traveling outside the hamlet boundaries to cultivate upland fields? The answer to this is somewhat complicated. It was observed during fieldwork that some agriculture fields used by people from Tat hamlet are on the border, and slightly over the border, of the hamlet. However, it was also observed that people from other villages and hamlets were cultivating fields on the border with Tat hamlet, and even inside of the hamlet's boundaries. Thus, it is hard to quantify how large an area of upland fields is cultivated by Tat inhabitants outside of the hamlet's boundaries. However, given that the system is reciprocal with other villages and hamlets, it appears that the result may even out over time and does not compromise the analysis done here.

Another possibility is that the agriculture land is producing more per hectare. This is certainly the case for the paddy field areas. Besides the small increase in the paddy area, the yields have also increased quite significantly due to improved varieties being grown, expansion of double cropping, and increased

Table 16.5 Comparison of general land cover in Tat hamlet from 1952 to 2003[a]

CLASS	1952 AREA (HA) (POP = 50)	1995 AREA (HA) (POP = 402 (1996))	1998 AREA (HA) (POP = 432)	2003 AREA (HA) (POP = 476)
Paddy field	7.7 (1.1)	14.7 (2)	15.3 (2.1)	17.7 (2.4)
Swidden	47.7 (6.6)	136.1 (18.7)	74.3 (10.2)	79.7 (11)
Secondary Regrowth (Total)	672.3 (92.4)	578 (79.4)	638 (87.7)	630.3 (86.6)
Grass, small bamboo, shrubs	201.1	483.3	532.6	506.5
Open canopy (less dense tree cover)	391.1	94.7	105.4	123.8
Closed Canopy (dense tree cover)	80.1	0	0	0
Total	727.6	728.8	727.6	727.6

Note. a: No population figures are available for 2000

use of fertilizers (Nguyễn Thanh Lâm *et al.* 2004). This improvement may play a role in decreasing the population pressure on the upland areas. The increased yield from the paddy fields may be influencing the way the local people manage their other natural resources, such as their upland agriculture fields, water resources, and even forest resources.

Finally, another possibility is that the composite swiddening agriculture system's inherent flexibility may play a role in mitigating the effect of pressure from population growth on the natural resource base. To examine this possibility these results can be added to analysis that was done by Fox *et al.* (2000) for the same hamlet. In that article a comparison was made between the land cover for Tat hamlet in 1952 and the land cover in 1995. The analysis was more general than what we have done here, as only three categories of land cover were looked at: paddy field, swidden, and secondary regrowth (total). Table 16.5 shows the data presented in this report grouped into these more general categories and compared to the 1952 data.

When the size of the change in each class is considered with regards to the total area of the hamlet it appears that the area of the different general land cover classes is little changed. The biggest changes have been in the paddy fields, which have doubled in size, and active swidden fields, which have not quite doubled in size. Looking at this data in relation to the population growth in the hamlet further shows just how stable land cover has been in the hamlet. While population has increased over 852 percent (from 50 people in 1952 to 476 people in 2003), paddy field area has increased 130 percent, upland agriculture and residential areas (referred to as swidden in this table) have increased by 67 percent, and overall secondary regrowth has decreased by 6 percent.

Conclusions

Based on the analysis above and evidence presented in the results section, it is possible to answer the two sub-questions posed at the beginning of this paper: How has land cover changed in Tat hamlet since 1995? How has it changed since 1952?

Since 1995, there has been a net increase in the area under the more mature forest vegetation (tall trees/mature bamboo) and a net decrease in the area under upland agriculture fields. The paddy field area has increased. The area covered by grasses, bushes, bamboo, and small trees has roughly stayed the same. Since 1952, the paddy field area has increased from 7.6 ha to 17.7 ha (+130%), upland agriculture and residential area has increased from 48 ha to 80 ha (+67%), and overall secondary forest vegetation regrowth has decreased in area from 672 to 630 ha (-6%). It appears that, overall, the land cover in Tat is not changing permanently, but rather is being managed to provide a stable level of fallow regrowth land cover that can be cleared on a 5 to 10 year rotation for upland fields. Based on the evidence presented here this appears to be the case going back to the 1950s. The recent evidence also shows that it is possible that, as yields from the paddy field area have increased, the hamlet is now managing its fallow lands to allow more of the tall tree and bush land cover to mature into tall tree/ mature bamboo land cover. If this continues, there may be a net increase in the mature types of secondary regrowth land cover in Tat in the future.

Let us now turn to the main question that motivated this study: Does the land cover change that has been found to have occurred in Tat hamlet support the view that composite swiddening leads to deforestation?

Given the evidence from our research, as well as from Fox *et al.* (2000), it does not appear that composite swiddening as it has been practiced in Tat hamlet leads to deforestation. In fact, in recent years it appears that the opposite may be occurring as the area dominated by the most mature form of secondary regrowth, tall trees and mature bamboo, has actually increased.

Our evidence shows that the composite swidden farming system in Tat hamlet currently does not cause deforestation, as the proportions of different types of land cover have not changed significantly since 1995, and one could argue this is the case since 1952. Thus, the accepted FAO definition of deforestation does not fit the situation in Tat hamlet. At most, one can say that forest degradation has taken place since 1952. However, it is more likely that the definition of deforestation does not actually fit this case. The available evidence suggests that the vast majority of the land that is cleared for swidden cultivation each year has been previously used for swidden cultivation. This being the case, most of the land in Tat hamlet should not be considered 'forest' land as per the FAO definition, at least not when considered in the time-frame of the study. It does appear that since 1952 the extent of the tall tree/mature bamboo area has gotten smaller and most likely decreased in quality. Given this situation one could argue that a certain amount of degradation has taken place. However, the same cannot be said when looking at the area since 1995. In that case it appears that the area, and possibly the quality, of the tall tree/mature bamboo areas have increased.

Contrary to what many government officials may think regarding swidden and shifting cultivation, the composite swidden farming system has provided a sustainable way for the people of Tat hamlet to use and manage their land and natural resource assets as its population density has increased by over 850 percent in 50+ years. The key question that remains is whether the flexibility that appears to be inherent in the system can accommodate further increases in population density? The initial evidence from over the past few years is that it can, as population density has continued to increase since 1998, upland agriculture field area has decreased and the most mature form of secondary regrowth, tall tree/mature bamboo, has increased. However, if this stability is based on increased rice yields from the paddy field area, the question of whether those increased yields are sustainable has to be asked, and further research is needed in order to answer this question. Finally, the impact of the change in land tenure rules need to be taken into account and the influence that their implementation in Tat hamlet may have had on the decrease in the area of active swidden fields in the years since 1995 should be further investigated.

Notes

1 The totals in each of the comparison tables vary slightly. The reason for this has to do with the problem of overlaying raster images, unless each raster image's individual cells completely coincide with the other raster image's cells, there will be a slight area along the border were the cells do not overlay completely. This area will not be used in the comparison (it will be classified in one image but not in the other) and correspondingly the totals vary by a few hectares when the maps that result from each image classification are compared.

2 The decrease in this area to 20 hectares in 2003 is attributed to a classification issue. In 1995, 1998, and 2000 the image acquisition date (June, November and September respectively) showed all cultivation areas when they were bare ground. In 2003 the image was acquired in early April when grass and stubble from the previous year would have been on the fields. In this case areas that were previously classified as residence and near residence

cropping areas (e.g., gardens) were misclassified as grass and grass and bush.

3 Population figures for 1996 are used since there are no population figures for 1995.

PART V
THE COMPOSITE SWIDDEN AGROECOSYSTEM OF TAT HAMLET IN WIDER PERSPECTIVE

Introduction

The three chapters in this part expand the focus of investigation beyond the borders of Tat hamlet to look at composite swidden agriculture in the wider context of Vietnam's Northern Mountain Region (NMR) and, even, all of montane mainland Southeast Asia.

Chapter 17 examines the extent to which composite swiddening occurs in other parts of the NMR. It attempts to answer the question repeatedly asked of us over the years by the donors who supported our research: "Is the composite swidden agroecosystem of Tat hamlet a unique case, or is this system found elsewhere in the northern mountains?" The underlying rationale for this question was, of course, the concern that the system might be such a specialized adaptation to the specific conditions in Tat hamlet as to have little applicability elsewhere in the NMR. If that should be the case, scaling-up results of our research would be difficult, if not impossible, so that there would be no justification for investing scarce resources in further in-depth investigations in Tat hamlet.

These concerns are effectively laid to rest by Chapter 17, which uses several ingenious indicators to map the spatial extent of major farming systems, including composite swiddening, found in five randomly chosen districts in the northwestern part of the NMR. It shows that, far from being an agroecosystem unique to Tat hamlet, composite swiddening is quite widespread elsewhere in the NMR.

Chapter 18 compares Tat hamlet with four other upland farming communities in the northern mountains in terms of their respective development statuses and environmental and social situations. These communities were selected to represent a wide range of ecological settings and agricultural adaptations within the region. The concept of emergent properties is used to compare Tat hamlet with the other communities in terms of productivity, diversity, sustainability, equitability, integration, and solidarity.

In terms of its overall development status, Tat hamlet scores relatively high in comparison to the other communities. That it performs as well as it does is somewhat surprising given its relatively remote location and lack of good agricultural land. Its performance in terms of the various emergent properties is quite a mixed one, however. It ranks high for diversity and integration, has medium rankings for productivity and equitability, but has low rankings for sustainability and solidarity. Its low ranking for sustainability is somewhat surprising, although it is in part an artifact of the nature of the sample of communities to which it is compared. None of the other communities are nearly as reliant on swidden agriculture as Tat, which inflates their ranking for sustainability. If Tat hamlet were to be compared to any of the many thousands of communities in the NMR that still depend on pure swiddening, it would certainly achieve a much higher relative ranking. Nevertheless, its low sustainability ranking does support our conclusion that the composite swidden agroecosystem is facing serious problems as the result of population pressure and scarcity of suitable land for swiddening.

Chapter 19 places composite swiddening in Tat hamlet into the broader context of the political ecology of shifting cultivation everywhere in montane mainland Southeast Asia. It points out that, since the colonial period, foresters, agronomists, and government officials throughout the region have been convinced that swidden farmers are willfully ignorant pyromaniacs who are responsible for most of the widespread deforestation and degradation of land resources in the uplands. The conventional view is not, however, supported by most recent scientific research, including our investigations in Tat hamlet. There is a large and growing body of literature that documents the highly developed agroecological knowledge of swidden farmers and demonstrates the sophistication of their locally adapted land and vegetation management techniques. Far from being a primitive and highly destructive form of agriculture, swidden agriculture can be, in appropriate contexts, a positive force for forest conservation.

This is not to say, however, that swidden agriculture does not leave any marks on the landscape. The long-term presence of swiddening has had major impacts on the vegetation cover throughout montane mainland Southeast Asia, which is now dominated by secondary forests in various successional stages. The inevitable consequence of this human disturbance of forests is habitat fragmentation. In Tat hamlet, as elsewhere in the region, what were once extensive tracts of dense mature forest have been converted into a mosaic of fields, pastures, and forest patches in varying successional stages. This fragmentation has important consequences for hydrology, soil erosion, and biodiversity. But the kinds of changes in vegetation cover resulting from swidden agriculture do not equate with 'deforestation' or 'land degradation.' Instead, it is precisely because farmers have continued to practice swidden agriculture that secondary vegetation, with the potential for regenerating into closed-canopy forests, still dominates the landscape of the region. The biggest threat to forests is not swidden agriculture, but the conversion of swidden fields to permanent commercial agriculture. This suggests that the best means of preserving forest may be to preserve and support swidden agriculture.

17
Distribution of Composite Swiddening Agroecosystems in Vietnam's Northern Mountain Region

Stephen J. Leisz, Nguyễn Thị Thu Hà, and Nguyễn Thị Bích Yên

Introduction

As our research in Tat hamlet has shown, composite swidden agriculture is a relatively diverse, dynamic and flexible system, and seems to hold considerable potential for intensification. However, it is not known how widely spread this farming system is across the NMR. In order to help close this knowledge gap a study testing methodology for mapping the extent of seven pure and mixed farming systems, including composite swiddening, in the NMR was carried out (Leisz *et al.* 2005). The study showed that it is possible to use available spatial information to identify the spatial extent of the major farming systems in the NMR of Vietnam. This chapter builds on these earlier results to derive a set of indicators that can be used to identify communes where composite swidden agriculture is found. The chapter concludes by providing an example of how these indicators are already being used by CARES to identify new study sites for studying composite swidden agriculture in the NMR.

Research Design and Methods

Leisz *et al.* (2005) used several geographic indicators to successfully identify where different farming systems are located in five randomly chosen districts in the northwestern NMR. Figure 17.1 shows the location of these districts.

It was found that, while permanent farming systems (defined as systems that make use of the same field for growing crops on a yearly basis) seem to be widespread in the five randomly chosen districts, systems that include composite swidden agriculture are also surprisingly widespread. Of the 124

Figure 17.1 Districts included in the study done by Leisz et al. (2005)

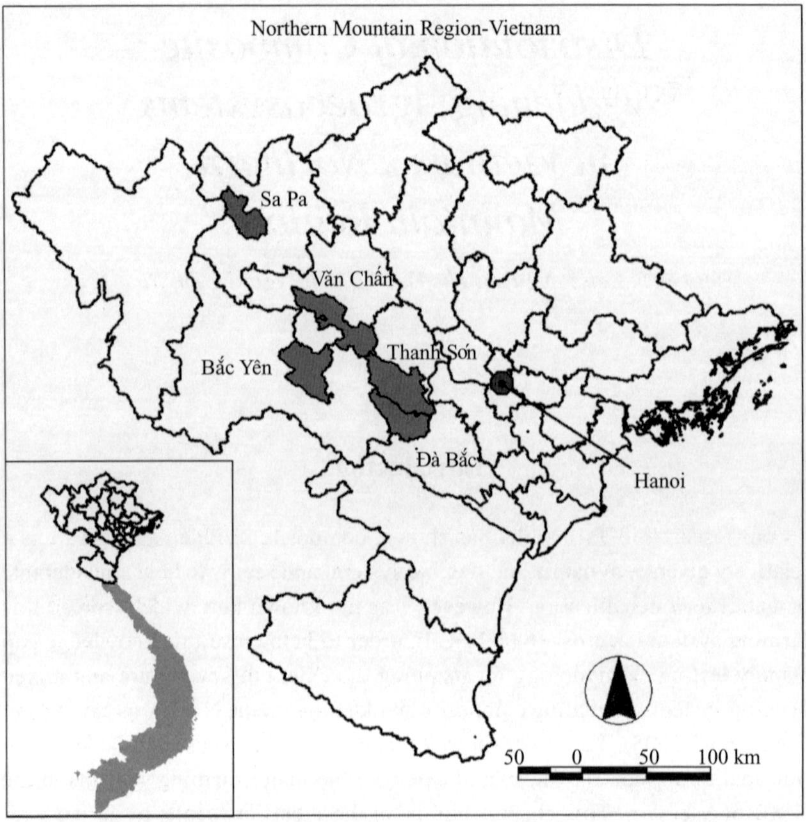

communes studied, 63 of them (50%) were classified as having composite swidden agriculture. Table 17.1 shows the overall classification of the farming systems by commune for the five districts, and Appended Table 17.A1 shows the geographic indicators used to identify the farming systems.

Based on these results, an analysis of the 63 communes that have composite swidden agriculture versus the 61 communes where no composite swidden agriculture is found was done. A number of features that distinguish the composite swidden agriculture communes from the non-composite swidden agriculture communes were identified.

Results

Figure 17.2 shows the farming system classification results from Leisz *et al.* (2005). Four of the five districts have communes where composite swidden

Table 17.1 Distribution of farming system types in the five districts (2005)

GENERAL FARMING SYSTEM TYPE	NUMBER OF COMMUNES	PERCENT OF TOTAL
1. 'Pure' general farming system type	*12*	*10%*
1.1 Rotational swidden agriculture	2	2%
1.2 Permanent paddy agriculture	5	4%
1.3 Permanent upland agriculture	5	4%
2. 'Mixed' general farming system type	*49*	*40%*
2.1 Mixed permanent paddy and upland agriculture	45	37%
2.2 Mixed permanent upland agriculture and rotational swidden agriculture	4	3%
3. Composite swidden agriculture	*63*	*50%*
3.1 Mixture of: permanent upland, paddy field, and composite swidden agriculture	44	35%
3.2 Only composite swidden agriculture	19	15%
Total	124	100%

Source: Adapted from Leisz et al. (2005)

agriculture is found. From an analysis of this information the rules for identifying communes with composite swidden agriculture are derived.

The most important aspects for identifying where composite swidden agriculture could possibly be found in the NMR were thought to include land cover; landscape pattern (fragmentation); topography (presence of flat valley bottom and sloping hills/mountainsides); hydrology; and population density. The following are the results of examining each of these variables in connection with the presence of composite swidden agriculture in a commune:

Land cover

The ratio of active agriculture land cover (e.g., cleared upland fields, paddy fields) to non-agriculture land cover (e.g., bush, trees, grass) appears to be a distinguishing factor with regards to whether composite swidden agriculture is found in a commune or not. If a commune has more than 40 percent of its territory devoted to agriculture, then composite swidden agriculture is never present. If a commune has between 25 percent and 40 percent of its area under agriculture, then it appears that composite swidden agriculture may be present in conjunction with other, permanent, forms of agriculture. If a commune has less than 25 percent of its area devoted to agriculture, and the rest is covered with vegetation found in fallow areas, then composite swidden agriculture can be found there, although other forms of swidden may also be found.

Figure 17.2 Communes with composite swidden agriculture in 2000 (adapted from Leisz et al. 2005)

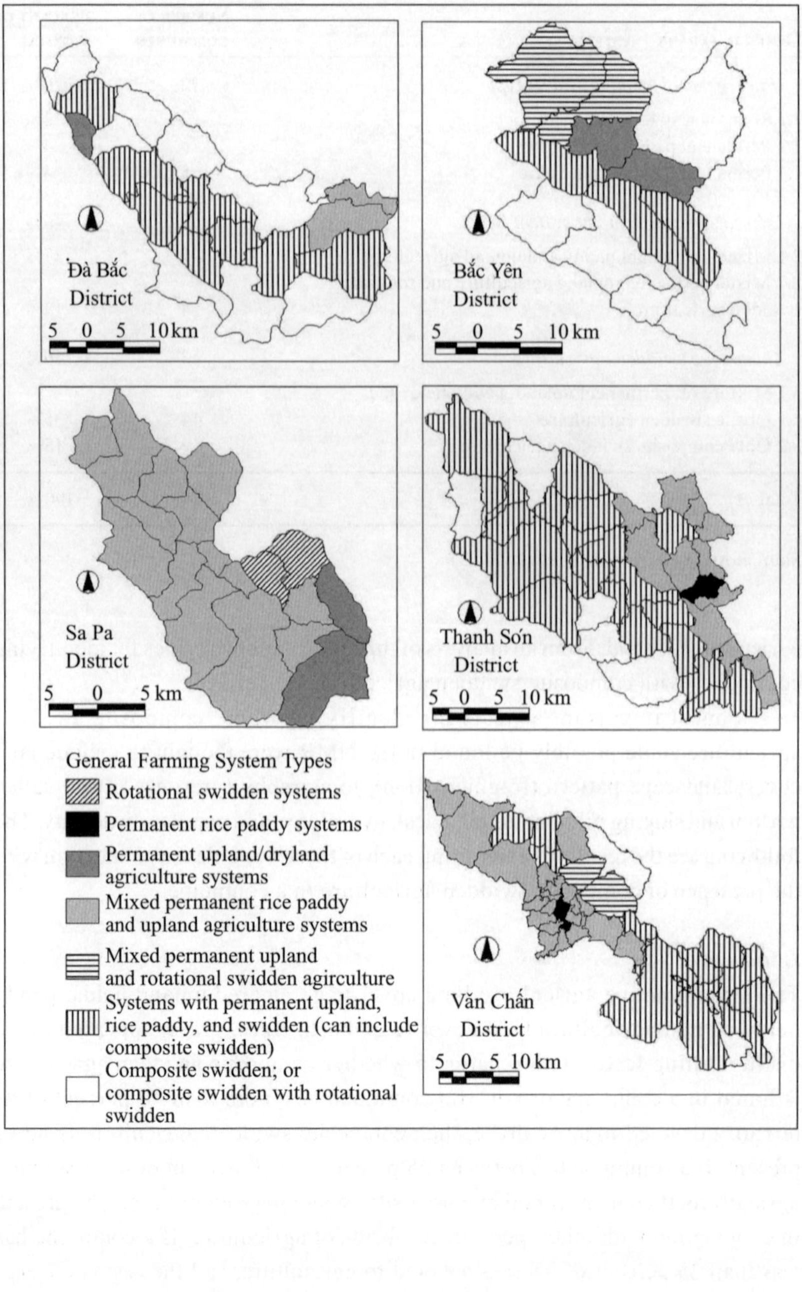

Landscape fragmentation

The patterns on the landscape and the way in which the landscape is fragmented are both indications of whether composite swidden agriculture is present or not. Appended Table 17.A1 provides both landscape fragmentation metrics (in the form of normalized landscape shape index) and visual pattern interpretations that can be used as indicators of whether or not composite swidden agriculture is present in a commune. In general, if the patches of upland agriculture areas are 0.3 to 10 ha in size and interspersed with other land cover (trees, bush, grass) then some form of swidden is indicated. If these patches are in close proximity (500 m) of small to medium size paddy fields then the area most likely supports composite swidden agriculture. If this pattern is found in a commune that also has large patches (> 15 ha) of paddy fields and/or large patches of upland agriculture fields, then the commune has permanent farming systems in addition to composite swidden agriculture. If rotational swidden fields are not in close proximity to paddy, then the areas have only rotational swidden farming systems.

Topography

Topography is a limiting factor for composite swidden agriculture. Composite swidden agriculture is only found in areas where there are small to medium size valley bottoms surrounded by hills and mountains with slopes of between 10 degrees and 45 degrees. (Small to medium size means the valley bottoms may range in size from 0.5 hectare to 15 hectares. It should be noted that swiddening is sometimes done on slopes steeper than 45 degrees, but this is not common.) The flat valley bottom areas must be wide enough to allow for most, if not all, the farmers in the community to cultivate paddy. Thus, total valley bottom size should allow for at least a 1:1 ratio of paddy cultivation to hillside cultivation. If valley bottoms are not wide enough then composite swidden agriculture will not be found. If valley bottoms are very wide, and there is a large amount of flat valley land, that is suitably irrigated, then permanent paddy field cultivation may be the main farming system, and little, if any, composite swidden agriculture will be found.

Water resources

It is important that there are streams and rivers that run through the flat valley bottoms in the commune. Without these water sources paddy field cultivation cannot be done and composite swidden agriculture will not exist.

Soil

Suitable soil characteristics are important (e.g., soils found in areas where limestone is prevalent may be too porous to support paddy cultivation, and there are situations where soil nutrient availability is a limiting factor for swidden cultivation). However, accurate soil maps at a suitable scale for this analysis are not available for the NMR. Therefore, soil properties as a distinguishing indicator for composite swidden agriculture could not be examined.

Population density

Before starting this work it was hypothesized, based on farming system literature, that population density plays an important role in determining whether swidden systems are found and by extension whether composite swidden agriculture is found. In order to test this, the population density in each of the communes of the five districts was calculated. The average population densities for each of the farming system types were then calculated (Appended Table 17.A2). While it is true that the highest population densities are found in communes where pure paddy systems and mixed permanent upland field and paddy field systems are found, the population densities for systems that include swidden were about equal. Therefore, it is not possible to use population density as a distinguishing factor, at the commune level, to identify potential composite swidden agriculture areas from other potential farming systems.

Table 17.2 shows the resulting indicators for identifying potential communes where composite swidden agriculture can be found in the NMR.

Conclusions

This chapter employs selected geographic indicators that can be used to identify communes in the NMR where composite swidden agriculture is potentially practiced. While this example does not conclusively prove that the indicators will be correct every time they are used, it does validate the belief that these geographic indicators can predict where composite swidden agriculture is likely to be found. More importantly, these geographic indicators can be used to identify areas where composite swidden agriculture can potentially be practiced, if it is decided as a policy matter to promote the extension of composite swidden agriculture to other rural communities in the NMR.

An example of how these indicators can be used operationally to identify where composite swidden agriculture is found is seen in the results of another study carried out by CARES as part of the University Support to Environmental Planning and Management Project (USEPAM) that has been implemented in conjunction with the Institute of Geography, University of Copenhagen. In this study, the indicators were used to predict where composite swidden agriculture should be found in two districts in Nghệ An Province. For USEPAM it was

Table 17.2 Indicators of composite swidden agriculture in a commune

Composite swidden agriculture status	% of landscape under agriculture	Ratio of upland ag. to paddy fields	NLSI upland	NLSI paddy	Water resources	Topography	Landscape pattern guide for visual interpretation
Only composite swidden agriculture	Less than 25%	Greater than 1.5:1 [a]	> 0.27	> 0.27	Small to medium streams running into and out of valley bottoms.	Small to medium areas of flat valley bottoms suitable for paddy fields; surrounding mountain/ hill slopes between 10 and 45 degrees	Patches of upland ag. 0.3 to 10 ha (swidden) interspersed with other land cover (regrowth and trees); in close proximity (500m) of small to medium size paddy fields.
Composite swidden agriculture with other farming systems in the same commune	Between 25% and (+/–) 40%	Usually roughly equal; can be more 3:1 or 1:3	Between 0.20 and 0.28 (can be > 0.28; determined by visual analysis)	> 0.28	Small to medium streams running into and out of valley bottoms; larger streams/rivers in large valley bottoms.	Small areas of flat valley bottoms suitable for paddy and large valley bottom areas in the same commune; surrounding mountain/hill slopes between 10 and 45 degrees	Large continuous upland ag. found in parts of commune; large continuous paddy found in parts of commune; patches of upland ag. (swidden) interspersed with other land cover in in close proximity (500m) of small and/or medium size paddy fields.

Note. a: It may be that the ratio is 1:1 or that more paddy than swidden is found. Visual pattern analysis should be used to supplement

necessary to choose study sites that represent the different possible farming systems found in the districts of Con Cuông and Tương Dương. A total of nine study sites were chosen, one has a permanent paddy farming system, six have rotational swidden systems, and three have composite swidden agriculture systems. At the time the sites were chosen it was not known if composite swidden agriculture was present in the area. Three of the nine sites chosen had the indicators noted in Table 17.2 for composite swidden agriculture. These three sites are Lưu Phong village in Lưu Kiền Commune (Tương Dương), Can village in Tam Thai Commune (Tương Dương), and Quẹ village in Bình Chuẩn Commune (Con Cuông). In all of the cases the indicators proved correct, and composite swidden agriculture was found in each of the villages. In both Lưu Phong and Quẹ, it is the main type of farming system present. In Can there are also households that rely only on permanent paddy farming, but in the past these households relied on composite swidden agriculture. In 2000 a project in the area promoted both permanent upland cropping and expanded the paddy areas, and some families stopped practicing composite swidden agriculture.

A final conclusion to this chapter should note that during our research it became clear that composite swiddening has historically been widespread in the northern mountains of Vietnam and indications are that it is still widespread today. This study randomly chose five districts in five different provinces and in four of these composite swidden agriculture is practiced. Furthermore, using the indicators from the study, three communes in two districts were identified in a sixth province, Nghệ An, where composite swidden is practiced. In all of the locations where composite swiddening is practiced today, and where field work has been done, local informants note that composite swiddening has deep historical roots within the community and the communities plan to continue practicing composite swiddening in the future. There are many other villages in the mountains of northern and north-central Vietnam where conditions are suitable for composite swidden agriculture. Based on our results it is reasonable to assume that composite swidden agriculture is practiced in many of them.

18
Tat Hamlet in Comparative Perspective

A. Terry Rambo, Stephen J. Leisz, Trần Đức Viên, Đào Minh Trường, Đào Trọng Hưng, Lê Quang Trung, and Nghiêm Phương Tuyến

Introduction

In this chapter, we look at Tat hamlet in comparison to four other rural communities in the Northern Mountain Region (NMR). These communities were selected to represent a wide range of ecological and social situations and agricultural adaptations within the region.

The NMR has extremely high levels of ecological and social diversity. It is also undergoing extremely rapid but unevenly distributed economic change (Rambo 1997a; Jamieson *et al.* 1998). The region presents many difficult development challenges because of its rugged terrain, poorly developed transportation and communications infrastructure, high concentrations of ethnic minority peoples, low educational levels, high rates of population growth, low levels of agricultural production, and weakly developed market system (Rambo 1997a). Income levels are far below the national average and the percentage of households living in absolute poverty is the highest of any region in the country (National Centre for Social Sciences and the Humanities 2001: 91). Good agricultural land is scarce and the natural environment has already been severely degraded by widespread deforestation, serious soil erosion, and loss of biodiversity (Rambo and Jamieson 2003). However, conditions vary greatly from place to place within this vast region, so that different local communities display very different ecological and social characteristics and have achieved very different levels of development.

Profiles of the five communities are first presented followed by a quantified assessment of their development status. The concept of emergent properties of ecological and social systems is introduced and Tat hamlet's performance in terms of productivity, diversity, sustainability, equitability, integration, and solidarity is compared to that of the other sample communities.

Figure 18.1 Location of study communities in the Northern Mountain Region

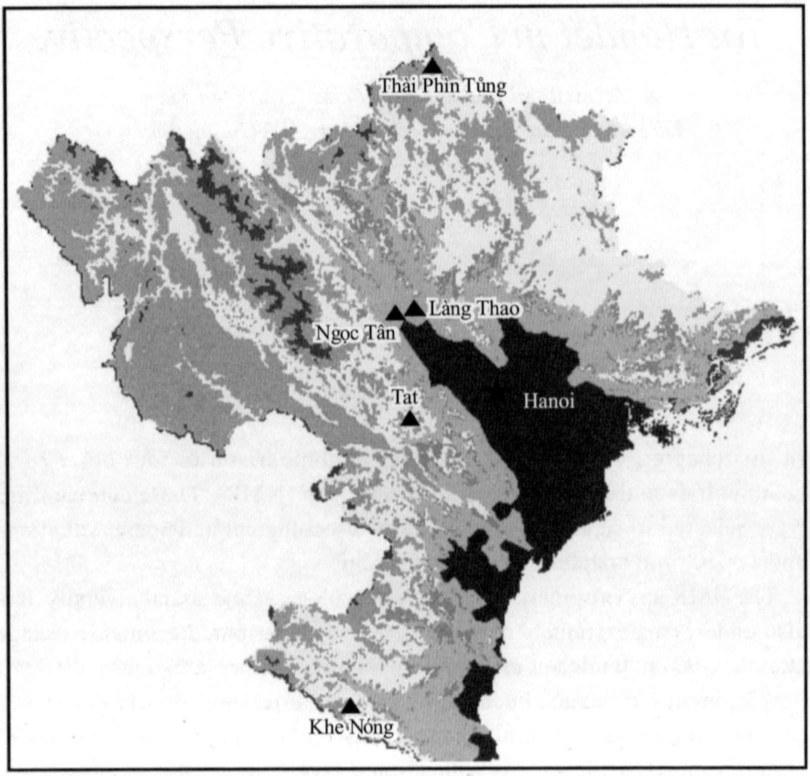

Study Design and Description of the Case Study Communities

The analysis in this chapter employs data collected in 1998-2000 in Tat hamlet and four other communities in different parts of the NMR. Figure 18.1 is a map showing the locations of the communities.

The five communities were selected for study as part of a long-term project on monitoring development trends in the northern mountains. The sample communities represent the most important types of development situations in the NMR. They differ in terms of ethnicity, topography, population density, accessibility, and the balances between subsistence-orientation and market-orientation (Table 18.1).

The five communities studied were:

- *Khe Nóng* Hamlet, Châu Khê Village, Con Cuông District, Nghệ An Province (Đan Lai ethnic minority). Khe Nóng is located in the forested

Table 18.1 Key characteristics of the five NMR communities

	Khe Nóng Hamlet	Thài Phìn Tủng Village	Tat Hamlet	Ngọc Tân Hamlet	Làng Thao Hamlet
Ethnic group	Đan Lai	White H'Mông	Đà Bắc Tày	Cao Lan	Kinh
Topography	Mountain-valley	High mountains	Mountain-valley	Hill and valley	Hill and valley
Area (ha)	1,370 (estimated)	2,484	737	122	227
No. of households	21	427	86	118	287
Population	113	2,505	432	532	1,232
Population density (persons/km^2)	8	101	59	437	543
Accessibility	Very low	Low	Medium	High	Very high
Mode of adaptation	Swidden farming, collection of forest products	Dry land farming, animal husbandry, distilling alcohol	Composite swiddening, animal husbandry, collection of forest products	Wet rice farming, agroforestry, animal husbandry, wage labor	Wet rice farming, agroforestry, animal husbandry, wage labor
Extent of market orientation	Subsistence oriented	Largely subsistence oriented with some market involvement	Mixed subsistence and market oriented	Market oriented	Market oriented

Source: Lê Trọng Cúc and Rambo 2001: 46, Table 2-1; 48, Table 2-2

hills of the Pù Mát Nature Reserve buffer zone. It is reachable only by walking for almost eight hours over forest trails so accessibility is rated as very low. Its economy is strongly subsistence-oriented. The Đan Lai engage in rotational swidden agriculture and collection of forest products and also cultivate a few small wet rice fields. Although the settlement is surrounded by primary forest, the villagers are prohibited from clearing new swiddens or exploiting forest resources by the nature reserve authorities. Khe Nóng is not officially recognized as a settlement so its boundaries have not been delineated.

- *Thài Phìn Tủng* Village, Đồng Văn District, Hà Giang Province (White H'Mong ethnic minority). Thài Phìn Tủng is a large, long settled village located in the high limestone mountains near Vietnam's northern border with China. Although an all-weather road passes through the village center, to reach the road from most houses requires at least a one hour walk over steep trails, and the nearest large urban center is more than one day's drive away so accessibility is ranked as low. Its White H'Mong inhabitants grow maize by means of 'rock pocket agriculture' for subsistence use and raise livestock and make alcohol for sale in the market. Population density is high, arable land scarce, and surface water unavailable in the long dry season, so that conditions of life are extremely difficult and the potential for development very limited.
- *Tat* Hamlet, Tân Minh Village, Đà Bắc District, Hòa Bình Province (Đà Bắc Tay ethnic minority). Tat is located in the mountain-valley realm in the Northwestern Mountains. It is served by an all weather road and is within 50 km of a major market center so is rated as having medium accessibility. Its ethnic minority inhabitants practice composite swidden farming. Subsistence production is still predominant but involvement in the market is rapidly increasing.
- *Ngọc Tân* Hamlet, Ngọc Quan Village, Đoan Hùng District, Phú Thọ Province (Cao Lan ethnic minority). Ngọc Tân is located in the rolling hills and valleys of the Midlands. Houses in the hamlet are connected to an all weather highway by laterite roads accessible to motor vehicles in all except the heaviest rains so accessibility is rated as high. The Cao Lan are a T'ai-speaking group who settled the Midlands shortly before the onset of massive Kinh in-migration (Gillogly and Nghiêm Phương Tuyến 1992). Ngọc Tân is now a small Cao Lan enclave surrounded by Kinh settlements. Its inhabitants engage in diversified agriculture including wet rice cultivation, animal husbandry, growing of timber and industrial tree crops, and services and wage labor.

- *Làng Thao* Hamlet, Ngọc Quan Village, Đoan Hùng District, Phú Thọ Province (Kinh ethnic group). This hamlet is located in the rolling hills and flat valleys of the midlands. It is set immediately next to a hard surfaced highway so is considered to have a very high level of accessibility. Its Kinh inhabitants, who migrated to Làng Thao from the Red River Delta in the early 1960s, engage in diversified agriculture, including wet rice growing, production of cash crops and tree crops such as tea, animal husbandry, and services and wage labor.

Methodology

A multi-community research approach was chosen for this study because the traditional area studies method of intensively studying a single community would not have yielded the kinds of information needed to make generalizations about the impact of development on the more than 2,000 villages in the NMR. On the other hand, extensive survey research in a very large sample of communities would not have provided sufficient understanding of locality-specific contexts to permit analysis of complex interactions between environmental and social factors within communities. Thus, a research strategy that represented a compromise between the depth of a single community study and the breadth of a survey was employed. The time spent in field research in each community was quite short, ranging from seven to fifteen days, but collection of data was assigned to several teams of specialists, each of which focused on particular types of information, so that a large quantity of data could be collected in a short period of time. The research teams employed several different methodologies to collect data on social and environmental conditions. A detailed description of data collection methods employed in this project is provided in Lê Trọng Cúc and Rambo (2001: 32-42).

Social and economic data

A large number of social and economic variables were described for each community. These included social organization and institutions; population size, composition and rate of increase; incidence of diseases and access to health care; food security and incidence of hunger; education levels, rates of school attendance, and literacy rates; sources and levels of household income and incidence of households living in poverty; access to land and other productive resources; housing and household possessions; access to safe water and electricity; access to transportation and communications; gender equity and ethnic relations; occurrence of social problems; and people's self perceptions of their development situation. A standardized household survey questionnaire was used to collect data on economics and social organization from the heads of households included in the sample. In the cases of Khe Nóng, Tat, Ngọc Tân, and Làng Thao, random samples of all households classified as permanent residents of these communities were employed. The samples were quite large with 20 out of 21 households interviewed in Khe Nóng, 42 out of 86 in Tat, 40 out of 122 in Ngọc Tân, and 40 out of 227 in Làng Thao. In Thài Phìn Tủng, a very large village with 427 households living in 10 hamlets, the sample of 40 households was selected from the 96 households living in three hamlets (Hapuda, Nhu Sa, and Thài Phìn Tủng).

Environmental data

Environmental variables described for each community were the extent and composition of vegetation cover, patterns of land use and exploitation of natural resources, structure of agroecosystems, and biodiversity. Two complementary research methodologies were employed to document environmental conditions: description of ecological transects and preparation of land use and vegetation cover maps from ground surveys and satellite image interpretation.

Comparison of the Development Status of the Communities

While we recognize that the concept of development itself is highly problematic, for purposes of this analysis we have employed the official Vietnamese definition of development as the provision of infrastructure and achievement of socioeconomic goals. Twelve standard indicators were used to assess the overall development status of the five communities. Community performance for each indicator was rated on a five-point scale running from 1 (very low) to 5 (very high). These scales are relative ones with the top point set according to the highest performance on that indicator of any of the five communities and the bottom point set according to the lowest performance level. An overall development status score was calculated for each community by averaging its scores for all 12 indicators. Table 18.2 presents the development status rankings of the communities.

The sample communities display three distinctive patterns of development: Khe Nóng and Thài Phìn Tủng rank low on virtually all development indicators; Tat displays mid-range rankings for most variables; and Ngọc Tân and Làng Thao rank relatively high on all measures. The communities in this sample thus appear to fully capture the range of variation in development levels of villages in the NMR as a whole. However, it needs to be emphasized that these are relative ranks that are only intended to allow comparison among communities in the NMR. If an absolute ranking scale were employed, even the most developed communities in the NMR would rank quite low if compared to those in developed countries, although Ngọc Tân and Làng Thao do quite well when compared to national averages for various development indicators in Vietnam.

Figure 18.2 presents a graphical comparison of the development status of the different communities.

This figure employs a display technique first used in a report on social and economic conditions in sample villages in the South Vietnam in the late 1950s (Mission 'Economie et Humanisme' 1959). It can be seen that the development status of Tat, especially in comparison to Khe Nóng and Thài Phìn Tủng, is relatively well balanced in terms of the different indicators.

The Concept of Emergent Properties

Human ecology is concerned with analysis of interactions between social systems and ecosystems (Rambo and Sajise 1984: 2-6). These systems can be defined at different levels in a hierarchical scale ranging from a single paddy field, to a farm, to a village and up to a whole region (e.g., the Red River Delta). In this paper the focus is on the local community level (hamlet or village). The main objectives of human ecology research on agricultural systems are to: (1) describe the structure and functioning of social systems and their agroecosystems, and (2) assess the performance of these systems in terms of emergent properties (Conway 1984: 37-38; Marten and Rambo 1988).

An emergent property can be defined as 'a characteristic of a system as a whole that comes into existence from the organization of the system's parts rather than from the characteristics of any of the parts themselves' (Marten 2001: 218). Emergent properties are generated by the complex interactions among the multiple different components making up the total system. Thus, for example, the yield of rice from a paddy field cannot be explained by simply looking at a single rice plant in isolation. Instead, one has to look at the interactions that occur among all of the rice plants, the soil, the water, the weeds, the pests, and, of course, the farmer's management practices. The actual yield of grain 'emerges' as the outcome of all these interactions within the paddy ecosystem (Conway 1984).

Key emergent properties that are considered in this paper are productivity, sustainability, diversity, equitability, integration, and solidarity:

Table 18.2 Development status of the five NMR communities (1998–2000)

DEVELOPMENT STATUS INDICATOR	KHE NÓNG	THÀI PHÌN TỦNG	TAT	NGỌC TÂN	LÀNG THAO
1. Cash income (percentage of households with annual incomes below 1 million VND)	Very Low (100 %)	Very Low (93 %)	Medium (43 %)	High (23%)	Very High (15%)
2. Food security (percentage of households with sufficient food during whole year)	Very Low (0.0%)	Low (48 %)	High (88 %)	High (88%)	Very High (90%)
3. Education (percentage of people aged 7 or older who have attended school)	Very Low (12 %)	Low (35 %)	High (77 %)	Very High (98%)	Very High (95%)
4. Health status (average no. of diseases/ household)	Very Low (5.1)	Low (3.67)	Low (3.98)	Medium (3.4)	Medium (3.18)
5. Access to health care (percentage of households using clinic and hospital)	Medium (45 %)	Low (23 %)	High (69 %)	Medium (50%)	Medium (41%)
6. Adoption of family planning (percentage of women currently using birth control)	Low (30 %)	Medium (50 %)	High (79 %)	Medium (50%)	Medium (52%)
7. Access to safe water (percentage of households with safe water in dry season)	Very Low (0.0 %)	Very Low (10 %)	High (77 %)	Very High (85%)	Very High (90%)
8. Sanitation (percentage of households with latrine)	Medium (45 %)	Very Low (2 %)	High (79 %)	Very High (80%)	Very High (80%)
9. Access to electricity (percentage of households connected to grid or with own micro-generator	Very Low (0.0)	Very Low (0.0)	High (67%)	Very High (83%)	Very High (95%)
10. Access to information (percentage of households listening to radio or TV)	Very Low (20 %)	Medium (65 %)	High (86 %)	High (88%)	High (88%)
11. Access to vehicular transportation (percentage of households owning bicycles and motorbikes)	Very Low (0 %)	Very Low (0 %)	Medium (17 %)	Very High (93%)	Very High (95%)

12. Travel outside village (percentage of households visiting district town in preceding year)	Low (40 %)	Very High (100 %)	High (81 %)	High (82%)	High (85%)
Total no. of development "points"	18	24	44	50	52
Average development status (total development points divided by 12 indicators)	1.50	2.00	3.67	4.17	4.33

Source: Lê Trọng Cúc and Rambo (2001:23, Table 12)

Figure 18.2 Development status diagrams of the five communities

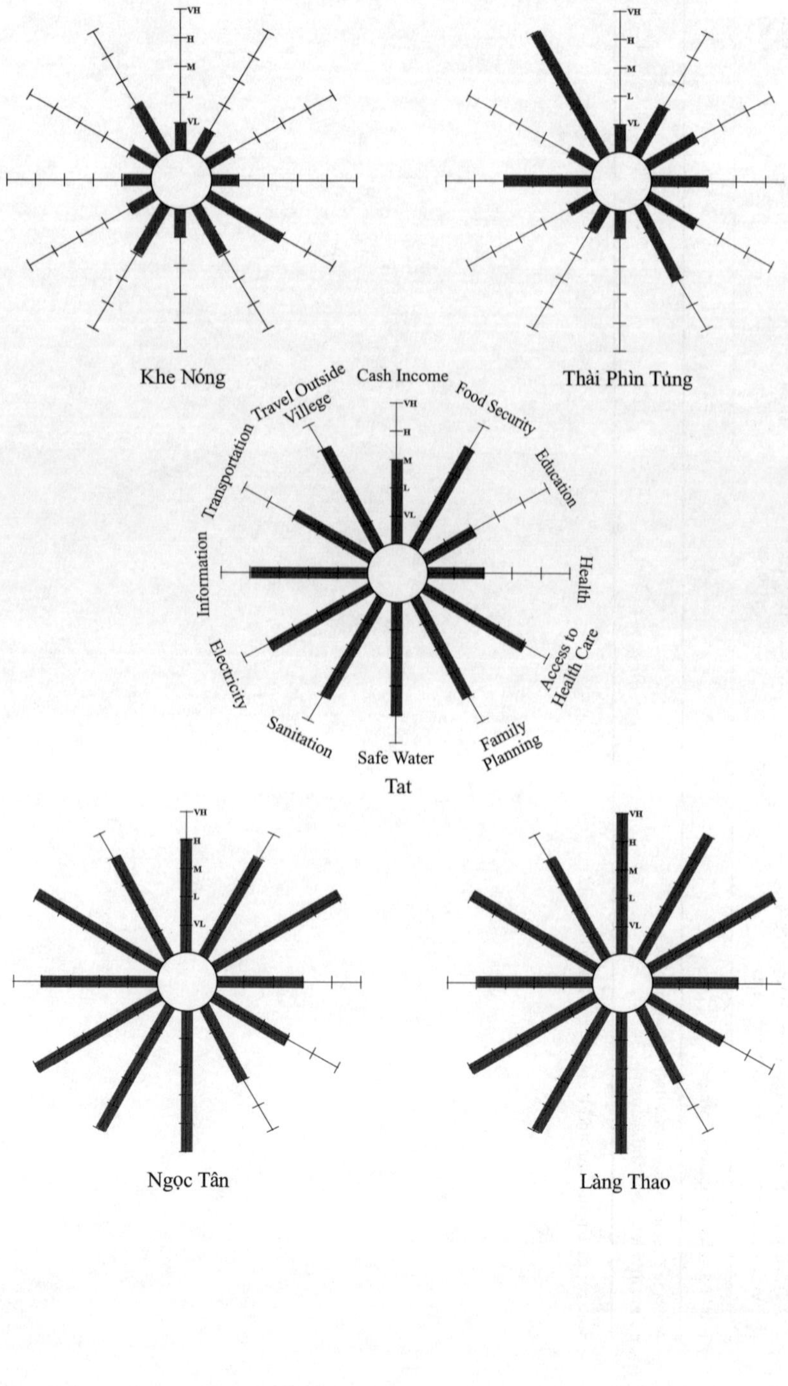

- *Productivity* is defined as '...the desired output of the system, measured in terms, say, of crop yield or net income' (Conway 1984: 37) or as 'the net increment in valued product per unit of resource (land, labour, energy, or capital). It is commonly measured as annual yield or net income per hectare or man-hour or unit of energy or investment' (Conway 1986: 23). Obviously, productivity is an emergent property of great concern to both the farmers and development policymakers.
- *Sustainability* is defined as the ability of a system to maintain an acceptable level of productivity for an extended period of time in the future without causing serious degradation of the resource base (Marten and Rambo 1988: 276). Since the 1992 Rio Earth Summit, sustainable development has become an explicit policy goal of all internationally supported rural development efforts but changes in actual practice have failed to match the change in official rhetoric (Rambo 1997b).
- *Diversity* is defined as the amount of ecological variability in the agricultural ecosystem. It can be measured at several different levels: genetic diversity within populations of individual species, species diversity within each ecological community, and diversity of ecological communities within the whole agricultural ecosystem.
- *Equitability* is a measure of how evenly the benefits and costs of agricultural production are distributed among its human participants (Conway 1986: 23; Marten and Rambo 1988: 282). Equitability can be measured at different hierarchical levels ranging from individuals to the nation. In this paper, distributional equity among households in the village is the focus.
- *Integration* is defined as the extent to which the local community is involved in larger social and economic systems. This property has often been called *autonomy*, which is, in effect, the opposite state from integration (Marten and Rambo 1988: 279). No human community is wholly autonomous but contemporary subsistence-oriented cultivators in the uplands of Southeast Asia are much less tightly integrated into larger systems than lowland farmers who are heavily dependent on the market system for their survival.
- *Solidarity* is a measure of the level of social cohesion among individuals and households making up a local community. Cohesive communities are characterized by abundant stocks of 'social capital,' the existence of which determines the ability of the community to make and enforce decisions about management of its agroecosystem (Rambo and Trần Đức Viên 2001a).

Emergent properties offer a useful way to measure the performance of village-level agroecosystems in the context of development. It should be remembered, however, that properties are intellectual constructs, not real 'things' and they are descriptive measures and not normative ones. A system may be more or less productive, more or less sustainable, more or less integrated into larger systems, etc. Whether or not having high levels for specific properties is good or not is a normative judgment that is made by human managers of the system. Different managers with different goals may evaluate such questions very differently depending on what their policy goals happen to be. For example, agricultural communities in Vietnam during the cooperative period were characterized by very high equitability but this was accompanied by extremely low productivity. After the *đổi mới* reforms returned land to peasant households to manage, however, productivity greatly increased but equitability declined. Clearly, trade-offs must be made. It is impossible to simultaneously maximize the performance of an agroecosystem for all emergent properties. And it is dangerous to concentrate on maximizing only the property of productivity, as has commonly been the case in rural development programs. Instead, it is necessary to keep the performance of the system as a whole in view in order to avoid undesirable outcomes of development interventions.

Comparison of the NMR Communities in Terms of Emergent Properties

In this section we present an analysis of each of the five communities in terms of their performance levels for each of the six emergent properties. For each property, the performance level displayed by the community is ranked as being high, medium, or low. Such a quantified approach is necessary to permit comparison among the communities. Although such quantification is essential if comparisons are to be made among the five communities, too much faith should not be placed in the rankings for any specific indicator. The data used to assess agroecosystem performance are often quite crude and not highly reliable and the assignment of rankings necessarily somewhat subjective. Use of multiple indicators to assess performance for each property increases the confidence that one can place in the overall rankings but the reader is asked to keep in mind that this is an experimental first effort to quantify agroecosystem performance.

It should also be emphasized that the ranking system is a relative one since it is based on the specific range of values displayed by these five NMR communities. Therefore, it is applicable only to rural communities in that region, and cannot be used to compare them with communities in other regions of Vietnam, or in other countries. Cross-regional or cross-national comparisons would require creation of an absolute ranking system but the data needed to create such a system are not currently available.

Each property was assessed using three or four separate indicators. For example, diversity was assessed using three indicators: plant species diversity, overall biodiversity, and the complexity of the agricultural ecosystem. For each indicator, a three-point ranking scale ranging from 1 (low) to 3 (high) was employed. A complete description of the quantitative rankings for each indicator for each of the six properties is presented in Rambo (2002). The absolute values for the scale are determined by the highest and lowest levels of performance displayed by the agroecosystems of the five communities for that indicator. An overall combined ranking for the performance of each agroecosystem for each of the emergent properties was then calculated by averaging its rankings for all the individual indicators, with values on the ranking scale ranging from a low of 1 to a high of 3. The performance of the agroecosystems in terms of productivity, diversity, sustainability, equitability, integration, and solidarity will be described in that order.

Productivity

Productivity is measured using three indicators: (1) the total yield per hectare of annual crops; (2) the total yield per ha of forest products from both natural and planted forests; and (3) the total yield per ha of livestock. Each of these indicators is described below.

1. *Productivity of annual crops*: This indicator is based on the yield of annual crops including wet rice raised in irrigated fields, maize and cassava raised on permanently cultivated dryland fields, and rice, maize, cassava, and other swidden crops.
2. *Productivity of forestland*: This indicator is based on the cash value of all timber and non-timber forest products derived from land covered with trees, including both natural and planted forests and tea gardens.
3. *Productivity of livestock*: This indicator is based on the sale price of all livestock, including cattle, buffalo, pigs, and chickens.

Because each component of the agroecosystem yields different products measured in different units it was necessary to convert all products into a common unit of value, in this case, their rice equivalent value (Leisz *et al.* 2001, Appendix 4D). Staple crops (rice, maize, cassava) were recorded by weight. Industrial crops (e.g. tea), agroforestry products, timber and non-timber forest products, and livestock were recorded according to their cash values. Maize and cassava were converted into rice equivalent values using Vietnamese Government conversion rates. The income earned from sale of cash crops, timber and NTFPs, and livestock was converted to rice equivalent value by dividing it by the market price of unhusked rice (1,500 VND/kg). The yields, in rice equivalent values (kg/ha), of all of the components were then summed to measure the total production of each of the three types of land use. These figures were then divided by the total number of hectares devoted to that land use within the community in order to determine the mean yield per ha. Table 18.3 presents the rankings for each type of land use and the combined productivity rankings for all the communities.

Khe Nóng and Thài Phìn Tủng have very low productivity, Tat has medium productivity, and Ngọc Tân and Làng Thao have very high productivity. The three indicators of productivity show a strong tendency to co-vary so that a community (e.g., Khe Nóng) that has low crop productivity also displays low productivity for forests and livestock while a community with high crop productivity (e.g., Làng Thao) also has high productivity for forests and livestock.

Diversity

Diversity of the agroecosystems is assessed using three indicators: (1) plant species diversity, (2) overall biodiversity, and (3) the number of components or sub-systems of which the agroecosystem is composed.

1. *Plant species diversity*: All wild and domesticated plant species found growing along three transects in each of the communities were enumerated and diversity indexes calculated using Simpson's formula (Southwich 1976). A higher index number indicates higher plant species diversity in the agroecosystem.
2. *Overall biodiversity*: Relative ratings of the overall biodiversity of each of the agroecosystems were assigned by a panel of three ecologists and natural resource management specialists with expert knowledge of the region. They took into account the number and rarity of plant and animal species in each vegetation community and the size of the area covered by each vegetation community. The rating scale ranges from 1 to 5, with a higher number indicating higher overall biodiversity.
3. *Agricultural ecosystem complexity*: The different agricultural ecosystems vary in their complexity. The agricultural ecosystems of Tat and Khe Nóng include many different components or subsystems whereas Thài Phìn Tủng, Ngọc Tân, and Làng Thao have simpler systems. The five communities were ranked according to the number of components, both natural and human influenced or created, of their agroecosystems. The larger the number of components, the higher the diversity ranking that is assigned to the community.

Table 18. 4 presents the diversity rankings for the communities.

Thài Phìn Tủng and Làng Thao have low diversity rankings, Ngọc Tân a medium ranking, and Khe Nóng and Tat have the highest rankings. The high

Table 18.3 Productivity rankings for the five NMR communities

	KHE NÓNG	THÀI PHÌN TỦNG	TAT	NGỌC TÂN	LÀNG THAO
Crop productivity (mean yield per ha, 1000 kg rice equivalent value)	Low (1.8)	Low (1.1)	Medium (2.6)	High (6.3)	High (4.6)
Forest productivity (mean yield per ha, 1000 kg rice equivalent value)	Low (0.004)	Low (0.0)	Medium (0.19)	High (3.2)	High (4.5)
Livestock productivity (mean yield per ha, 1000 kg rice equivalent value)	Low (0.014)	Low (0.132)	Low (0.057)	Medium (1.3)	High (2.1)
Combined productivity ranking	Low (1)	Low (1)	Medium (1.7)	High (2.7)	High (3)

Notes:
Crop productivity ranking scale: Low < 2,000 kg; Medium = 2,000–4,000 kg; High > 4,000 kg
Forest productivity ranking scale: Low < 100 kg; Medium = 100–2,000 kg; High > 2,000 kg
Livestock productivity ranking scale: Low < 1,000 kg; Medium = 1,000–2,000 kg; High > 2,000 kg
Combined productivity ranking scale: Low < 1.5; Medium = 1.5–2.5; High > 2.5

Table 18.4 Diversity rankings of the five NMR communities

	KHE NÓNG	THÀI PHÌN TỦNG	TAT	NGỌC TÂN	LÀNG THAO
Plant species diversity (Simpson's diversity index)	High (0.96)	Medium (0.87)	High (0.93)	Medium (0.71)	Low (0.45)
Overall biodiversity (judgmental ratings by ecologists)	High (4.70)	Low (1.67)	Medium (2.82)	Low (1.16)	Low (1.02)
Complexity of agroecosystem (number of components)	Medium (10)	Low (5)	High (14)	Medium (7)	Medium (7)
Combined diversity ranking	High (2.7)	Low (1.3)	High (2.7)	Medium (1.7)	Low (1.3)

Notes:
Plant species diversity index ranking scale: Low < 0.50; Medium = 0.51–0.90; High > 0.90
Overall biodiversity ranking scale: Low < 2.00; Medium = 2.01–4.00; High > 4.00
Agroecosystem complexity ranking scale: Low = 1–6; Medium = 7–10; High > 10
Combined diversity ranking scale: Low < 1.5; Medium = 1.5–2.5; High = 2.5–3.0

rankings for the latter two communities reflect the fact that they still have appreciable areas of natural forest, which has higher diversity than any other ecosystem in the region, and also because their agroecosystems incorporate more components than those of the other communities. There is a reasonable level of agreement between the three diversity indicators for each community.

Sustainability

Four indicators of sustainability are employed: (1) the extent to which vegetation cover and land use practices protect the soil from erosion; (2) the share of total production that is obtained from land uses that are known to be highly sustainable in the uplands; (3) local people's perceptions of environmental changes in their community during the past five years; and (4) local people's perceptions of declining soil fertility in their fields.

1. *Protection of soil*: A panel of three ecologists and natural resource management specialists estimated the extent to which different types of vegetative cover and land use practices protect the soil from erosion in each of the communities and assigned a soil protection rating of from 1 (the least protection) to 5 (the greatest protection) to each type of land use. These ratings were then multiplied by the percentage of the total land area covered by each type (based on vegetation and land use maps prepared from satellite images). These figures were then added together to calculate the mean soil protection rating of each agroecosystem
2. *Production from relatively sustainable land uses*: Another indicator of sustainability is the percentage of total agroecosystem production (in rice equivalent values) that is derived from wet rice fields (or, in Thài Phìn Tủng, cultivation of maize in rock pockets) and agroforestry land uses. These production systems are recognized to have high sustainability in comparison to most systems of annual dryland crop cultivation or logging of forests.
3. *Perceptions of environmental change*: The perception of local people of change in the condition of their local environment over time is another indicator of sustainability. After the interviewer read them a brief definition of the term 'environment' to make sure that they all understood the meaning of this term in the same way, respondents were asked to rate the present and past conditions of the natural environment in their community on a 10-point ladder scale. Mean values were calculated for past and present ratings in order to determine the direction and magnitude of change. People in communities with negative scores perceived that the local environment had deteriorated while those in communities with positive scores perceived that the state of their environment had improved.
4. *Perceptions of declining soil fertility*: Respondents were presented with a standardized list of environmental problems, including declining soil fertility, and asked to indicate those that occurred in their community. Communities with lower percentages of respondents reporting declines in soil fertility are ranked higher for sustainability than those with higher percentages.

Table 18.5 Sustainability rankings for the five NMR communities

	KHE NÓNG	THÀI PHÌN TỦNG	TAT	NGỌC TÂN	LÀNG THAO
Soil protection (judgmental ratings by ecologists)	High (4.72)	Low (2.05)	Medium (2.88)	Medium (3.02)	Medium (2.62)
Land use sustainability (percentage of total production derived from sustainable land uses)	Low (39.6)	Medium (63.8)	Low (21.9)	Medium (67.9)	High (83.4)
Environmental change ranking (perceived change from past to present)	Low (–0.28)	Medium (–0.14)	Low (–0.25)	High (+1.33)	High (+1.13)
Soil fertility decline ranking (percentage of respondents perceiving decline)	High (65.0)	Medium (73.3)	Low (82.5)	High (60.0)	Medium (75.0)
Combined sustainability ranking	Medium (2.0)	Medium (1.8)	Low (1.3)	High (2.5)	High (2.5)

Notes:
Soil protection ranking scale: Low < 2.5; Medium = 2.5–4.0; High > 4.0
Land use sustainability ranking scale: Low < 50%; Medium = 50–70%; High > 70%
Environmental change ranking scale: Low = –0.20 to –0.30; Medium = –0.20 to +1.00; High > +1.00
Sustainability ranking scale for decline in soil fertility: Low = 100–76%; Medium = 75–64%; High = 65% and lower
Combined sustainability ranking scale: Low < 1.8; Medium = 1.8–2.4; High > 2.4

The rankings of the communities for sustainability are presented in Table 18.5.

Tat has the lowest combined sustainability ranking, reflecting continued heavy reliance on swidden cultivation in a situation of increasing population pressure and land scarcity. Khe Nóng and Thài Phìn Tủng both have medium levels of sustainability. In the case of Khe Nóng, this is a reflection of its very low population density, while in Thài Phìn Tủng the predominant method of cultivation provides excellent protection against soil erosion and the limestone derived soils maintain their fertility for extended periods of cultivation. The high sustainability rankings of Ngọc Tân and Làng Thao are reflections of the fact that they have large areas of wet rice fields and have recently reforested most of the formerly barren hill slopes.

Equitability

Equitability is measured using four indicators: Gini coefficients for differences among households in (1) income, (2) landholdings, (3) ownership of total physical assets; and (4) the gap in income between the top and bottom quintiles of households. The Gini coefficient is a commonly used measure of economic inequality. It ranges from 0 to 1 with 0 representing perfect equality and 1 indicating perfect inequality (Keister 2000: 63). In a community with a Gini coefficient of 1, a single household would own everything. Because a high Gini coefficient indicates that the distribution of income or assets is highly skewed, communities with the lowest Gini coefficients (e.g., Khe Nóng, Thài Phìn Tủng, and Ngọc Tân) receive higher equitability rankings than communities (e.g., Làng Thao) with the highest Gini coefficients.

1. *Income differences*: This indicator uses the Gini coefficient to measure the equitability with which per capita cash income is distributed among households in the communities.
2. *Differences in landholdings*: The equitability of distribution among households of per capita area of landholdings is measured using the Gini coefficient.
3. *Differences in total physical assets*: The estimated cash values of all land, livestock, houses, and valuable possessions, such as tools and consumer durables, possessed by households are added together to calculate their total physical assets. The Gini coefficients for distribution of total physical assets among households are used to determine equitability for this indicator.
4. *Income gap between rich and poor households*: The gap between the mean cash incomes of the wealthiest and poorest quintiles of households in the communities is used as a fourth indicator of equitability. A larger gap indicates lower equitability.

Table 18.6 presents the combined community rankings for the four indicators of equitability.

All of the communities, except Làng Thao, display high or medium overall rankings for equitability. Thài Phìn Tủng has the highest combined equitability ranking. This may be a reflection of the tribal characteristics of H'Mông society where the strong lineage structure serves as a mechanism for ensuring that member households have relatively equal access to resources. Khe Nóng and Ngọc Tân also display high overall equitability while Tat has medium equitability. The low ranking of Làng Thao may reflect its unique history as a recent settlement made up of unrelated households that migrated there from the Red River Delta so that effective redistributive mechanisms may be lacking.

The rankings for each of the four indicators of equitability for each community also appear to be highly correlated, with the exception of Tat, which ranks low on two indicators and high on two others.

Table 18.6 Equitability rankings for the five NMR communities

	KHE NÓNG	THÀI PHÌN TỦNG	TAT	NGỌC TÂN	LÀNG THAO
Cash income (Gini coefficient)	Medium (0.31)	Medium (0.33)	High (0.26)	Medium (0.33)	Low (0.38)
Landholding (Gini coefficient)	Medium (0.38)	High (0.35)	Low (0.49)	High (0.34)	Low (0.45)
Total physical assets (Gini coefficient)	High (0.25)	Medium (0.27)	Low (0.31)	Medium (0.27)	Low (0.31)
Income gap (times)	Medium (11.0)	High (6.8)	High (6.1)	Medium (11.0)	Low (14.0)
Combined equitability ranking	High (2.3)	High (2.5)	Medium (2.0)	High (2.3)	Low (1.0)

Notes:
Income equitability ranking scale: Low > 0.35; Medium = 0.35–0.30; High < 0.30
Landholding equitability ranking scale: Low > 0.40; Medium = 0.40–0.36; High < 0.36
Physical assets equitability ranking scale: Low > 0.30; Medium = 0.30–0.26; High < 0.26
Income gap ranking scale: Low > 12; Medium = 8–12; High < 8
Combined equitability ranking scale: Low < 1.5; Medium = 1.5–2.0; High > 2.0

Integration

Integration refers to the extent to which local communities are involved in larger social and economic systems. This property is assessed using four indicators: (1) the literacy rate for the population; (2) the extent to which household economies are subsistence-oriented; (3) the percentage of households with members engaged in non-farm occupations; and (4) the percentage of households having cash income derived from government transfer payments (salaries and pensions for cadre, development assistance).

1. *Literacy rate*: Ability to read and write is a necessary condition for individuals to participate fully in the larger social system so that the literacy rate is a useful indicator of integration. The rates are calculated for members of the population aged seven-years or older. Communities with low literacy rates are ranked low for integration.
2. *Extent of subsistence orientation of household economies*: The ratio of in-kind to cash income is an indicator of the extent of community involvement in the market economy. A high ratio indicates that the community is heavily subsistence-oriented whereas a low ratio indicates stronger market involvement and, thus, greater integration into larger systems.
3. *Non-farm employment*: The share of households having at least one member engaged in full-time non-farm employment is an indicator of the

extent to which the economy of the community is integrated into the larger economic system

4. *Government transfer payments*: The number of households that obtain part of their income in the form of transfer payments from the government in the form of salaries paid to local cadre, pensions paid to retired cadre and veterans, and assistance provided by various development projects, is another indicator of involvement in larger systems. The percentage of households receiving government transfer payments is the indicator of the extent of integration.

Table 18.7 presents the combined integration rankings for the five communities.

Khe Nóng has the lowest level of integration into larger systems, Thài Phìn Tủng has a somewhat greater degree of integration, while Tat, Ngọc Tân, and Làng Thao all display high levels of integration. Correlation among the several indicators for each community is reasonably strong with the exception of Thài Phìn Tủng.

Solidarity

Solidarity is measured in terms of three indicators: (1) the extent to which inter-household relations within the communities are friendly and cooperative; (2) the occurrence of quarrels among neighbors; and (3) the occurrence of fights among community members.

1. *Friendly and cooperative relations among households*: Respondents were asked to state whether they thought that relations among households in their community were friendly and cooperative, friendly but not cooperative, or neither friendly nor cooperative. Communities with a high percentage of respondents who stated that inter-household relations in their community were friendly and cooperative are ranked as having high solidarity.
2. *Quarrels among neighbors*: Respondents were asked if neighbors ever quarreled in their community. Communities with a higher percentage of respondents stating that quarrels occur among neighbors are rated lower for solidarity.
3. *Fighting among members of the community*: Respondents were asked if physical fights ever occurred in their community. Communities having higher percentages of affirmative responses are rated as lower in solidarity.

Table 18.8 presents the combined solidarity ratings for the five communities.

Solidarity is very high in Khe Nóng and Thài Phìn Tủng, somewhat lower in Ngọc Tân, and very low in Tat and Làng Thao. The low solidarity found in Tat and Làng Thao may be explained by their unique historical experiences. Tat is a Tày settlement that was subject to the despotic rule of a Mường

Table 18.7 Integration rankings for the five NMR communities

	KHE NÓNG	THÀI PHÌN TỦNG	TAT	NGỌC TÂN	LÀNG THAO
Literacy rate (%)	Low (13.0)	Low (13.0)	High (92.0)	High (97.0)	High (97.0)
Ratio of in-kind to cash income	Low (1.8)	High (0.3)	Medium (0.9)	High (0.6)	High (0.3)
Non-farm employment (Percentage of households with at least 1 member engaged in full-time non-farm employment)	Low (0.0)	Medium (10.0)	High (26.2)	High (30.0)	High (20.0)
Government transfer payments (Percentage of households receiving payments)	Low (10.0)	Low (20.0)	Medium (33.0)	Medium (30.0)	High (57.5)
Combined integration ranking	Low (1.0)	Medium (1.8)	High (2.5)	High (2.8)	High (3.0)

Notes:
Literacy ranking scale: Low < 50%; Medium = 50–80%; High > 80%
Ratio of in-kind to cash income ranking scale: Low > 1:1; Medium = 1:1–0.7:1; High < 0.7:1
Non-farm employment ranking scale: Low < 10% Medium = 10–19%; High > 19%
Transfer payments ranking scale: Low < 29%; Medium = 30–50%; High > 50%
Combined integration ranking scale: Low < 1.5; Medium = 1.5–2.4; High > 2.4

Table 18.8 Solidarity rankings for the five NMR communities

	KHE NÓNG	THÀI PHÌN TỦNG	TAT	NGỌC TÂN	LÀNG THAO
Inter-household relations friendly and cooperative (Percentage of respondents agreeing)	High (90.0)	High (92.0)	Medium (83.0)	High (88.0)	Low (62.0)
Quarrels with neighbors (Percentage of respondents reporting occurrence of quarrels)	Medium (20.0)	High (5.0)	Low (33.0)	Medium (20.0)	Low (23.0)
Fighting inside community (Percentage of respondents reporting occurrence of fights)	High (5.0)	High (0.0)	Low (30.0)	Medium (28.0)	Low (35.0)
Combined solidarity ranking	High (2.7)	High (3.0)	Low (1.3)	Medium (2.3)	Low (1.0)

Notes:
Inter-household relations ranking scale: Low < 70%; Medium = 70–85%; High > 85%
Quarrels with neighbors ranking scale: Low > 20.0%; Medium = 10.0–20.0%; High < 10.0%
Fighting ranking scale: Low = 30.0% +; Medium = 10.0–29.0%; High < 10.0%
Combined solidarity ranking scale: Low < 2.0; Medium = 2.0–2.5; High > 2.5

overlord during the colonial period so that it failed to develop effective internal structures of governance (Rambo and Trần Đức Viên 2001a, Chapter 11 in this volume) while Làng Thao is a recently-formed migrant settlement made up of unrelated households originating from different places in the Red River Delta. Làng Thao is also more subject to strains resulting from rapid social change due to its tight integration into the larger economy and society. Khe Nóng, Thài Phìn Tủng, and Ngọc Tân, on the other hand, are long-settled communities whose inhabitants are bound together by strong ties of kinship and co-residence.

Tat Hamlet in Comparative Perspective

The rankings achieved by the five communities for all six of the emergent properties are presented in Table 18.9. Figure 18.3 presents a graphic comparison of the performances of the communities.

Tat hamlet achieves high rankings for diversity and integration, medium rankings for productivity and equitability, and low rankings for sustainability and solidarity. The low ranking for solidarity is a reflection of the particular historical experience of the people of Tat hamlet and is not attributable to their reliance on composite swiddening. Tat's low ranking for sustainability, however, is somewhat surprising in the light of some of the findings about the composite swidden agroecosystem presented in preceding chapters. But it should be recognized that this low ranking is in part an artifact of the limited number of types of agroecosystems that are included in the small sample. Tat and Khe Nóng are the only communities in the sample that are heavily dependent on swidden agriculture. In the case of Khe Nóng, which has a medium ranking for sustainability, the area of swidden fields is tightly restricted by the Pù Mát Nature Reserve authorities so that most of the territory of the hamlet remains under undisturbed forest cover, which explains why it achieves higher scores than Tat on most of the sustainability indicators. Thài Phìn Tủng, Ngọc Tân, and Làng Thao all have permanent forms of cultivation that minimize soil erosion and so also achieve higher rankings. Thus, in comparison to these four communities, Tat hamlet does indeed have the lowest level of sustainability. Unfortunately, however, our sample did not include any communities wholly dependent on pure swidden agriculture. It would be necessary to compare Tat hamlet to such pure swiddening communities in order to be able to assess the relative sustainability of composite swiddening versus pure swiddening agroecosystems.

Table 18.9 Combined rankings for the emergent properties of the five NMR communities

EMERGENT PROPERTIES	KHE NÓNG	THÀI PHÌN TỦNG	TAT	NGỌC TÂN	LÀNG THAO
Productivity	Low (1.0)	Low (1.0)	Medium (1.7)	High (2.7)	High (3.0)
Diversity	High (2.7)	Low (1.3)	High (2.7)	Medium (1.7)	Low (1.3)
Sustainability	Medium (2.0)	Medium (1.8)	Low (1.3)	High (2.5)	High (2.5)
Equitability	High (2.3)	High (2.5)	Medium (2.0)	High (2.3)	Low (1.0)
Integration	Low (1.0)	Medium (1.8)	High (2.5)	High (2.8)	High (3.0)
Solidarity	High (2.7)	High (3.0)	Low (1.3)	Medium (2.3)	Low (1.0)

Figure 18.3 Profiles of emergent property rankings of the five NMR communities

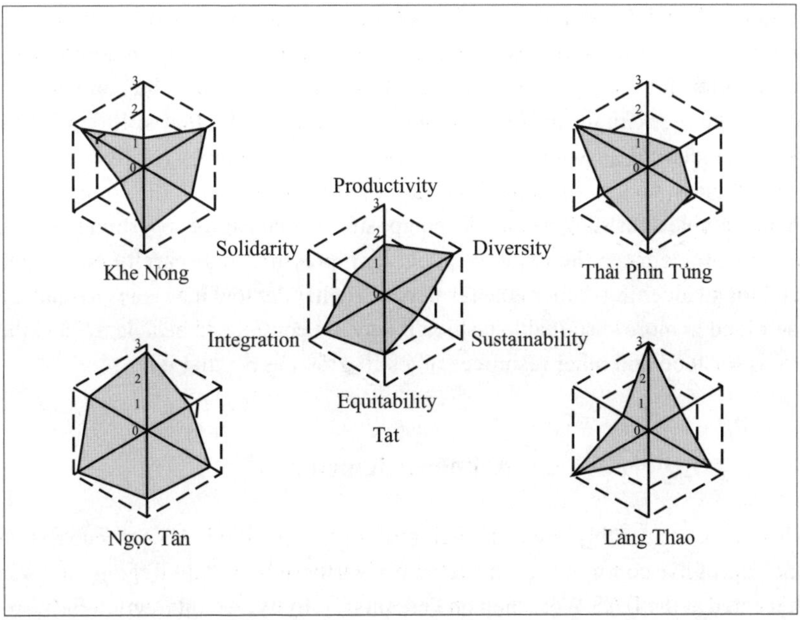

Conclusions

In terms of its overall development status, Tat hamlet scores relatively high in comparison to the other communities. It achieves much higher ranking on most development indicators than Khe Nóng and Thài Phìn Tủng, and only lags somewhat behind top-ranking Ngọc Tân and Làng Thao. It performs very well in terms of food security, education, access to health care, adoption of family planning, access to safe water, sanitation, access to electricity, and travel outside of the village, but does not perform very well with regard to health status, cash income, and ownership of bicycles or motorbikes. That it performs as well as it does is somewhat surprising given its relatively remote location and lack of good agricultural land.

Its performance in terms of the various emergent properties is quite a mixed one. It ranks high for diversity and integration, has medium rankings for productivity and equitability, but has low rankings for sustainability and solidarity. Its low ranking for productivity is not surprising, given the land-extensive nature of composite swidden agriculture. Tat's low ranking for solidarity is probably an accurate reflection of the character of its social system. Few other communities in the NMR are so weakly organized and so lacking in internal social capital. Its low ranking for sustainability, however, is more problematic. As was already noted, this ranking is valid only in comparison to the other four sample communities. If Tat hamlet were to be compared to any of the many thousands of communities in the NMR that still depend on pure swiddening, it would certainly achieve a much higher relative ranking. Nevertheless, its low sustainability ranking in comparison to the communities in our sample does suggest that the composite swidden agroecosystem is facing serious problems as the result of population pressure and scarcity of suitable land for swiddening. Unless the farmers of Tat hamlet find new ways to manage their land in more sustainable fashion it may not continue to be able to meet the needs for food and other resources of a still growing population.

Acknowledgments

This is a considerably revised version of a paper entitled 'An agroecosystem analysis of five communities in Vietnam's Northern Mountain Region' that was presented at the IIAS Workshop on Peasants' Activity: An Interaction Between Culture and Nature held in Leiden, 28-30 August 2002 (Rambo 2002).

Funding for this research was provided by grants to the Center for Natural Resources and Environmental Studies (CRES) of the Vietnam National

University, Hanoi, from the Swedish International Development Cooperation Agency (SIDA) and the Ford Foundation and to the East-West Center from SIDA, the Ford Foundation, the John D. and Catherine T. MacArthur Foundation, and the Keidanren Nature Conservation Fund. Field research was carried out between December 1998 and March 2000. Detailed findings are presented in the project final report (Lê Trọng Cúc and Rambo 2001).

Terry Grandstaff, Neil Jamieson, Gerald Marten, Gary Murfin, Thomas Sikor, Tanaka Koji, Patma Vityakon, and Yanigasawa Masayuki made many helpful criticisms and suggestions for improvement of earlier drafts of this paper. Đào Minh Trường prepared the maps and diagrams. The views expressed in this paper are those of the authors and are not necessarily shared by any of the above named organizations and individuals.

19
Forest Interrupted: Tat Hamlet and the Political Ecology of Swidden Agriculture

Jefferson Fox

Introduction

The hills along the road to Tat hamlet are steep and covered with a mosaic of forest patches, agricultural fields, and a variety of shrub, bamboo, and grass plots. The product of a long history of swidden agriculture, this landscape resembles that in many parts of montane mainland Southeast Asia, an extensive region that covers the western half of Vietnam, most of Laos, and large areas of Northern Thailand, Southern Yunnan, and Eastern Myanmar (Figure 19.1).

Since the colonial period, agronomists and government officials throughout Asia have been convinced that swidden cultivators are irresponsible pyromaniacs destroying forests and endangering soil fertility (see Jewitt 1995; Bryant 1997; Schmidt-Vogt 1998). Captain P. Cupet, a member of the Pavie Mission to Indochina between 1879 and 1895, for example, wrote,

> These savages are the greatest destroyers of forests I know. One never meets them except in the vicinity of the highest crests, on elongated hilltops offering a great surface of arable land.... The terrain is almost everywhere deforested and covered with rai [mountain rice swiddens] (Cupet 2000: 35-40).

For decades, government officials and foresters have cursed swidden farmers for destroying valuable forest cover (Myers 1984). Nature conservationists have campaigned for the eviction of mountain dwellers whose land-use practices they claim destroy tree cover, degrade soils, increase runoff, and cause downstream floods (ASB 1994; Myers 1992). Recently, swidden agriculture has been blamed not only for destroying biodiversity and soil fertility, but also for global warming (Bandy, Garrity, and Sanchez 1993; Adger and Brown 1994).

Figure 19.1 Map of Montane Mainland Southeast Asia

The scientific knowledge base used to make these claims, however, is coming under increasingly critical scrutiny. Many of these claims are based on environmental 'myths' or orthodoxies that are highly overstated. This is particularly true in relation to forest ecology, soil erosion and degradation, hydrology, and fire impacts. Many claims by enthusiastic foresters and conservationists are often poorly supported by evidence and often have negative consequences for the people and ecosystems involved (Hamilton 1988; Rocheleau and Ross 1995; Fairhead and Leach 1998; Calder 1999; Lindert 2000). Recent research points to insufficient empirical data in many cases to support specific scenarios of degradation, in particular, an absence of long-term data needed to determine directional trends, a failure to separate the impacts of naturally occurring processes from those induced by human activities, and a failure to distinguish seasonal from permanent changes in vegetation cover (Hamilton 1987; Ives and Messereli 1989; Agarwal and Chak 1991; Saberwal 1998; Fox *et al.* 2000).

Rambo (1998) defines composite swiddening as an agricultural system that integrates permanent wet rice fields and rotating swidden plots into a single household resource system. In the eyes of most people, however, composite swiddening and swiddening (without wet rice fields) look identical. Experts estimate that swidden cultivation (both composite and non-composite) is practiced on about 30 percent of the world's exploitable soil and roughly one half of the total land area in the tropics (Andriesse and Schelhaas 1987). The total number of people engaged in some form of swidden agriculture has been loosely estimated to be about one billion people. Approximately 22 percent of the population of the developing world in tropical and subtropical countries relies directly or indirectly on some form of swidden cultivation. This final chapter seeks to explore the environmental myths related to swidden agriculture and to place this ancient land-use practice (in both its composite and non-composite forms) into a richer and more nuanced context.

Forsyth (2003) and others (Fairhead and Leach 1996; Braun 2002) suggest that environmental orthodoxies are not solely the result of scientific errors or of technological progress in various scientific fields producing new 'truths'. These authors suggest that 'truth' or scientific 'laws' only hold true under given epistemic boundaries, and that these 'truths' may be contradicted when empiricism on the same topics is conducted under a different set of boundaries. 'Alternative framings of environmental problems represent a further insight into biophysical reality from a new perspective, and can provide a more accurate (and usually more socially relevant) explanation of change than the larger-scale generalization' (Forsyth 2003: 217). Many environmental orthodoxies or 'factual' explanations of environmental problems are highly problematic. They underestimate biophysical uncertainties and often fail to recognize how people value environmental change. This does not deny that environmental problems exist, nor legitimize destructive practices, but rather it seeks to understand the complex social and political influences upon how we perceive and explain environmental problems, and then accept such explanations.

The research described in this book sought to engage natural and social scientists in seeking an understanding of the human ecology of composite swidden agriculture, especially its impacts on the natural environment, in a single hamlet in the mountains of northern Vietnam. More broadly these authors were interested in the social dynamics of this indigenous agricultural system, in its impacts on land-use and vegetation cover, and in how different inhabitants understood this process and responded to it technically and socially. A fundamental element of this study has been to establish with as much precision as possible the course, pattern, and causes of vegetation change in Tat hamlet's forest mosaic. Clearly, this first involves attention to evidence of vegetation change, which researchers compiled from documentary, photographic, and oral accounts during the last half century. It also involves attention to the current conditions on the ground, to specific land management techniques, grounded in farmers' ecological knowledge, and to the social and economic relations that have conditioned land use.

Swidden Landscapes and Socialized Nature

This book adds to a now large and growing body of literature that documents the agroecological knowledge of Asian farmers. This literature also demonstrates the sophistication of local land, soil and vegetation management techniques and the wisdom and creativity of farmers (Conklin 1957; Kunstadter, Chapman, and Sabhasri 1978; Spencer 1966; Schmidt-Vogt 1998; Padoch, Harwell, and Susanto 1998; Dove 1985; Nibbering 1991; Ramakrishnan 1992). When considering the ecological concepts and explanations employed by the residents of Tat hamlet, however, it has not always been straightforward to understand these in terms familiar to western scientific literature on forest ecology. This is due in large part to the different assumptions made by scientists and local inhabitants concerning the relationship between social and ecological processes (Fairhead and Leach 1996).

Since the Enlightenment western scientists have assumed that nature can be investigated as something separate from human society. 'Nature' thus became the essential reference point against which to discuss human impact. Scientists and policy makers hence view swidden systems and their landscapes of successional vegetation as 'degraded,' and closed-canopy forests as the natural 'climax' vegetation within an otherwise abused landscape. Swidden cultivators are widely perceived to be primitive, backwards, unproductive, wasteful, and exploitative and destructive of natural resources and the environment. These beliefs led to many current national laws and policies that are antagonistic toward swidden cultivators and aim to replace swidden cultivation with forms of sedentary agriculture considered to be more modern. Negative attitudes toward shifting cultivators are also prevalent in many research and development institutions in both hemispheres.

In contrast, the arguments in this book turn on the concept of *social nature*—a term that indicates the inevitable intertwining of society and nature in any and all social and ecological projects (Braun 2002; Harvey 1996), rejecting the view that nature is something separate from humans. What we see as 'natural' internalizes not only ecological relations but social relations too. The idea that nature is separate from human society leads to arguments for limits to growth and carrying capacities (Ehrlich 1968; Devall and Sessions 1985) and the rejection of swidden agriculture (Myers 1984). Lost in this language of limits are the social and political causes of scarcity, as well as humanity's creative capacity to adapt to nature (Braun 2002). Some environmentalists reject the idea that nature is socially constructed because they fear that it undermines efforts to 'save' nature, and will give free rein to the transformation of nature by humans (Soule and Lease 1995). Against this view, we argue that seeing the world in terms of nature-culture 'hybrids,' presents the best way to understand present-day social and environmental conditions and can lead us towards a more ecologically sustainable and socially just future.

At the same time as social scientists have been rethinking the separation of nature and society, and developing the concept of 'socialized nature', ecologists have undergone a shift in fundamental assumptions about the orderly functioning of biophysical systems and systems ecology. New attention is now given to issues of inherent instability, non- and disequilibria, chaotic fluctuations and the role of specific catastrophic events in ecosystems. This new interpretation of nature undermines classical representations of ecosystems in terms of tendency to equilibrium, and of universal laws—such as of plant succession to climax vegetation. Sometimes called 'new' or 'non-equilibrium' ecology (Hollings 1973, 1979; Wiens 1977; Botkin 1990; Zimmerer 2000), this new view of nature still perceives the world within the traditional nature-society dichotomy. Nevertheless, it denies the vision of society and ecology as in functional equilibrium. Correspondingly, it compromises managerial perspectives that assume that society can predict and manage ecological relations for stability. New ecology, however, provides conceptual space for reevaluating vegetation dynamics in landscapes dominated by swidden systems. When freed from its nature-versus-society straitjacket, this approach enables a better understanding of local reasoning and decision-making, and leads to a reappraisal of people's roles in managing environmental stability or directing change.

Political Ecology of Land Management

To understand historical patterns of landscape change, it is necessary to pay attention not only to the technical aspects of land management but also to land-use practices within the broader social and political relations that influence how people deal with land and vegetation, and, more broadly, how regional economic, political and demographic changes shape these local dynamics. Here, we draw on a large and useful body of literature on agrarian evolution in Asia which investigates how land use is shaped by the dynamics of resource access, use and control, setting these firmly within the context of social form and process (Agrawal and Ostrom 2001; Brosius and Tsing 1998; Ganjanapan 1994; Johnson and Forsyth 2002; Li 2002; Peluso 1992; Vandergeest and Peluso 1995).

Tenure, labor, and resource control patterns can shift over time, along with broader social institutional arrangements, with major effects on material, land and vegetation use. Shifts can result from changing demographic conditions, or changing local and regional political and economic relationships. Considering how regional, national or international processes interact with local ones over time is also a shared feature of various different approaches in political ecology (Rigg and Sakunee 2001; Hirsch 1993; Li 2002; Robbins 1998).

In Tat hamlet, the past half-century has seen major changes in the wider political economy. As Rambo and Trần Đức Viên point out in Chapter 11, within the span of a single life, the villagers have experienced four radically different forms of social organization. These periods are the feudal period, the resistance war period, the cooperative period, and the current period in which the household is the key unit for management of land and natural resources. In each of these periods the balance of power between the individual households, local community governance institutions, and extra-local governance institutions has varied with consequent differences in the way in which resources have been managed.

The 1993 Land Law and subsequent implementing laws provided a legal framework for land tenure, use, and transactions in Vietnam that atttempted to facilitate more intensive land use. The law affirms that while land is the property of the state, it may be leased to economic organizations, the armed forces, state offices, polticial and social organziations, and to households and individuals for stable long-term use. Annual cropland is allocated for 20 years and perennial cropland (forests) for 50 years. Households or individuals receiving land are given rights to exchange, transfer, lease, mortgage, and pass on the land for inheritance. The state, however, retained the right to recover land if it is not used according to its assigned purpose or is needed for some purpose of national interest. For many upland farmers in Tat hamlet and elsewhere, the Land Law may have provided increased tenure security but it constrained their right to continue practicing swidden agriculture.

Forest Fragmentation and its Consequences

With a long history of swidden agriculture, the land surface of montane mainland Southeast Asia is dominated by secondary forest across the full spectrum of the various stages of forest growth. Brown and Lugo (1990) define secondary forests as those resulting from human disturbance of forest vegetation (e.g., logging, swiddening, forest products collecting). Gomez-Pompa and Vasquez-Yanes (1974) refer to the present as the 'era of secondary vegetation' because most tropical countries have larger areas of secondary forests than primary ones. Brown and Lugo (1990) state that secondary forests cover more than 600 million ha in the tropics, and that managers and the public often disregarded them as useless brush. Potter, Brookfield, and Byron (1994) suggest that as much as 26 percent of all land in Southeast Asia falls into the 'other' category where other is defined as shrub, brush, pasture, waste and other land-use categories, many of which are actually different successional stages of secondary forests.

The almost inevitable consequence of human disturbance and a common characteristic of secondary forest is habitat fragmentation. Laurance and Bierregaard (1997) suggest that the fragmented landscape is becoming one of the most ubiquitous features of the tropical world. FAO (1993) points out that the pattern of forest fragmentation varies in different regions of the world. Forest conversion and fragmentation in the Brazilian Amazon, for example, has developed along a network of roads resulting in a fishbone pattern (Skole and Tucker 1993; Whitmore 1997). In Southeast Asia, patterns of change in land cover differ significantly from those of tropical America, although many of the underlying social and economic factors are very similar. In Vietnam, remaining forests increasingly persist as islands in a 'sea' of agricultural land (Collins, Sayer, and Whitmore 1991). In northern Thailand, Fox *et al.* (1995) found that between 1954 and 1976 the number of forest patches in a 10,000 square hectare area increased from 33 patches (with mean patch size of 219 ha) to 64 patches (with mean patch size of 84 ha).

The phenomenon of forest fragmentation has been studied in some detail with regard to the ecological effect of such change, especially on biodiversity and species composition of forest ecosystems (e.g., Schelhas and Greenberg 1996; Laurance and Bierregaard 1997). Forest fragments often provide a wide array of benefits to rural people who in turn value, protect, and frequently manage these forest patches in very productive ways (Schelhas and Greenberg 1996).

Nevertheless, the economic role of forest fallows has often been overlooked by social scientists and until recently has rarely been taken into consideration in rural development programs (Alcorn 1996; Schelhas 1996; Browder 1996; Lebbie and Schoonmaker Freudenberger 1996; Pinedo-Vasquez and Padoch 1996). Similarly, comparatively little work has yet been done to extend these investigations to an analysis of the effects, direct and indirect, on the hydrological functions of forests. In northern Vietnam it is clear that what were once dense forests have been converted into a mosaic of fields, pastures, and forest patches, but important questions remain concerning how and to what extent these changes in land cover have affected watershed functions.

Hydrological effects of forest fragmentation depend on the size of the fragments, the degree of fragmentation, and the distribution of different types of fragments on the landscape. For a given area of fragments, the smaller the fragments, the greater the fraction of remaining forest area in close proximity to a forest edge. The discontinuity in land surface properties across a forest edge gives rise to microclimatic effects extending toward the interior of a forest fragment (Collins and Pickett 1987; Murcia 1995). This edge effect acts to enhance the impacts attributed to land-cover change on climatic and biological processes, inflating the effective areal extent of deforestation. Furthermore, the spatial heterogeneity of land surface characteristics resulting from forest fragmentation can induce changes in mesoscale air circulation, convection, cloud cover, and rainfall (Anthes 1984; Yan and Anthes 1987; Segal *et al.* 1988; Giorgi 1989; Henderson-Sellers *et al.* 1993). Microclimate variations in vapor pressure deficit, photosynthetically active radiation, and air temperature have been found to extend 30 to 60 meters into forest fragments (Kapos 1989; Williams-Linera 1990; Matlack 1993; Young and Mitchell 1994). The edge-effect impacts vary with fragment shape (Laurance and Yensen 1991), topography, latitude, and time of year (Kapos 1989; Matlack 1993; Young and Mitchell 1994).

The effects of land-cover fragmentation on surface hydrology are dependent on the relationship between the spatial land-cover pattern and the topography. In a fragmented landscape, the delivery efficiency of Horton overland flow (HOF) may be reduced wherever a low-infiltration (HOF-generating) land-cover patch lays directly upslope of a high-infiltration patch. Some or all of the runoff (and sediment) generated on the upslope patch may be infiltrated (and deposited) in the lower patch. The basin-wide effect of this process depends on the frequency with which HOF-generating patches lie upslope of high-infiltration patches. This is a function of both the degree of fragmentation and the location of fragments within the watershed. For example, in a landscape with the lowest degree of fragmentation, each land class would be confined to a single contiguous patch. In comparison with a highly fragmented landscape, the contiguous patch would have relatively short upslope or downslope boundaries with contrasting land cover classes. Two key questions arise from this hypothesis: are land-cover patches related to topography, and if so, does the resulting distribution enhance or reduce the buffer effect?

Methods for Diversifying and Localizing Environmental Science

A fundamental contradiction faces researchers interested in both the construction of social and environmental reality and the material conditions to which they are linked. Robbins (1998: 71) suggests:

> That if metaphors underlie the landscapes as texts, then an interrogation of these is prerequisite to understanding the landscape. If on the other land, land provides the resources for production, then measurement of material land degradation is necessary for understanding the workings of the social and political economy.

How can an empirical study of landscape change do both simultaneously? How does an analysis of human/environment interaction proceed if it wishes to embrace claims to the historicity of scientific categories, including those of ecological science, while retaining the 'observation language' of ecology that might be abandoned in 'theoretical anarchism' (Robbins 1998: 71)?

Numerous studies have recognized the need for integrating biophysical explanations with social and political contextualization of environmental problems. Fairhead and Leach (1996), for example, compared a variety of satellite images with local oral histories to question orthodox beliefs about the prevalence and causes of deforestation in West Africa. Robbins (1998) developed a GIS database to compare the implications of different forest classification techniques on the identification of 'forest' and 'forestland in northwest India. Forsyth (1996) used a variety of sources including oral histories, historical aerial photographs, GIS, and historic measurements of soil erosion to follow what happened after the establishment of a village in a mountainous region of northern Thailand. The aim of these 'hybrid' studies (Forsyth 2003) is not to uncover biophysical change in a final, complete, and realistic manner. It is rather to reveal how scientific discussions of land change parallel the experiences of local people and their landscape and what this tells us about the multiple causes and drivers of land-use and land-cover change. To promote the economic and social conditions conducive to ecologically sound development, public policy must be based on a better understanding of what drives changes in land use.

The research in this book sought to integrate biophysical explanations with social, political and economic explanations of environmental change in Tat hamlet. Throughout the study researchers faced tension between conducting an analysis that scrutinized the specific historical roots of categories used to describe the natural world, and an analysis that simultaneously quantified these same categories. This approach draws into question the assumption of an unproblematic linkage between land use and land cover in global change models (Geist and Lambin 2002). Tensions also exist between this approach and those outlined by Rindfuss *et al.* (2003) for linking micro-level human behavior and remotely sensed data in a primarily quantitative manner.

Like other studies of swidden cultivation in Southeast Asia, researchers in this book conclude that the moist tropical forests of this region are impressively resilient. Under the right conditions, these forests regenerate and gradually recover after both human induced disturbances (swidden cultivation, clear-cutting, conversion to plantation) and natural events (hurricanes, landslides, lightening) (e.g., Schmidt-Vogt 1998; Fox et. al. 1995; Cannon, Peart, and Leighton 1998; Lawrence, Peart, Leighton 1998). Ironically, it has been precisely because farmers have continued to practice swidden agriculture during this time period that secondary vegetation, with the potential for regenerating into closed-canopy forests, still dominates the landscape of the region. *The biggest threat to forests is not swidden agriculture, but the conversion of swidden fields to commercial agriculture. This suggests, again ironically, that the best means of preserving forest is to preserve and support swidden agriculture.*

Appendix

- Appended Table 5.A1: List of plants identified in fallow fields in Tat hamlet with information on their utilization by the villagers
- Appended Table 5.A2: Genus and species of plants in Tat hamlet fallow fields according to family names (Alphabetic order)
- Appended Table 5.A3: Plant species found in fields in different fallow stages
- Appended Table 8.A1: Useful plants species grown in home and tree gardens in Tat hamlet in 1998 (n = 15) and 2004 (n = 30)
- Appended Table 9.A1: Checklist of plant species in secondary-evergreen broad leaf forest of medium to tall trees (compiled by Le Tran Chan)
- Table 9.A2: Checklist of plant species in bamboo forest mixed with woody trees (compiled by Le Tran Chan)
- Appended Table 15.A1: Nutrient balance of the swidden fields (kg/ha/year)
- Appended Table 15.A2: Nutrient balances (kg/0.102 ha/year) in paddy fields adjacent to upland stream outlet
- Appended Table 17.A1: Rules used for classifying general farming system type by commune
- Appended Table 17.A2: Land cover, landscape, topographic and population density information for each of the general farming system types identified in the NMR

Appended Table 5.A1 List of plants identified in fallow fields in Tat hamlet with information on their utilization by the villagers

No.	Scientific name	Vietnamese name	Đà Bắc Tày name	Function
	I. Psilotophyta	**Ngành Khuyết lá Thông**		
	Psilotaceae Kainitz			
1.	*Psilotum nudum* (L.) P. Beauv.[a]	Không hạt lá thông		The whole plant is used for treating injury; to reduce rheumatism
	II. Lycopodiophyta	**Ngành Thông Đất**		
	Lycopodiaceae P.Beauv. ex Mirb.			
2.	*Lycopodiella cernua* (L.) Pie	Thông đất	Nha mung ước	Ornamental plant
	Selaginellaceae Mett.			
3.	*Selaginella argentea* (Hook. et Grev.) Spring	Quyển bá bạc		Ornamental plant
4.	*Selaginella biformis* A.Braun. ex Kuhn	Quyển bá hai dạng		Ornamental plant; treating burns
5.	*Selaginella dolichoclada* Alston	Quyển bá nhánh dài		
6.	*Selaginella pennata* (D.Don) Spring			
7.	*Selaginella picta* A.Braun. ex Baker	Quyển bá đốm		The whole plant is used for treating haemoptysis (a disease that symptoms include coughing up blood from the lungs)
8.	*Selaginella rolandi-principis* Alston	Quyển bá hoa đá		
9.	*Selaginella uncinata* (Desv.) Spring	Quyển bá móc		Ornamental plant; whole plant is used to treat hepatitis
	III. Polipodiophyta	**Ngành Dương xỉ**		
	Adiantaceae (C. presi) Ching			
10.	*Adiantum capillus-veneris* L.	Ráng vệ nữ		Whole plant is used for treating coughs, improving flow of milk for lactating women, and hemorrhages
11.	*Adiantum capillus-veneris var. laciniata* C.Chr.	Ráng		

12.	*Adiantum soboliferum* Wall. ex Hook.	Ráng vệ nữ có chồi		
13.	*Cheilanthes chusana* Hook.	Ráng có môi chu		Whole plant is used for treating dysentery, painful urination and pimples
14.	*Cheilanthes subrufa* Baker	Ráng có môi hung		
15.	*Doryopteris ludens* (Hook.) J.Sm.	Ráng đinh ba thường		
16.	*Onychium lucidum* Spring	Ráng đề xi sáng		
	Aspleniaceae Mett.ex A.B. Frank			
17.	*Asplenium normale* D.Don	Tổ điểu thường		Whole plant is used to treat injury, dysentery, snakebite
18.	*Asplenium sublaserpitiifolium* Ching ex Blume et Ching	Tổ điểu lá gần thìa là		
19.	*Asplenium tenerum* Forst.	Tổ điểu nhỏ		
20.	*Diplazium megaphyllum* (Bak.) Christ.	Ráng song bào lá to		
	Azollaceae Wettst.			
21.	*Azolla caroliana* Willd.	Bèo hoa dâu Carolin		
22.	*Azolla pinata* R.Br.	Bèo hoa dâu		
	Cyatheaceae Kaulf.			
23.	*Cyathea podophylla* (Hook.) Copel.	Ráng gỗ dầy		Tender leaves are used as vegetable
	Gleicheniaceae (R.Br.) C. Presl			
24.	*Dicranopteris dichotoma* (Thunb.) Benth.	Guột		Whole plant is used for treating kidney diseases
25.	*Dicranopteris linearis* (Burm.) Underw.	Tế thường		Root is used for preventing parasitic worms
26.	*Dicranopteris splendida* (Hand.-Mazzer.) Tagawa	Tế lộng lẫy		
	Lindsaeaceae			
27.	*Lindsaea orbiculata* (Lam.) Mett. ex Kuhn		Quạt xoè tròn	Leaves and stems are used for treating dysentery

Appended Table 5.A1 Continued

No.	Scientific name	Vietnamese name	Đà Bắc Tày name	Function
	Marsileaceae Mirb.			
28.	*Marsilea crenata* J.Presl	Rau bợ răng		
29.	*Marsilea quadrifolia* L.	Rau bợ thường		Whole plant is used for treating leucorrhoea (commonly known as white discharge from the vagina, is due to excessive production of normal vaginal secretions)
	Polypodiaceae Bercht. & J. presl			
30.	*Drynaria bonii* C.Chr.	Tắc kè đá bon	Co cút cam	Stumps and roots are used to reduce backaches and rheumatism
31.	*Drynaria fortunei* (Mett.) J.Sm.[a]	Tắc kè đá fóoctan	Co cút cam	Stumps and roots are used to reduce backaches and rheumatism
	Pteridaceae Spreng ex Jamson			
32.	*Pteris biaurita* L.	Ráng sẹo gà hai tai		Whole plant is used for treating pimples and injury
33.	*Pteris ensiformis* Burm.f.	Ráng sẹo gà hình gươm		Whole plant is used for treating dysentery, malaria, and as diuretic
34.	*Pteris longipes* D.Don	Ráng sẹo gà dài		
35.	*Pteris semipinnata* L.	Ráng sẹo gà nửa lông chim		Whole plants is used as antiseptic, detoxicant, and to treat itch
36.	*Pteris linearis* Poir	Ráng sẹo gà dải		Whole plant is used to treat rheumatism
	Salviniaceae T. Lestib.			
37.	*Salvinia cucullata* Roxb.	Bèo tai chuột		
	Schizeaceae Kaulf.			
38.	*Lygodium flexuosum* (L.) Sw.	Bòng bong lắt léo		Stem is used for haematuria (the appearance of blood in the urine) and as a diuretic
39.	*Lygodium japonicum* (Thunb.) Sw.	Bòng bong Nhật		Leaves are used for treating kidney stones, encephalitis
40.	*Lygodium polystachyum* Wall. ex Moore	Bòng bong nhiều nhánh		Whole plant is used as a diuretic and antipyretic

	Thyrsopteridaceae			
41.	*Cibotium barometz* (L.) J.Sm.[a]	Cầu tích	Cút cu li	Stem and root are used for treating backaches. Hair on the stem is used as hemostatic
42.	*Thelypteris triphylla* (Sw.) Iwats.			
	Thelypteridaceae			
	IV. Pinophyta	**Ngành Thông**		
	Gnetaceae Lindl.			
43.	*Gnetum montanum* Markgr.	Dây gắm	Chưa muổi	Stem is used for making rope, crossbow strings, and for treating rheumatism; Edible fruit
44.	*Gnetum latifolium* Blume	Dây gắm lá rộng	Chưa muổi	Stem is used for making rope, crossbow strings, and for treating rheumatism; Edible fruit
	Pinaceae Lindl			
45.	*Pinus merkusiana* Cool et Gaussen	Thông hai lá		Timber
	V. Magnoliophyta	**Ngành Ngọc lan**		
	A. Magnoliopsida	**Lớp Hai lá mầm**		
	Acanthaceae Juss.			
46.	*Strobilanthes cusia* (Ness) Kuntze	Chàm mèo		Leaf is used for dyeing cloth
	Actinidiaceae			
47.	*Saurauia tristyla* DC.	Nóng	Co mác mầu	Edible fruit
	Alangiaceae			
48.	*Alangium chinese* Lour.	Thôi ba		Roots and leaves are used for treating rheumatism
	Amaranthaceae			
49.	*Amaranthus spinosus* L.	Rau dền gai	Phắc hom khao	Edible vegetable; Leaf is used to treat pimples
50.	*Amaranthus tricolos* L.	Rau dền canh	Phắc hom đeng	Edible vegetable (Leaf is used for treating dysentery; seed is good for eyesight)

Appended Table 5.A1 Continued

No.	Scientific name	Vietnamese name	Đà Bắc Tày name	Function
51.	*Amaranthus viridis* L.	Rau dền cơm		Edible vegetable; Root is used for treating dysentery
	Anacardiaceae			
52.	*Dracontomelum mangiferum* BI	Sấu	Mác chủ	Edible fruit
53.	*Mangifera indica* L.	Xoài		Edible fruit; Skin of fruit is used for toothache treatment
54.	*Rhus chinensis* Mill.	Muối	Mạy pật	Wood is used for making gunpowder; Fruit is used as spice
55.	*Rhus javanica* L	Muối	Mạy pật	
56.	*Gliota megalocarpe* (Evad.) Tardieu	Sơn		Root is used for treating hepatitis and asthma. Its resin causes an allergy
57.	*Spondias lakoensis* (Pierre). Stapf	Dâu da xoan	Co mác lỏ	Edible fruit
	Annonaceae			
58.	*Desmos chinensis* Lour.	Hoa giẻ thơm		Cooked flower is used for women who have difficulty giving birth
59.	*Desmos cochinchinensis* Lour.	Giẻ Nam bộ		Cooked root is used for women who have difficulty giving birth
60.	*Goniothalamus sp.*			
61.	*Fissistigma polyanthoides* (DC.) Merr.	Lạc nôm mō	Chưa mac vi	Wood is used as restorative and to treat diarrhea
62.	*Orophea tonkinensis* Fine et Gagnep.			
63.	*Polyalthia cerasoides* (Roxb.) Bedd.	Đuōi trân		Root is used for antipyretic
64.	*Polyalthia parviflora* Ridl.			
	Apiaceae			
65.	*Centella asiatica* (L.) Urb.	Rau má	Phắc chèn	Whole plant is used as antipyretic
66.	*Coriandrum sativum* L.	Rau mùi	Phắc chi'	Fruit is used for stimulating digestion
67.	*Eryngium foetidum* L.	Rau mùi tàu	Phắc hom chi lao	Leaf is used as spice, to reduce digestive disorder, to treat cold

.	**Apocynaceae**			
68.	*Alstonia scholaris* (L.) R.Br.	Sữa	Co tin pết	Bark and stem are used for treating malaria and used as restorative
69.	*Tabernaemontana buffalina* Lour.	Lánkỏ		
70.	*Wrightia lecomtei* Pit.	Mức Lecomte		
71.	*Wrightia pubescens* R.Br.	Mức lông mềm		Roots, barks, and leaves are used for treating scrofula and arthritis
72.	*Wrightia macrocarpca* Pit.	Thừng mực		
	Aquifoliaceae			
73.	*Ilex loeseneri* Tardieu	Bùi		
	Araliaceae			
74.	*Aralia vietnamensis* Ha	Cuồng	Phắc nam	Young leaves are used as vegetable; roots are used to treat hepatitis, sore throat, rheumatism
75.	*Aralidium pinnatifidum* Miq.			
76.	*Polyscias fruticosa* (L.) Harms	Đinh lăng		Root and leaves are used as restorative for anemic persons and women after giving birth. Stem is used for treating rheumatism and backaches
77.	*Schefflera heptaphylla* (Lour.) Harms	Chân chim	Co tảng mạy	Bark used for restorative medicine, and to strengthen tendons and bones
78.	*Schefflera tonkinensis* R.Vig.	Chân chim Bắc bộ	Co tảng mạy	Bark is used as restorative and to strengthen tendons and bones
79.	*Trevesia sphaerocarpa* Grushv. & Skvorts.			
80.	*Macropanax membranifolium* Shang	Đại đinh lá mỏng		
	Asclepiadaceae			
81.	*Dischidia acuminata* Cost.	Tai chuột, Dây hạt bí		Whole plant is used for treating leucorrhoea (commonly known as white discharge from the vagina, is due to excessive production of normal vaginal secretions)
82.	*Cryptolepis buchanani* Roem. Et Sch.	Dây càng cua		

Appended Table 5.A1 Continued

No.	Scientific name	Vietnamese name	Đà Bắc Tày name	Function
	Asteraceae			
83.	*Artemisia vulgaris* L.	Ngải cứu	Nhả ngài	Leaves are used to induce menstrual flow, treat coughs and influenza
84.	*Blumea hieracifolia* (D.Don) DC.	Cúc lông bạc		Whole plant used for diuretic
85.	*Dichrocephala integrifolia* (L.f.) Kuntze	Rau chuối		Leaves used for diuretic
86.	*Emilia sonchifolia* (L.) DC.	Rau má lá rau muống		Leaves are used as detoxicant, antiseptic
87.	*Erechtiles hieracifolia* (L.) Raf.	Hoàng thất lá hẹp		Young shoots and buds are used as vegetable
88.	*Erigeron crispus* Pourr.			
89.	*Chromolaena odorata* (L.) r. King et H. Rob.	Cỏ lào	Nhả nhật	Leaves are used for treating dysentery and as haemostatic
90.	*Lactuca indica* L.	Bồ công anh Việt Nam		Leaves are edible, used as restorative, and used for treating furuncle, breast abscess, improving milk flow for lactating women, and cleaning blood
91.	*Lactuca parishii* Craib			
92.	*Pluchea indica* (L.) Less.	Cúc tần		Leaf is used for spicy vegetable, treating dermatological diseases, root is used for antipyretic
93.	*Spilanthes iabadacensis* A.H.Moore	Cúc áo, Kim hoa		Whole plant used for diuretic
94.	*Xanthium Strumarium* L.	Ké đầu ngựa	Mác nháng	Fruit is used for treating rheumatism, pimples
95.	*Wedelia calendulacea* Less.	Sài đất		Leaves are used for treating venereal diseases
96.	*Gynura crepioides* Benth.		Rau tàu bay	
97.	*Eupatorium reevesii* Wall.			
98.	*Hemistepta lyrata* Bunge			
	Balanophoraceae			

99.	*Balanophora fungosa* J.R.et G.Forst.	Dơ đất	(Cuchó)	Whole plant is used as restorative and for reducing backaches
100.	**Begoniaceae**			
101.	*Begonia bonii* Gagnep.	Chua khan		Leaf is used for cooking sour soup
	Bignoniaceae			
102.	*Oroxylum indicum* (L.)Vent.	Núc nác	Phắc ca	Edible young fruit; bark is used for treating dermatological diseases
	Bombacaceae			
103.	*Bombax malabarica* (DC.) Merr	Gạo	Co ngịu	Bark is used for treating burns
104.	*Ceiba pentandra* (L.) Gaertn.	Gòn		Boiled seeds are eaten
	Brassicaceae			
105.	*Brassica juncea* (L.) Czern.	Cải canh	Phắc cát	Edible vegetable
106.	*Raphanus sativus var. longipinnatus* Baill.	Cải củ		Edible vegetable; seeds are used for treating cough
107.	*Rorippa dubia* (Pers.) Haroon	Cải đất núi		Whole plant is used for treating cold and hepatitis
108.	*Rorippa indica* (L.) Hiern	Cải hoang		Edible vegetable, used to treat fever and hepatitis
	Burseraceae			
109.	*Canarium album* (Lour.) Raeusch. ex DC.	Trám trắng	Co cườm	Fruit is used as vegetable
	Caricaceae			
110.	*Carica papaya* L.	Đu đủ	Mác lầu	Edible fruit, stalks of leaves can be used to treat earaches
	Celastraceaee			
111.	*Celastrus sp.*			
	Ceratophyllaceae			
112.	*Ceratophyllum demersum* L.	Rong đuôi chó		Whole plant is used as fodder for pig
	Chenopodiaceae			
113.	*Chenopodium ambrosioides* L.	Dầu giun		Extract from whole plant is used to treat ascarid worms and hookworms

Appended Table 5.A1 Continued

No.	Scientific name	Vietnamese name	Đà Bắc Tày name	Function
114.	*Chenopodium album* L.	Rau muối		Young stalks are edible and are used for treating dysentery
	Combretaceae			
115.	*Terminalia catappa* L.	Bàng		Leaf is used for treating fever, dysentery; edible fruit
116.	*Terminalia myriocarpa van* Heurck et Muell.-Argent	Chiêu liêu xanh		Leaf is used for treating asthma, timber is used for house construction and domestic use
	Convolvulaceae			
117.	*Argyreia capitata* (Vahl) Choisy	Bạc thau hoa đầu	Co pác hắp	Leaf is used for treating heavy menstrual bleeding, setting bone fractures
118.	*Ipomoea aquatica* Forssk.	Rau muống	Phặc búng boong	Leaf and top of plant are used as vegetable, for treating food poisoning, and stopping nosebleeds
119.	*Ipomoea batatas* Lam.	Khoai lang	Phặc búng	Leaves, tops of plant, tubers are used for pig fodder
120.	*Ipomoea purpurea* Kunth	Bìm bìm tía		Seeds are used for edema, and constipation
121.	*Ipomoea rubro- coerulea* L.			
122.	*Merremia bimbim van* (Gagnep.) Ooststr.	Bìm bìm đất		Seeds are used as uretic
	Crassulaceae			
123.	*Kalanchoe blossfieldiana*	Thuốc bỏng		Leaf is used for treating burns; ornamental plants
	Cucurbitaceae			
124.	*Cucumis sativus* L.	Dưa chuột	Qua	Edible fruit
125.	*Cucurbita maxima* Duch. ex Lam.	Bí đỏ	Phặc đeng	Edible fruit, edible stalk of vegetable, seeds are used for treating intestinal flat worms
126.	*Luffa acutangula* (L.) Roxb.	Mướp tàu	Co mac bốp	Edible fruit
127.	*Momordica charantia* L.	Mướp đắng	Co mác khôm	Edible fruit

128.	*Momordica cochinchinensis* (Lour.) Spreng.	Gấc	Co mác cước	Edible fruit
129.	*Sechium edule* (Jacq) Swartz	Su su		Edible fruit
	Dilleniaceae			
130.	*Dillenia hookeri* Pierre	Sổ hooker	Co tum	Edible fruit
131.	*Dillenia indica* L.	Sổ bà	Co mác thản	Edible fruit
132.	*Tetracera sarmentosa* (L.) Vahl	Chặc chìu	Co thiểu thản	Stem is used for strengthening tendons and bones
	Dipterocarpaceae			
133.	*Dipterocarpus retusus* Blume	Chò nâu		Trunks are used for house posts and beams; resin is used to make torches
134.	*Vatica tonkinensis* A. Chev.	Táu mật		Trunk is used for house posts
	Ebenaceae			
135.	*Diospyros kaki* L.f.	Hồng	Co mác hông	Fruit is used for treating hiccups of children
136.	Diospyros sp.			
	Elaeocarpaceae			
137.	*Elaeocarpus balansae* A.DC.	Côm Balansa	Co phôm phan	Wood is used as substrate for growing mushrooms
138.	*Elaeocarpus grandiflorus* A.J.E.Sm.	Côm hoa lớn	Co phôm phan	Wood is used as substrate for growing mushrooms. Fruit is used for antipyretic
139.	*Elaeocarpus petiolatus* (Jacq.) Wall.	Côm		
	Euphorbiaceae			
140.	*Mallotus microcarpus* Pax & Hoffm.	Bùm bụp trái rho’		
141.	*Croton agyrata* Blume			
142.	*Alchornea rugosa* (Lour.) Muell.-Argent	Bọ nẹt		Leaf and root are used to treat malaria. Seeds are used as purgative medicine
143.	*Alchornea trewioides* (Benth.) Muell.-Argent	Vông đo’	Co táu	

Appended Table 5.A1 Continued

No.	Scientific name	Vietnamese name	Đà Bắc Tày name	Function
144.	*Aleurites molucana* (L.) Willd.	Lai		Seeds is source of oil and used as purge
145.	*Antidesma bunins* (L.) Spreng.	Chòi mòi tía	Co mác mâu	Edible fruit; root is used to induce menstrual flow
146.	*Antidesma fordii* Hemsl.	Chòi mòi lông vàng	Co mác mâu	Edible fruit; timber
147.	*Baccaurea ramiflora* Lour.	Dâu gia	Mác phầy	Edible fruit; leaf is used for treating pimples and allergy
148.	*Bischofia trigoliata*	Nhội	Co phát	Young leaves are used as spicy vegetable; for treating itching pudenda
149.	*Breynia fruticosa* (L.) Hook.f.	Bồ cu vẽ	Co ót đông	Leaf is used for treating snake bites
150.	*Breynia minutiflora* Hook.f.			
151.	*Endospermum sinensis*	Vạng trứng		Timber tree
152.	*Euphorbia antiquorum* L.	Xương rồng ông		Stem is used for treating toothaches and pimples
153.	*Euphorbia hirta* L.	Cỏ sữa lá lớn		Whole plant is used for treating dysentery
154.	*Euphorbia indica* Lam.	Cỏ sữa Ấn Độ		Whole plant is used for treating dysentery
155.	*Glochidion littorale* Blume	Bọt ếch biển	Co ản mật	Leaf and root are used for treating intestinal pain
156.	*Macaranga denticulata* (Blume) Muell.-Argent	Ba soi	Co tạu	Leaf is used for treating pimples
157.	*Mallotus paniculatus* (Lam.) Muell.-Argent	Bụp bạc	Co tạu	Root is used as bandage to cover injury so that it is speedily recovered
158.	*Manihot esculenta* Crantz	Sắn	Mằn mạy	Edible tuber
159.	*Phyllanthus reticulatus* Poir.	Phèn đen	Co mác cảng pa	Leaf is used for treating scabies and dysentery
160.	*Phyllanthus urinaria* L.	Chó đẻ		Whole plant is used for treating hepatitis
161.	*Ricinus communis* L.	Thầu dầu	Co nga váng	Seeds are used for purge
162.	*Sapium discolor* (Benth.) Muell.-Argent	Sòi tía		Bark and root are used for treating constipation, oedema

163.	*Sauropus androgynus* (L.) Merr.	Bô ngót	Co ót	Edible vegetable; leaf is used for treating placenta pain after giving birth
164.	*Vercinia montana* Lour.	Trẩu		Attar is extracted from seeds and it is used for painting. Seeds can be used for treating pimples
165.	**Fabaceae**			
166.	*Millettia pachyloba* Drake			
167.	*Entada phaseoloides* (L.) Merr.	Bàm bàm		
168.	*Bauhinia blakeana*	Móng bò		
169.	*A. magnum* Willd.			
170.	*Arachis hypogea* L.	Lạc	Thua đin	Industrial crop; seeds are edible and delicious
171.	*Archidendron chevalieri* (Kost.) I.C.Nielsen	Mán đỉa Chevalier		Timber tree
172.	*Archidendron clypearia* (Jack.) I.C.Nielsen	Mán đỉa		Boiled leaf is used for bathing to treat scabies and injury on the human skin. Timber tree
173.	*Archidendron lucidum* (Benth.) I.C.Nielsen	Mán đỉa trâu		Timber tree; leaf is used for treating diarrhea
174.	*Bauhinia acumilata* L.	Móng bò trắng		Seeds are used for treating stomachache
175.	*Casia tora* L.	Thảo quyết minh	Nhả cooc bẻ	Seeds are used to make tea substitute to improve eyesight
176.	*Dalbergia ovata* Grah.	Trắc trứng		Timber tree
177.	*Delonix regia* (Hook.) Raf.	Phượng vĩ		Bark is used for treating fever, dyspepsia, and for reducing blood pressure
178.	*Desmodium microphyllum* (Murr) DC.	Thóc lép lá nhỏ	Co khẩu mẩu	Whole plant is used for inducing menstrual flow and treating nephritis, gastritis, headache, and toothache
179.	*Desmodium strigillosum* Schindl.	Thóc lép lông ráp	Co khẩu mẩu	Roots are used to treating swelling associated with injuries
180.	*Gleditsia fera* (Lour.) Merr.	Bồ kết	Co kết	Leaf is used as purgative medicine
181.	*Glycine max* (L.) Merr.	Đậu tương		Seeds are used to make soya curd
182.	*Lablab purpureus* (L.) Sweet *subsp. Purpureus*	Đậu ván		Seeds used for restorative stomach

Appended Table 5.A1 Continued

No.	Scientific name	Vietnamese name	Đà Bắc Tày name	Function
183.	*Ormosia balansae* Drake	Ràng ràng mít		Trees are used for timber
184.	*Ormosia inflata* Merr. et Chun	Ràng ràng mọng		Trees are used for timber
185.	*Pachyrrhizus erosus* (L.) Urb.	Củ đậu		Edible root. Toxic seeds are used for treating scabies
186.	*Pelthophorum dasyrrachis* (Miq.) Kurz	Hoàng linh		Trees are used for timber. Bark is used to treat cough
187.	*Phaseolus vulgaris* L.	Đậu cô ve		Edible vegetable
188.	*Psophocarpus tetragonolobus* (L.) A.DC.	Đậu rồng		Edible vegetable
189.	*Pueraria thomsoni* Benth.	Sắn dây		Root is used for antipyretic
190.	*Vigna radiata* (L.) Wilczek	Đậu xanh	Thua kheo	Pods are eaten as vegetable
191.	*Vigna unguiculata* (L.) Walp. *subsp. sesquipedalis* (L.) Verdc.	Đậu đũa	Thua thăng	Edible vegetable
192.	*Vigna unguiculata* (L.) Walp. *subsp. Unguiculata*	Đậu trắng	Thua khao	Edible vegetable
193.	*Zenia insignis* Chun[a]	Mûông linh		Trees are used for timber
	Fagaceae			
194.	*Castanopsis indica* (Roxb.) A.DC.	Cà ổi Ấn Độ	Co mác có	Trees are used for timber
195.	*Lithocarpus sp.*	Dẻ gai		
	Clusiaceae			
196.	*Cratoxylon formosum* (Jack.) Dyer	Thành ngạnh va'ng		
197.	*Garcinia cochinchinensis* (Lour.) Choisy	Bứa nhà	Co mác lưới	Edible fruit. Leaf is used for treating phagedaena (a rapidly spreading ulcer); bark is used to treat allergies
198.	*Garcinia cowa* Roxb.	Tai chua		Edible fruit
199.	*Garcinia multiflora* Champ. ex Benth.	Dọc	Mác lộc	Edible fruit
200.	*Garcinia tinctoria* (DC.) Wight	Bứa nhuộm	Co mác lưới	Leaf is used to treat beriberi and stomachache

	Hamamelidaceae			
201	*Liquidambar formosana* Hance	Sau sau		Leaf is used for dyeing sticky rice to become black
	Icacinaceae			
202.	*Gonocaryum lobbianum* (Miers) Kurz	Quỳnh lam		Leaves are eaten as vegetable, and used for making alcohol yeast Root is used as antipyretic and detoxicant
203.	*Gomphandra mollis* Merr.			
	Juglandaceae			
204.	*Engelhardtia spicata* Blume	Chẹo lông		Timber tree
	Lamiaceae			
205.	*Ensholtzia ciliata* Hyland	Kinh giới		Edible vegetable can be used for treating cough
206.	*Mentha arvensis* L. *var. javanica* Hook.f.	Bạc hà		Leaf is used as edible vegetable and it is used for treating fever, headache, and gastritis
207.	*Ocimum basilicum* L.	Húng quế		Leaf is used as spice
208.	*Ocimum gratissimum* L.	Hương nhu trắng	Co cốc thân	Leaves are used for treating cough, extracting attar
209.	*Perilla frutescens* Britto *var. frutescens*	Tía tô	Co hom tô đeng	Leaf, fruit, and stem are used for treating cough, fever and sweat
	Lauraceae			
210.	*Cinnamomum balansae* Lecomte[a]	Vù hương		Trees are used for timber, attar can be extracted from its leaves, barks and roots
211.	*Cinnamomum sericans* Hance	Ô phát		
212.	*Litsea cubeba* (Lour.) Pers.	Màng tang	Co mạy chang	Attar is extracted from fruit
213.	*Litsea pierrei* Lecomte			
214.	*Machilus robusta* W.W.Smith			
	Loganiaceae			
215.	*Gelsemium elegans* (Gardn.et Champ.) Benth	Lá ngón		Toxic leaves; whole plant can be used for treating eczema, fungi, pimples on the human skin

Appended Table 5.A1 Continued

No.	Scientific name	Vietnamese name	Đà Bắc Tày name	Function
	Magnoliaceae			
216.	*Manglietia hainamensis* Dandy	Mỡ Hải Nam		Timber tree
	Malvaceae			
217.	*Abelmoschus moschatus* Medicus	Vông vang		Seeds are used for treating snakebites
218.	*Hibiscus rosa-siensis* L.	Dâm bụt		Ornamental plant; Young bud is used for treating pimples
219.	*Sida subcordata* Span.	Bái lá to		
	Melastomaceae			
220.	*Medinilla assamica* (C.B.Clarke) Chen	Mua leo		Young leaves are used for cooking sour soup, and treating injuries and discolored bruises
221.	*M. subsessilis* (Craib) Nayar			
222.	*Melastoma bauchei* Guill.	Mua Bauche	Co á	
223.	*Melastoma imbricatum* Wall. ex Clarke	Mua ông	Co á	
224	*Melastoma normale* D.Don	Mua thường	Co á	Whole plant is used for treating dysentery, indigestion, external injury
225.	*Melastoma septemnervium* (Lour.) Merr.	Mua trắng	Co á	Leaf is used for treating fever, reducing blood pressure
226.	*Pseudodissochaeta septentrionalis* (W.W.Sm.) Nayar			
227.	*Sonerila annamica* Guill.	Sơn linh Trung bộ		
228.	*Sonerila cantonensis* Stapf	Sơn linh Quảng Đông		
	Meliaceae			
229.	*Chukrasia tabularis* A.Juss.[a]	Lát hoa		Timber; bark is used for treating diarrhea
230.	*Heynia trijuga* Roxb.			

231.	*Melia azedarach* L.	Xoan		Timber; its toxic bark is used for treating worms
232.	*Melia azedarach var toosendan*	Xoan		
233.	*Toona sureni* (Blume) Merr.	Xương mộc		Timber; bark is used for treating malaria, dysentery; flower is used for inducing menstrual flows
234.	*Walsura bonii* Pell.	Lòng tong		Timber; edible fruit
	Menispermaceae			
235.	*Cyclea tonkinensis* Gagnep.	Sâm nama đaỏ		Leaf is used for treating snakebites, rheumatism
236.	*Pericampilus glaucus* (Lam.) Merr.	Dây chõa ỏ		Leaf is used for stopping bleeding; better digestion, treating stomachache; root is used for treating poisonous snakebites
237.	*Tinospora sp.*			
	Moraceae			
238.	*Artocarpus gomezianus* Wall.	Chay		Edible fruit
239.	*Artocarpus heterophyllus* Lam.	Mít	Mác mị	Timber, edible fruit. Leaf is used for increasing amount of milk of mothers
240.	*Ficus altissima* Blume	Đa tía		Root is used as antipyretic, detoxicant, and reliever
241.	*Ficus auriculata* Leur.	Vả	Co mác ngoa	Edible fruit
242.	*Ficus hirta var. roxburghii* (Miq.) King	Ngái lông dầy	Co mác đớ	Whole plant used to treat rheumatism
243.	*Ficus langkokensis* Drake	Sung làng cốc		
244.	*Ficus pisocarpa* Blume	Rau sộp		Fruit and young leaves are used as vegetable
245.	*Ficus racemosa* L.	Sung	Co mác đớ	Edible fruit
246.	*Ficus subtecta* Corner	Sung phù		
247.	*Morus nigra* L.	Dâu quả đen	Co mạy mòn	Edible fruit; fruit is used to make drinks; root is used for treating toothache; bark is used as a purgative medicine and for killing parasitic worms

Appended Table 5.A1 Continued

No.	Scientific name	Vietnamese name	Đà Bắc Tày name	Function
248.	*Ficus vasculosa* Wall. ex Miq.			
249.	*Artocarpus styracifolius* Pierre	Vỏ khoai		Chewed with betel
250.	*Streblus ilicifolius* (Vidal) Corner	Ruối ô rô		Wood is used for hoe handles. Bark is used to treat pimples and as detoxicant
	Myristicaceae			
251.	*Knema corticosa* Leur.	Máu chó lá nhỏ		Seeds are used for treating scabies and dermatological diseases
252.	*Knema petelotii* Merr.	Máu chó Petelot		
	Myrsinaceae			
253.	*Maesa montana* A.DC.	Đơn nem núi		
254.	*Ardisia sp.*	Cơm nguội		
255.	*Embelia ribes* Burm.f.	Chua ngút	Tẩm cùi	Edible fruit. Leaf is used for cooking sour soup. Seeds are used for treating intestinal flat worms
256.	*Maesa balansae* Mez	Đơn trâu	Co chặp	Leaves are used to treat allergies and scabies
257.	*Maesa membranacea* A.DC.	Đơn màng	Co chặp	
258.	*Maesa parvifolia* A.DC.	Đơn trà	Co chặp	Leaves are used for drinking as tea
259.	*Maesa perlarius* (Lour.) Merr.	Đơn nem	Co chặp	Leaves are used to treat pimples, asthma, cough, and drink as tea
	Myrtaceae			
260.	*Eucalyptus camaldulensis* Dehnh.	Trắng		Tree is used for timber, attar is extracted its leaves, young branches
261.	*Psidium gujava* L.	Ổi	M'c éi	Edible fruit. Leaf is used for treating diarrhea
262.	*Rhodomyrtus tomentosa* (Aiton) Hassk.	Sim	Mác nim	Edible fruit, making drinks; leaf used to treat diarrhea

263.	*Syzygium jambos* (L.) Alston	Gioi		Edible fruit; leaf can be used to stop diarrhea and dysentery, and to treat sore eyes
264.	*Syzygium samarangense* (Blume) Merr. et Perry	Gioi	Co khi xả	Edible fruit; Leaf is used for antiseptic
	Onagraceae			
265.	*Ludwigia octovalvis* (Jacq.) Raven *ssp. Octovalvis*	Rau mương lông		Whole plant is used to treat dysentery, diuretic, cough, and pimples
	Oxalidaceae			
266.	*Averrhoa carambola* L.	Khế	Mác phường	Edible fruit; Leaf is used for taking bath to treat skin allergy
267.	*Oxalis corniculata* L.	Chua me đất hoa vàng		Leaf is used for cooking sour soup, medicine for treating dysentery
	Passifloraceae			
268.	*Passiflora foetida* L.	Lạc tiên	Co tào tin	Stem and leaf are used for treating insomnia
	Pedaliaceae			
269.	*Sesamum orientale* L.	Cây vừng		Edible seeds; attar can be extracted from seeds
	Piperaceae			
270.	*Piper betle* L.	Trầu không		Leaf is chewed with betel to help strenthen teeth, and to treat trachoma
271.	*Piper lolot* C.DC.	Lá lốt	Phắc pát	Leaf is used as vegetable and spice; root is used to treat rheumatism
	Plantaginaceae			
272.	*Plantago major* L.	Mã đề	Co mạ đê	Whole plant is used for diuretic
	Polygalaceae			
273.	*Polygala tonkinensis* Chodat			
	Polygonaceae			
274.	*Polygonum chinensis* L.	Thồm lồm	Co thút thến	Whole plant is used as antipyretic, detoxicant, and to treat pimples, otorrhoea (ear-disease), hepatitis, and snakebites

Appended Table 5.A1 Continued

No.	Scientific name	Vietnamese name	Đà Bắc Tày name	Function
275.	*Polygonum flaccidum* (Meisn.) Steud.	Nghể mềm		Whole plant is used to treat bacillary dysentery, enteritis, gastritis, rheumatism, snakebites
276.	*Polygonum odoratum* Lour.	Rau răm	Co hom phết	Leaf is used as spice and to treat snakebites
277.	*Polygonum orientale* L.	Nghể đông		Leaf is used to treat stomachache, as diuretic, and for improving eyesight
	Portulacaceae			
278.	*Portulacca pilosa* L.	Sam lông		Leaf is used as diuretic, and for inducing menstrual flow, reducing inflammation and external injury
Rhamnaceae				
279.	*Zyzyphus mauritiana* Lam.	Táo dây		Edible fruit
	Rosaceae			
280.	*Prunus mume* Sieb. et Zucc.	Hoa mai		Ornamental plant
281.	*Prunus persica* (L.) Bartsch *var. persica*	Đào	Co mác tao	Ornamental plant, edible fruit, seeds are used to treat coughs
282.	*Prunus salicina* Lindl. *var. salicina*	Mận	Co mac man	Edible fruit; seeds are used to treat bone ache; bark is used to treat cough, dysentery
283.	*Rosa chinensis* Jacq.	Hoa hồng		Ornamental plant
284.	*Rubus alceaefolius* Poir.	Mâm xôi	Co hủ	Leaf is used for better digestion, for drinking as tea, root is used for treating hepatitis
285.	*Rubus leucanthus* Hance	Ngấy trâu	Co hủ	Edible fruit; leaf is used for better digestion
286.	*Rubus moluccanus* L.	Đum đảo Molacca		Edible fruit; fresh leaf is used to treat bed-wetting
	Rubiaceae			
287.	*Urophyllum* sp.			
288.	*Lasianthus tonkinensis* (Drake) Pit.	Xú Bắc bộ		

289.	*Spermacoce articularis* (L.f.)	Ruột gà		Root is used to treat fever
290.	*Cephalanthus tettrandra* (Roxb.) Ridd.	Gáo nước		
291.	*Dentella repens* J.R. et C.F.Forst.	Cỏ màn màn		Whole plant is used as bandage to cover injuries on the skin and speed up recovery
292.	*Hedyotis hispida* Retz.	Bòi ngòi lông		Whole plant is used to treat measles, and the body pains
293.	*Lasianthus langkokensis* Pit.	Xú làng kốc		
294.	*Mussaenda dehisens* Craib.	Bướm bạc		
295.	*Neonauclea calycina* Merr.		Co xạn xạn	
296.	*Paederia scandens* (Lour.) Merr.	Mơ lông		Whole plant is used to treat rheumatism, cough, pneumonia
297.	*Psychotria rubra* (Lour.) Poir.	Lấu		
298.	*Randia spinosa* Blume	Găng trâu	Mạy căng	Fruit is used as substitute for soap for washing clothes
299.	*Uncaria lecomtei* Pit.	Câu đằng Lecomte	Co cáu	
300.	*Wendlandia paniculata* A. DC.	Hoắc quang	Cọ cao	Leaf is used to treats injuries
	Rutaceae			
301.	*Citrus amblycarpa* (Hassk.) Ochse	Quýt tà		
302.	*Citrus aurantifolia* (C.Christm.) Sw.	Chanh giấy		Fruit is used as antipyretic, for stimulating digestion, and helping phlegm run out
303.	*Citrus grandis* (L.) Osb.	Bưởi	Co mác pục	Edible fruit; attar can be extracted from flower; bark of fruit is used to treat coughs, asthma; leaf is used to treat fever
304.	*Citrus medica* L.	Phật thủ		Edible fruit; fruit is used to improve digestion, and to treat anorexia and cough
305.	*Citrus sinensis* (L.) Osb.	Cam	Co mác cam	Edible fruit; bark of fruit is used to improve digestion and reduce burping
306.	*Clausena lansium* (Lour.) Skeels	Hồng bì	Co mác mặt	Edible fruit; root is used to treat stomachaches and rheumatism
	Sapindaceae			

Appended Table 5.A1 Continued

No.	Scientific name	Vietnamese name	Đà Bắc Tày name	Function
307.	*Dimocarpus longan* Lour.	Nhãn	Mác ngận	Edible fruit; fruit is used as restorative
308.	*Litchi sinensis* (Vahl) Radlk.	Vải	Mác pái	Edible fruit; seeds are used to treat stomachache, intestinal pain, and orchitis.
309.	*Nephelium lappaceum* L.	Vaĩ rừng	Co nghiêu bân	Edible fruit
	Sapotaceae			
310.	*Madhuca pasquieri* (Dub.) H.J.Lam.[a]	Sến mật		Timber tree; leaf is used for treating burns
311.	*Madhuca floribunda* (Pierre ex dubard) H.J.Lam	Sến trắng		
312.	*Manilkara achras* (Mill.) Fosberg	Hồng xiêm		Edible fruit; bark is used for treating diarrhea. Seeds are used as diuretic
313.	*Magnolia coco* (Lour.) DC.	Trứng gà		Edible fruit
	Saururaceae			
314.	*Gymnotheca chinensis* Decne	Dièp cá suối		Whole plant is used to treat stomachache, rheumatism, fever, and pimples
315.	*Houttuynia cordata* Thunb.	Dièp cá		Edible vegetable; leaf is used as antipyretic, and to treat enteritis
	Scrophulariaceae			
316.	*Acrocephalus indicus* (Burm.f) Küntze	Nhân trần	Co thâm nà	Whole plant is used to treat liver and to improve digestion
317.	*Limnophila ruellioides* Pennell			
318.	*Mimulus strictus* Benth.	Mị thảo hẹp		
319.	*Paulownia fargesii* Franch.			
320.	*Scoparia dulcis* L.	Cam thảo đất		Whole plant used to treat cough
	Solanaceae			
321.	*Capsicum frutescens* L.	Ớt		Fruit used as spice

322.	*Lycopersicon esculentum*	Cà chua		Edible fruit
323.	*Solanum erianthum* D.Don	Cà hôi		Leaf is used to treat hemorrhoids. Root is used to treat stomachaches and rheumatism
324.	*Solanum melongena* L.	Cà		Edible fruit
325.	*Solanum torvum* Sw.	Cà dại hoa trắng		Root is used for treat stomachache, toothache, and cough
326.	*Solanum tuberosum* L.	Khoai tây		Edible root
327.	*Solanum undatum* Poir.	Cà pháo		Edible fruit
	Sonneratiaceae			
328.	*Duabanga sonneratioides* Ham.	Phay		Timber tree
	Sterculiaceae			
329.	*Pterosperrmum heterophyllum* Hance			
330.	*Pterospermum pierrei* Hance	Lòng mang đài tua		Timber tree
331.	*Sterculia lanceolata* Cav.	Sảng	Co phắc lịn	Bark is used to treat pimples, leucorrhoea (commonly known as white discharge from the vagina, is due to excessive production of normal vaginal secretions)
332.	*Sterculia principis* Gagnep.	Trôm canh		
333.	*Sterculia tonkinense* A.DC.	Trôm Bắc bộ		Edible seeds
	Styracaceae			
334.	*Styrax tonkinensis* (Pierre) Craib ex Hardw.	Bồ đề		Timber tree; resin is used to treat cough
	Theaceae			
335.	*Camelia sinensis* (L.) Kuntze	Chè		Leaves and buds are used as drink
336.	*Eurya quinquelocularis* Kobuski	Súm năm ô		
	Thymeleaceae			
337.	*Daphne bholua* Ham. ex G.Don	Dó kép		

Appended Table 5.A1 Continued

No.	Scientific name	Vietnamese name	Da Bac Tay name	Function
	Tiliaceae			
338.	*Corchorus capsularis* L.	Đay quả tròn		Leaf is used as vegetable, to treat cough and dysentery. Bark is used for fiber
	Ulmaceae			
339.	*Gironniera subaequalis* Planch.	Ngát		Timber tree
	Urticaceae			
340.	*Elatostemma dissectum* Wedd.			
	Verbenaceae			
341.	*Callicarpa nudiflora* Hook.et Arn.	Tử châu hoa trần	Co phạ pằm	Root is used to treat stomachache, nosebleed, and hepatitis
342.	*Callicarpa rubella* Lindl.	Tu hú quả tím	Co phạ pằm	Root is used to treat rheumatism; cooked leaves are used for bathing to treat scabies
343.	*Clerodendrum cyrtophyllum* Tucrz.	Bọ mẩy		Edible young leaf; root is used as restorative
344.	*Clerodendrum kaempferi* (Jack) Sieb.	Mò	Co poong pi	Root is used to treat leucorrhoea (commonly known as white discharge from the vagina, is due to excessive production of normal vaginal secretions)
345.	*Stachytarpheta jamaicensis* (L.) Vahl	Đuôi chuột	Nha ha ngụ	Leaf used to treat urinary infections, diarrhea, coughs
	Vitaceae			
346.	*Cayratia ceratophora* Gagnep.	Vác có sừng		
347.	*Cayratia japonica* (Thunb.) Gagnep.	Vác nhật		Whole plant is used as detoxicant, diuretic, and to treat leucorrhoea and hepatitis
348.	*Tetrastigma eberhardtii* Gagnep.	Tứ thư eberhardt	Chưa ện on	
	B. Liliopsida	**Lớp một lá mầm**		
	Agavaceae			

349.	*Cordyline fruticosa* (L.) Goepp.	Huyết dụ		Ornamental plant; flower, root and leaf are used as haemostatic (for stopping bleeding)
	Alliaceae			
350.	*Allium ascalonium* L.	Hành ta		Leaves are used for spice
351.	*Allium fistulosum* L.	Hành hương		Leaves are used for spice
352.	*Allium sativum* L.	Tỏi		Leaves are used for spice and to treat fever
	Araceae			
353.	*Acorus gramineus* Soland.	Thạch xương bồ	Co lạ khao nặm	Root is used to treat rheumatism
354.	*Alocasia odora* (Roxb.) C. Koch	Môn bạc hà	Phước	Bottom leaves are used as vegetable
355.	*A. tonkinensis* Engl. et Gehrm.			
356.	*Colocasia antiquorum* Schott	Khoai sọ	Phước	Edible root
357.	*Colocasia esculenta* (L.) Schott	Khoai sọ	Phước	Edible root
358.	*Homalomena occulta* (lour.) Schott.	Thiên niên kiện	Co vạt vẹo	Root is used to treat rheumatism
359.	*Pistia stratiotes* L.	Bèo cái		Whole plant is used for treating pimples
	Arecaceae			
360.	*Acera catechu* L.	Cau	Mác lang	Fruit used for chewing betel; seeds used to treat intestinal flat worms
361.	*Arenga pinnata* (Wurmb) Merr.	Báng	Co tao	Stem used to make flour and alcohol, young stem used for vegetable; fruit used to treat aches and pains
362.	*Calamus petreus* Lour.	Mây đá		Stem is used to make baskets, pillows, tables, chairs, etc.
363.	*Calamus tetradactylus* Hance	Mây nếp	Co vai	Stem is used to make baskets, pillows, tables, chairs, etc.
364.	*Calamus tonkinensis* Becc.	Mây bắc bộ	Co vai	Stem is used to make baskets, pillows, tables, chairs, etc.
365.	*Caryota monostachya* Becc.	Móc bông đơn		Edible stem
366.	*Cocos nucifera* L.	Dừa		Edible fruit

Appended Table 5.A1 Continued

No.	Scientific name	Vietnamese name	Đà Bắc Tày name	Function
367.	*Livistona chinensis* (Jacq.) R.Br. ex Mart.	Cọ		Leaf is used for house roofs
368.	*Pinanga baviensis* Becc.	Cau dại Ba Vì		Seeds are used to treat intestinal flat worms; bark used as diuretic
	Bromeliaceae			
369.	*Ananas comosus* (L.) Merr.	Dứa		Edible fruit
	Cannaceae			
370.	*Canna edulis* Ker	Dong riềng		Edible root
	Costaceae			
371.	*Costus tonkinensis* Gagnep.	Mía dò hoa gốc	Co ỏi ưởng	Root used as diuretic and to treat urine retention
	Cyperaceae			
372.	*Cyperus pygmaeus* Rottb.	Cú lùn		
373.	*Cyperus tonkinensis* Clarke	U du Bắc bộ		
374.	*Fimbristylis gracilenta* Hance	Mao thư mảnh		
375.	*Fimbristylis quinquangularis* (Vahl) Kunth	Cỏ lông lợn	Nha lục nựn	
376.	*Kyllinga nemoralis* (Forst.et C.F.Forst.) Dandy ex Hutch. et Dalzell	Cỏ bạc đầu		Whole plant used to treat coughs, diarrhea, dysentery
377.	*Scleria tonkinensis* Clarke	Cương Bắc bộ		
	Dioscoreaceae			
378.	*Dioscorea cirrhosa* Prain et Burkill	Củ nâu	Chưa mác bau	Root eaten and used as dye, and to treat diarrhea, dysentery
379.	*Dioscorea esculenta* (Lour.) Burkill *var. fasciculata* Burkill	Củ từ	Chưa mằn khao	Tuber eaten and used as detoxicant
380.	*Dioscorea persimilis* Prain et Burkill	Củ mài	Chưa mằn kép	Root eaten and used as restorative

	Dracaenaceae			
381.	*Dracaena sp.*	Bồng bồng		
	Hypocynaceae			
382.	*Curculigo gracillis* (Kurz.) Wall. ex Hook.f.	Tam nănġ		
	Lemnaceae			
383.	*Lemna aequinoxialis* Welw.	Bèo tấm		
	Marantaceae			
384.	*Maranta arundinacea* L.	Củ dong		Edible root
385.	*Phrynium tonkinense* Gagnep.	Dong Bắc bộ	Co thoong chinh	Leaf is used for wrapping up traditional cakes. It is used as alcohol detoxicant
386.	*Phrynium dispermum* Gagnep.	Dong		
	Musaceae			
387.	*Musa acuminata* Colla	Chuối rừng	Co cuổi đông	Edible stem; root is used as tonic for pregnant women.
388.	*Musa paradisiaca* L.	Chuối tiêu	Co cuổi	Edible fruit; resin is used to treat ringworm
	Poaceae			
389.	*Chrysopogon aciculatus* (Retz.) Trin.	Cỏ may	Co nhả	Whole plant is used to treat jaundice
390.	*Coix lacrynma-jobi* L.*var. lacrynma-jobi*	Y dĩ	Co mác pắt	Fruit is used as restorative
391.	*Cymbopogon martinii* (Roxb.) Watson	Sả hoa hồng		Attar can be extracted from leaf; stem cooked with water is used to bath and stem is also used for treating cough
392.	*Cynodon dactylon* (L.) Pers. *var. dactylon*	Cỏ gà		Whole plant is used to treat urine retention and kidney stone
393.	*D. brandisii* (Munro) Kurz			
394.	*Digitaria violascens* Link	Cỏ chân nhện tím		
395.	*Dendrocalamus membranaceus* Munro	Luồng Thanh hoá		
396.	*Eleusine indica* (L.) Gaertn.	Cỏ mần trầu	Co nhả lút	Whole plant is used to treat high blood pressure, and enteritis

Appended Table 5.A1 Continued

No.	Scientific name	Vietnamese name	Đà Bắc Tày name	Function
397.	*Imperata cylindrica* (L.) P.Beauv. *var. major* (Nees) C.E.Hubb.	Cỏ tranh	Co khẹm kha	Stem is used as diuretic, and to treat urine retention
398.	*Miscanthus sinensis* Andres	Cỏ chè vè		Stem cooked with water is used to wash injuries on the human skin
399.	*Neyraudia reynaudiana* (Kunth) Keng ex Hitchc	Sậy khô		Stem and leaf are used to treat nephritis (inflammation of the kidney)
400.	*Paspalum conjugatum* Berg	Cỏ đắng đôi		
401.	*Paspalum scrobiculatum* L.	Cỏ đắng		
402.	*Saccharum officinarum* L.	Mía	Co ỏi	Stem is edible; stems, tops of plant and leaves are used for feeding livestock
403.	*Saccharum spontaneum* L.	Cỏ bông lau		Flower used as stuffing for pillows
404.	*Setaria glauca* (L.) P.Beauv.	Cỏ đuôi chó		Whole plant is used to treat ulcers, itchy skin
405.	*S. fertilis* (Steud.) Clayton			
406.	*Zea mays* L.	Ngô		Corn silk is used as diuretic
	Pontederiaceae			
407.	*Monochoria hastata* (L.) Solms	Rau mác thon		Leaf is used to absorb puss, carbuncle on the spine, and furuncle
	Smilacaceae			
408.	*Smilax glabra* Roxb.[a]	Thổ phục linh	Cẩu vài lèng	Stem and root are used to treat rheumatism, and used as restorative, and detoxicant
409.	*Smilax megacarpa* A. et C.DC.	Kim cang quả to		Stem and root are used to treat rheumatism, restorative medicine, antidotal
	Stemonaceae			
410.	*Stemona tuberosa* Lour.	Bách bộ	Co mằn tham thíp	Root is used to treat cough and typhus

	Zingiberaceae			
411.	*Alpinia galanga* (L.) Willd.	Riềng nếp	Co khá	Root is used as spice and digestion stimulant
412.	*Alpinia tonkinensis* Gagnep.	Riềng Bắc bộ	Co khá	Root is used as spice and digestion stimulant
413.	*Amomum maximum* Roxb.	Đậu khấu chín cánh		
414.	*Amomum thyrsoideum* Gagnep.	Sa nhân hoa thưa	Co mác néng	Root is used to treat malaria. Fruit is used to treat dyspepsia and foetal derangement
415.	*Curcuma aromatica* Salisb.	Nghệ trắng		Root is used to induce menstrual flow and to treat rheumatism
416.	*Curcuma domestica* Lour.	Nghệ	Co nghệ lương	Root is used as spice, to treat stomachache; root is cooked with sticky rice to improve women's health after giving birth
417.	*Curcuma zedoaria* (Berg) C.Christm.	Nghệ đen	Co nghệ đăm	Root is used to treat stomachache and induce menstrual flow
418.	*Kaempferia angustifolia* Roscoe	Địa liền lá hẹp		Root is used to treat stomachache and dyspepsia
419.	*Zingiber officinale* Roscoe	Gừng	Khinh	Root is used as spice and to treat stomachache and cold

Appended Table 5.A2 Genus and species of plants in Tat hamlet fallow fields according to family names (Alphabetic order)

No.	Family name	Number of genus	Genus index (%)	Number of species	Species index (%)
1.	Acanthaceae	1	0.35	2	0.48
2.	Actinidiaceae	1	0.35	1	0.24
3.	Adiantaceae	4	1.41	7	1.67
4.	Agavaceae	1	0.35	1	0.24
5.	Alangiaceae	1	0.35	1	0.24
6.	Alliaceae	1	0.35	3	0.72
7.	Amaranthaceae	1	0.35	3	0.72
8.	Anacardiaceae	5	1.77	6	1.43
9.	Annonaceae	5	1.77	7	1.67
10.	Apiaceae	3	1.06	3	0.72
11.	Apocynaceae	3	1.06	5	1.19
12.	Aquifoliaceae	1	0.35	1	0.24
13.	Araceae	5	1.77	7	1.67
14.	Araliaceae	6	2.12	7	1.67
15.	Arecaceae	7	2.47	9	2.15
16.	Asclepiadaceae	1	0.35	1	0.24
17.	Aspleniaceae	2	0.71	4	0.95
18.	Asteraceae	13	4.59	16	3.82
19.	Azollaceae	1	0.35	2	0.48
20.	Balanophoraceae	1	0.35	1	0.24
21.	Begoniaceae	1	0.35	1	0.24
22.	Bignoniaceae	1	0.35	1	0.24
23.	Bombacaceae	2	0.71	2	0.48
24.	Brassicaceae	3	1.06	4	0.95
25.	Bromeliaceae	1	0.35	1	0.24
26.	Burseraceae	1	0.35	1	0.24
27.	Cannaceae	1	0.35	1	0.24
28.	Caricaceae	1	0.35	1	0.24
29.	Celastraceaee	1	0.35	1	0.24
30.	Ceratophyllaceae	1	0.35	1	0.24
31.	Chenopodiaceae	1	0.35	2	0.48
32.	Combretaceae	1	0.35	2	0.48
33.	Convolvulaceae	3	1.06	6	1.43
34.	Costaceae	1	0.35	1	0.24

Appended Table 5.A2 Continued

No.	Family name	Number of genus	Genus index (%)	Number of species	Species index (%)
35.	Crassulaceae	1	0.35	1	0.24
36.	Cucurbitaceae	5	1.77	6	1.43
37.	Cyatheaceae	1	0.35	1	0.24
38.	Cyperaceae	4	1.41	6	1.43
39.	Dilleniaceae	2	0.71	3	0.72
40.	Dioscoreaceae	1	0.35	3	0.72
41.	Dipterocarpaceae	2	0.71	2	0.48
42.	Dracaenaceae	1	0.35	1	0.24
43.	Ebenaceae	1	0.35	2	0.48
44.	Elaeocarpaceae	1	0.35	3	0.72
45.	Euphorbiaceae	18	6.36	25	5.97
46.	Fabaceae	21	7.42	28	6.68
47.	Fagaceae	2	0.71	2	0.48
48.	Gleicheniaceae	1	0.35	3	0.72
49.	Gnetaceae	1	0.35	2	0.48
50.	Guttiferaceae	2	0.71	5	1.19
51.	Hamamelidaceae	1	0.35	1	0.24
52.	Hypocynaceae	1	0.35	1	0.24
53.	Icacinaceae	2	0.71	2	0.48
54.	Juglandaceae	1	0.35	1	0.24
55.	Lamiaceae	4	1.41	5	1.19
56.	Lauraceae	3	1.06	5	1.19
57.	Lemnaceae	1	0.35	1	0.24
58.	Lindsaeaceae	1	0.35	1	0.24
59.	Loganiaceae	1	0.35	1	0.24
60.	Lycopodiaceae	1	0.35	1	0.24
61.	Magnoliaceae	1	0.35	1	0.24
62.	Malvaceae	3	1.06	3	0.72
63.	Marantaceae	2	0.71	3	0.72
64.	Marsileaceae	1	0.35	2	0.48
65.	Melastomaceae	4	1.41	9	2.15
66.	Meliaceae	5	1.77	6	1.43
67.	Menispermaceae	3	1.06	3	0.72
68.	Moraceae	4	1.41	13	3.10
69.	Musaceae	1	0.35	2	0.48

Appended Table 5.A2 Continued

No.	Family name	Number of genus	Genus index (%)	Number of species	Species index (%)
70.	Myristicaceae	1	0.35	2	0.48
71.	Myrsinaceae	3	1.06	7	1.67
72.	Myrtaceae	4	1.41	5	1.19
73.	Onagraceae	1	0.35	1	0.24
74.	Oxalidaceae	2	0.71	2	0.48
75.	Passifloraceae	1	0.35	1	0.24
76.	Pedaliaceae	1	0.35	1	0.24
77.	Pinaceae	1	0.35	1	0.24
78.	Piperaceae	1	0.35	2	0.48
79.	Plantaginaceae	1	0.35	1	0.24
80.	Poaceae	14	4.95	18	4.30
81.	Polygalaceae	1	0.35	1	0.24
82.	Polygonaceae	1	0.35	4	0.95
83.	Polypodiaceae	1	0.35	2	0.48
84.	Pontederiaceae	1	0.35	1	0.24
85.	Portulacaceae	1	0.35	1	0.24
86.	Psilotaceae	1	0.35	1	0.24
87.	Pteridaceae	1	0.35	5	1.19
88.	Rhamnaceae	1	0.35	1	0.24
89.	Rosaceae	2	0.71	7	1.67
90.	Rubiaceae	13	4.59	15	3.58
91.	Rutaceae	2	0.71	6	1.43
92.	Salviniaceae	1	0.35	1	0.24
93.	Sapindaceae	3	1.06	4	0.95
94.	Sapotaceae	3	1.06	4	0.95
95.	Saururaceae	2	0.71	2	0.48
96.	Schizeaceae	1	0.35	3	0.72
97.	Scrophulariaceae	5	1.77	5	1.19
98.	Selaginellaceae	1	0.35	8	1.91
99.	Smilacaceae	1	0.35	2	0.48
100.	Solanaceae	3	1.06	7	1.67
101.	Sonneratiaceae	1	0.35	1	0.24
102.	Stemonaceae	1	0.35	1	0.24
103.	Sterculiaceae	2	0.71	5	1.19
104.	Styracaceae	1	0.35	1	0.24

Appended Table 5.A2 Continued

No.	Family name	Number of genus	Genus index (%)	Number of species	Species index (%)
105.	Theaceae	2	0.71	2	0.48
106.	Thelypteridaceae	1	0.35	1	0.24
107.	Thymeleaceae	1	0.35	1	0.24
108.	Tiliaceae	1	0.35	1	0.24
109.	Ulmaceae	1	0.35	1	0.24
110.	Urticaceae	1	0.35	1	0.24
111.	Verbenaceae	3	1.06	5	1.19
112.	Vitaceae	2	0.71	3	0.72
113.	Zingiberaceae	5	1.77	9	2.15
Total		283	100.00	419	100.00

Appended Table 5.A3 Plant species found in fields in different fallow stages

No.	Scientific name	Vietnamese name
	A. Second year fallow	
1.	*Adiantum soboliferum* Wall. ex Hook.	Ráng vệ nữ có chồi
2.	*Alchornea trewioides* (Benth.) Muell.-Argent	Vông đỏ
3.	*Embelia ribes* Burm.f.	Chua ngút
4.	*Litsea cubeba* (Lour.) Pers.	Mãng tang
5.	*Lygodium japonicum* (Thunb.) Sw.	Bòng bong Nhật
6.	*Melastoma imbricatum* Wall. ex Clarke	Mua ông
7.	*Melastoma septemnervium* (Lour.) Merr.	Mua trắng
8.	*Paspalum conjugatum* Berg	Cỏ đắng đôi
9.	*Pericampilus glaucus* (Lam.) Merr.	Châu đảo
10.	*Rhus chinensis* Mill.	Muối
11.	*Styrax tonkinensis* (Pierre) Craib ex Hardw.	Bồ đề
12.	*Tetrastigma eberhardtii* Gagnep.	Tứ theberhardt
	B. Fifth year fallow	
13.	*Alchornea trewioides* (Benth.) Muell.-Argent	Vông đỏ
14.	*Amomum maximum* Roxb.	Đậu khấu chín cánh
15.	*Ficus pisocarpa* Blume	Rau sộp
16.	*Fissistigma polyanthoides* (DC.) Merr.	Lạc nôm mõ
17.	*Imperata cylindrica* (L.) P.Beauv.*var. major* (Nees) C.E.Hubb.	Cỏ tranh
18.	*Maesa montana A.DC.*	Đơn nem núi
19.	*Melastoma septemnervium* (Lour.) Merr.	Mua trắng
20.	*Adina cordifolia* (Roxb.) Hook. f. ex Brandis	Gáo
21.	*Pericampilus glaucus* (Lam.) Merr.	Châu đảo
22.	*Pueraria montana* (Lour.) Merr.	Sắn dây
23.	*Saurauia tristyla* DC.	Nóng
24.	*Selaginella uncinata* (Desv.) Spring	Quyển bá móc
25.	*Dendrocalamus membranaceus* Munro	Luồng Thanh hoá
	C. Secondary forest after 10 years of fallow	
26.	*Adiantum capillus-veneris var. laciniata* C.Chr.	Ráng
27.	*Alchornea trewioides* (Benth.) Muell.-Argent	Vông đỏ
28.	*Antidesma burvius*	Thuộc Họ Thầu Dầu
29.	*Cryptolepis buchanani* Roem. Et Sch.	Dây cãng cua
30.	*Eupatorium odoratum* L.	Cỏ lào
31.	*Litsea cubeba* (Lour.) Pers.	Mãng tang
32.	*Lycopodiella cernua* (L.) Franco et Vasc.	Thông đất

Appended Table 5.A3 Continued

No.	Scientific name	Vietnamese name
33.	*Maesa montana A.DC.*	Đơn nem núi
34.	*Millettia pachyloba* Drake	Thuộc Họ Đậu
35.	*Ormosia balansae Drake*	Ràng ràng mít
36.	*Tabernaemontana buffalina* Lour.	Lài trâu Lả nhỏ
	D. Secondary forest after 20 years fallow	
37.	*Adiantum soboliferum* Wall. ex Hook.	Ráng vệ nữ có chồi
38.	*Alchornea trewioides* (Benth.) Muell.-Argent	Vông đỏ
39.	*Bauhinia penicilliloba* Pierre ex Gagnep.	Thuộc họ Đậu
40.	*Dicranopteris dichotoma* (Thunb.) Benth.	Guột
41.	*Dioscorea cirrhosa* Prain et Burkill	Củ nâu
42.	*Elatostemma dissectum* Wedd.	Thuộc họ Gai
43.	*Entada sp.*	Thuộc họ Đậu
44.	*Lasianthus tonkinensis* (Drake) Pit.	Thuộc họ Cà Phê
45.	*Livistona chinensis* (Jacq.) R.Br. ex Mart.	Cọ
46.	*Macropanax membranifolium* Shang	Thuộc họ Nhân sâm
47.	*Nephelium lappaceum* L. Munt.	Vải Sừng
48.	*Ormosia balansae* Drake	Ràng ràng mít
49.	*Sapium discolor* (Benth.) Muell.-Argent	Sòi tía
50.	*Tabernaemontana buffalina* Lour.	Lài trâu La ńhỏ
51.	*Tinospora sp.*	Họ Tiết dê
52.	*Dendrocalamus membranaceus* Munro.	Luồng Thanh hoá
	E. Other dominant species after 20 years fallow	
53.	*Cephalanthus tettrandra (*Roxb.) Ridd.	Gáo nước
54.	*Dillenia hookeri* Pierre	Sổ hooker
55.	*Diospyros sp.*	Thị rừng
56.	*Elaeocarpus petiolatus (*Jacq.) Wall.	Thuộc họ Côm
57.	*Garcinia cochinchinensis (*Lour.) Choisy	Bứa nhà
58.	*Lithocarpus sp.*	Thuộc họ Dẻ
59.	*Litsea cubeba (*Lour.) Pers.	Mãang tang
60.	*Macaranga denticulata (*Blume*)* Muell.-Argent	Ba soi

Appended Table 8.A1 Useful plants species grown in home and tree gardens in Tat hamlet in 1998 (n = 15) and 2004 (n = 30)

No	Latin name	Vietnamese name	Da Bac Tay name	Percentage of gardens where found		Uses
				1998	2004	
1	*Aerva sanguinolenta*	Hoa mào gà		27	3.33	6
2	*Allium fistulosum*	Hành	Họm tiêu	53	23.3	3, 4
3	*Allium sativum*	Tỏi	Họm bùa	33	10	3, 4
4	*Allium ramosum*	Hẹ		27		3, 4
5	*Alpinia officina*	Riềng	Cọ khả	60	30	3, 4
6	*Amarantus spinosus*	Dền gai	Pặc hổm	26		3
7	*Annanas comosus*	Dứa	Cọ hoài na	40	50	1, 3
8	*Annona squamosa*	Na	Mặc na	27		1
9	*Arachis hypogaea*	Lạc	Mặc họ	33		3
10	*Areca catechu*	Cau	Mặc lang	27	7	8
11	*Artemisia vulgaris*	Ngải cứu	Pặc lá ngải	40		3, 4
12	*Artocarpus heterophyllus*	Mít	Cọ mí	47	63.3	1
13	*Averrhoa carambala*	Khế	Mặc pương	13	13.3	1, 3
14	*Belamcandra chinensis*	Xạ can	Cọ han	27	17	4
15	*Benincasa hispida*	Bí xanh	Mặc pác	40		3
16	*Brassica gongiloides*	Su hào	Mặc su hào	13	27	3
17	*Brassica oleracea*	Bắp cải	Pặc cải bắp	40	10	3
18	*Brassica rapa*	Cải trắng	Pặc cạt	40		3
19	*Brassica juncea*	Cải dưa	Pặc cạt xôm	40	80	3
20	*Camelia sinensis*	Chè	Cọ chè	80	10	8
21	*Canna orientalis*	Dong riềng	Mặc đót	40	3.3	2, 3
22	*Capsicum annuum*	Ớt chỉ thiên	Mặc ớt chỉ thiên	33		3
23	*Capsicum frutescens*	Ớt nhà	Mặc ượt	60	13.3	3
24	*Carica papaya*	Đu đủ	Cọ mặc tàch	27	40	1, 3
25	*Cereus peruvianus*	Xương rồng	Cọ họt ngước	27		6
26	*Citrus limonia*	Chanh	Cọ lựu	47	30	1, 3
27	*Citrus grandis*	Bưởi	Cọ púc	40	40	1, 4
28	*Colocasia gigantea*	Dọc mùng	Pặc ọc	40	37	3, 4
29	*Colocasia antiquorum*	Khoai sọ	Mặc pược	53	7	3
30	*Corchorus capsularis*	Rau đay	Pặc rau đay	40		3
31	*Coriandrum sativum*	Rau mùi	Pạc hóm	60	-	3, 4
32	*Curcuma longa*	Nghệ	Cọ han lửa	40	20	3,4
33	*Cucurbita pepo*	Bí ngô	Cọ mặc úc	53		3
34	*Cymbopogon citratus*	Sả chanh	Pịnh chơ	33	23.3	3, 4

Appended Table 8.A1 Continued

No	Latin name	Vietnamese name	Da Bac Tay name	Percentage of gardens where found 1998	2004	Uses
35	*Dimocarpus longan*	Nhãn	Cọ nhàn	33	27	1
36	*Dolichos lablab*	Đậu ván	Mặc đậu ván	27	10	3
37	*Eleutherine subaphylla*	Sâm đại hành		13	10	4
38	*Eryngium foetidum*	Mùi tàu	Hợm plu lào	40		3
39	*Elsholtzia ciliata*	Kinh giới	Hợm kinh giới	40	3.3	3, 4
40	*Glycine soja*	Đậu tương	Mặc đậu tương	13		3
41	*Hibiscus rosa-sinensis*	Hoa dâm bụt	Cọ ải huôn	27		4, 6
42	*Hura crepitans*	Vông đồng	Cọ dầu	27		6
43	*Impomea batatas*	Khoai lang	Mặc on	33	10	2, 3
44	*Lactuca sativa*	Xà lách	Pặc xà lách	27		3
45	*Lilium longiflorum*	Hoa loa kèn	Cọ hoa kèn	13		6
46	*Litchi chinensis*	Vải	Cọ pặc vải	40	23.3	1
47	*Livistona cochinchinensis*	Cọ	Cọ cú	33	3.3	7
48	*Luffa cylindrica*	Mướp	Mặc cái	33	7	3
49	*Lycopersicon esculentum*	Cà chua	Mặc cà chua	27		3
50	*Manihot esculenta*	Sắn	Cọ mên cọ	47	3.3	2
51	*Maranta arundinacea*	Hoàng tinh	Mặc tọ	27		2
52	*Melia azedarach*	Xoan	Cọ hiên	27	20	5
53	*Mentha piperita*	Bạc hà		13		3, 4
54	*Momordica charantia*	Mướp đắng	Mặc cà noi	27		3, 4
55	*Momordica cochinchinensis*	Gấc	Mặc gấc	13	7	3, 4
56	*Morus alba*	Dâu tằm	Cọ mòn	13		4
57	*Musa paradisiacal*	Chuối	Cọ cuôi	73	90	1, 3
58	*Nicotiana tabacum*	Thuốc lá	Cọ thuốc lá	27		8
59	*Paederia foetida*	Lá mơ	Bợ mơ	27		3, 4
60	*Piper lolot*	Lá lốt	Pắc ơ lật	60	13.3	3
61	*Piper betle*	Trầu không	Bợ pu	27		8
62	*Plantago major*	Mã đề	Pặc lá mã đề	27		4
63	*Portulaca pilosa*	Hoa mười giờ		27		6
64	*Pouteria zapota*	Trứng gà	Pặc xay cày	13	27	1
65	*Prunus salicina*	Mận	Cọ mặc phục	53	23.3	1, 6
66	*Prunus persica*	Đào	Cọ mặc cai	40	3.3	1, 6
67	*Armenia vulgaris*	Mơ	Cọ mặc phục	60	27	1, 6
68	*Psidium guajava*	Ổi	Cọ ổi	33	47	1
69	*Pueraria thomsonii*	Sắn dây	Mịn cọ chừa	40		3, 4

Appended Table 8.A1 Continued

No	Latin name	Vietnamese name	Da Bac Tay name	Percentage of gardens where found 1998	2004	Uses
70	*Raphanus sativus*	Cải củ	Pặc cà cộ	40		3
71	*Rosa sp.*	Hồng đất	Cọ hồng đất	13		1
72	*Saccharum officinarum*	Mía	Cọ oai	47	17	3
73	*Sansevieria trifasciata*	Lưỡi hổ	Cọ lín sữa	13		6
74	*Sauropus androgynus*	Rau ngót	Pặc ót	40	60	3
75	*Solanum album*	Cà trắng	Mặc cựa	40	3.3	3
76	*Sopora japonica*	Hoa hoè		27		8, 4
77	*Syzigium jambos*	Cây doi		33	7	1
78	*Tagetes erecta*	Hoa cúc	Cọ hoa cúc	27		6
79	*Vigra unguiculata*	Đậu đen	Mặc đậu đen	40		3
80	*Zea mays*	Ngô	Klé	27		2
81	*Zingiber officinale*	Gừng	Cọ khịnh	60	23.3	3, 4
82	*Ziziphus mauritiana*	Táo ta	Cọ táo	27	13.3	1
New added plants/trees						
83	*Chukrasia tabularis*	Lát hoa	Cọ lát	0	13.33	5
84	*Liquidambar formosana Hance.*	Sau sau	Cọ sau sau	0	3.33	5
85	*Crinum giganteum Andr.*	Lá láng	Pạc lá láng	0	3.33	4
86	*Kalanchoe pinnata (Lamk.) Oken.*	Lá bỏng	Pạc lá bỏng	0	3.33	4
87	*Ocimum gratissimum L.*	Hương nhu trắng	Cọ hương nhu	0	3.33	4
88	*Basella rubra L.*	Mùng tơi	Pạc mùng tơi	0	3.3	3
89	*Ipomoea aquatic Forssk.*	Rau muống	Pạc bung	0	7	3
90	*Fagopyrum cymosum*	Rau chua	Pạc xồm	0	3.3	3
91	*Dichrocephala auriculata*	Rau chuôi	Pạc mược	0	3.3	3
92	*Perilla frutescens*	Tía tô	Tín đảnh	0	3.3	3
93	*Solanum melongena L.*	Cà tím	Mạc khựa	0	10	3
94	*Dillenia indica L.*	Sổ bã	Cọ sổ	0	3.3	5
95	*Canarium album (Lour.)*	Trám trắng	Cọ trám	0	3.3	5
96	*Clausena lansium (Lour.)*	Hồng bì	Cọ vòng	0	3.3	1
97	*Citrus medica var. sacrodactylis (Sieb.)*	Phật thủ	Dứt mư	0	3.3	1
98	*Dracuntomelon duperreanum Pierre.*	Sấu	Cọ sấu	0	7	1,5
99	*Citrus sinensis (L.) Osb.*	Cam	Cọ cam	0	3.3	1
100	Diospyros Kaki	Hồng	Cọ hồng	0	73.3	1
101	Mangifera indica L.	Xoài	Cọ xoài	0	50	1

Code of uses: 1—fruit, 2—staple food, 3—vegetable, 4—medicine, 5—timber, 6—ornamental, 7—palm trees, 8—stimulant

Appended Table 9.A1 Checklist of plant species in secondary-evergreen broad leaf forest of medium to tall trees (compiled by Le Tran Chan)

No.	Scientific name	Vietnamese name
	I. ANGIOSPERMAE	**NGÀNH HẠT KÍN**
	A. Dicotyledones	**Lớp Hai lá mầm**
	1. Acanthaceae	**Họ Ô rô**
1	*Thunbergia grandiflora* Roxb.	Dây bông báo
	2. Amaranthaceae	**Họ Rau dền**
2	*Achyranthes aspera* L.	Cỏ xước
3	*Amaranthus spinosus* L.	Rau dền gai
	3. Anacardiaceae	**Họ Đào lộn hột**
4	*Rhus chinensis* Mill.	Muối
5	*Toxicodendron succedanea* (L.) Mold.	Sơn rừng
	4. Ancistrocladaceae	**Họ Trung quân**
6	*Ancistrocladus tectorius* (Lour.) Merr.	Dây trung quân
	5. Annonaceae	**Họ Na**
7	*Alphonsea boniana* Fin. et Gagnep.	Thau lĩnh
8	*Desmos chinensis* Lour.	Hoa dẻ
9	*Fissistigma pallens* (Fin. et Gagnep.) Merr.	
10	*Orophea tonkinensis* Fin. Et Gagnep.	
	6. Apiaceae	**Họ Hoa tán**
11	*Eryngium foetidum* L.	Mùi tàu
	7. Apocynaceae	**Họ Trúc đào**
12	*Astonia scholaris* (L.) R.Br.	Sữa
13	*Ervatamia balansae* (Pitard) Pichon	Ớt lản Balansae
14	*E. hoabinhensis* Ly	Ớt lản Hoà Bình
15	*Melodinus tournieri* Pierre	Giom đỏn
16	*Strophanthus caudatus* (Burm.f.) Kurz	Sừng trâu
17	*Wrightia pubescens* R.Br.	Thừng mức lông mếm
	8. Araliaceae	**Họ ngũ gia bì**
18	*Acanthpanax trifoliatus* (L.) Merr.	Ngũ gia bì gai
19	*Heteropanax fragrans* (Roxb.) Seem.	Tung trắng
20	*Schefflera octophylla* (Lour.) Harms	Chân chim tám lá
21	*Trevesia palmata* (Roxb.) Vig.	Đu đủ rừng
	9. Asclepiadaceae	**Họ Thiên lý**
22	*Dischidia acuminata* Cost.	Tai chuột
23	*Hoya macrophylla* Blume	Hoa sao
24	*Streptocaulon griffithii* Hook.f.	Hà thủ ô trắng

Appended Table 9.A1 Continued

No.	Scientific name	Vietnamese name
	10. Asteraceae	**Họ Cúc**
25	*Ageratum conyzoides* L.	Cỏ aỉt lợn
26	*Blumea lacera* DC.	Hoàng đầu
27	*Crassocephalum crepidioides* (Benth.) S.Moore	Rau tàu bay
28	*Elephantopus mollis* H.B.K.	Cúc chân voi
	11. Araceae	**Họ ráy**
29	*Pothos grandis* S.Buchet	
30	*Pothos repens* (Lour.) Druce	
	12. Begoniaceae	**Họ Thu hải đường**
31	*Begonia chinensis* A.DC.	Thu hải đường tàu
	13. Burseraceae	**Họ Trám**
32	*Canarium album* (Lour.) Raeusch.	Trám trắng
33	*C. tramdenum* Dai et Jakovl.	Trám đen
	14. Caprifoliaceae	**Họ cơm cháy**
34	*Viburnum latescens* Blume	Cơm cháy
	15. Clusiaceae	**Họ bứa**
35	*Garcinia multiflora* Champ.	Dọc
36	*G. oblongifolia* Benth. ex Champ.	Bứa
	16. Hamamelidaceae	**Họ Sau sau**
37	*Liquidambar formosana* Hance	Sau sau
	17. Juglandaceae	**Họ óc chó**
38	*Engelhardtia Chrysolepsis Hance* Wall.	Chẹo tiá
	18. Lauraceae	**Họ Long não**
39	*Caryodaphnopsis tonkinensis* (Lecomte) Ridl.	Cà lồ
40	*Litsea monopetala* (Roxb.) Pers.	Bời lời lá tròn
41	*Phoebe kunstleri* Gamble	Lưỏi nai
	19. Leeaceae	**Họ Gối hạc**
42	*Leea bracteata* C.B.Clarke	Gối hạc
	20. Convolvulaceae	**Họ Bìm bìm**
43	*Argyreia acuta* Lour.	Bạc thau
44	*Lepistemon binectariferum* (Wall.) Kuntze	Lân hùng
	21. Cucurbitaceae	**Họ Bầu bí**
45	*Gymnopetalum cochinchinensis* (Lour.) Kurz	Cứt quạ
46	*Gymnostemma pentaphyllum* (Thunb.) Makino	Cổ yếm
47	*Hodsonia macrocarpa* (Blume) Cogn.	Đại hái
	22. Dilleniaceae	**Họ sổ**
48	*Dillenia indica* L.	Sổ bà
49	*D. ovata* Wall. Ex Hook.f. et Thomson	Sổ xoan

Appended Table 9.A1 Continued

No.	SCIENTIFIC NAME	VIETNAMESE NAME
	23. Dipterocarpaceae	**Họ Dầu**
50	*Dipterocarpus retusus* Blume	Chò nâu
	24. Eleocarpaceae	**Họ Côm**
51	*Elaeocarpus balansae* A.DC.	Côm balansae
52	*E. grandiflorus* Sm.	Côm hoa to
	25. Euphorbiaceae	**Họ thầu dầu**
53	*Aleurites montana* (Lour.) Wils.	Trẩu
54	*Baccaurea ramiflora* Lour.	Dâu da đất
55	*Biscofia javanica* Blume	Nhội tía
56	*Macaranga denticulata* (Blume) Muell.	Lá nến
57	*Ostodes paniculata* Blume	
58	*Sapium discolor* (Champ. ex Benth.) Muell. Arg.	Sòi tía
59	*S. rotundifolium* Hemsl.	Sòi lá tròn
	26. Fabaceae	**Họ Đậu**
60	*Archidendron clypearia* (Jack) I.Niels.	Mán đỉa
61	*Millettia ichthyochtona* Drake	Thàn mát
62	*Ormosia balansae* Drake	Ràng ràng mít
63	*Peltophorum dasyrrhachis* (Miq.) Kurz	Lim xẹt, Hoàng linh
64	*Saraca dives* Pierre	Vàng anh
65	*Pueraria phaseoloides* (Roxb.) Schindl.	Sắn dây dại
	27. Fagaceae	**Họ giẻ**
66	*Castanopsis indica* (Roxb.) A.DC.	Giẻ gai, Cã ổi Ấń Độ
67	*Lithocarpus balansae* Drake	Sồi đá, Gĩẻ lá mái
68	*Quercus chevalieri* Hickel et A.Camus	Sồi cau
	28. Loganiaceae	**Họ Mã Tiền**
69	*Strychnos angustifolia* Benth.	Hoàng nàn
	29. Loranthaceae	**Họ Tầm Gửi**
70	*Helixanthera parasitica* Lour.	Chùm gửi ký sinh
71	*Taxillus balansae* (Lecomte) Danser	
	30. Magnoliaceae	**Họ Ngọc lan**
72	*Michelia tonkinensis* A.Chev.	Giổi
	31. Melastomataceae	**Họ Mua**
73	*Medinilla petelotii* Merr.	Me mái
74	*Phyllagathis tonkinensis* (Cogn.) Stapf	Me bắc
	32. Meliaceae	**Họ Xoan**
75	*Aglaia gigantea* Pierre	Gội tiá
76	*Aphanamixis polystachya* R.Parker	Gội nước

Appended Table 9.A1 Continued

No.	Scientific name	Vietnamese name
	33. Moraceae	**Họ Dâu tằm**
77	*Ficus fulva* Reinw. ex Blume	Ngoã lông vàng
78	*F. heterophylla* L.f.	Vú bò
79	*F. variegata* Blume	Ngoã
	34. Myrsinaceae	**Họ Đơn nem**
80	*Ardisia sylvestris* Pit.	Lá khôi rừng
81	*Maesa acuminatissima* Merr.	Đơn nem
82	*M. balansae* Mez	Đơn nem balansae
	35. Myrtaceae	**Họ Sim**
83	*Syzygium tonkinens* (Gagnep.) Merr. et Perry	Trâm bắc
	36. Oleaceae	**Họ Nhài**
84	*Jasminum eberhardtii* Gagnep.	
	37. Proteaceae	**Họ Tung**
85	*Helicia petelotii* Merr.	Tung petelotii
	38. Rubiaceae	**Họ Cà phê**
86	*Ixora coccinea* L.	Đơn đỏ
87	*Lasianthus baviensis* Drake	Xú hương Ba Vì
88	*Mussaenda dehiscens* Craib	Bướm bạc
89	*Psychotria rubra* (Lour.) Poit.	Lấu đỏ
90	*Tarenna baviensis* Pit.	
	39. Rutaceae	**Họ Cam**
91	*Zanthoxylum scabrum* Guillaum	Sẻn vảy
	40. Sapindaceae	**Họ Bồ hòn**
92	*Allophyllus caudatus* Raddk.	
93	*Nephelium cuspidatum* Blume	Chôm chôm rừng
	41. Solanaceae	**Họ Cà**
94	*Lycianthes biflora* (Lour.) Bitter	Cà hai hoa
95	*Solanum spirale* Roxb.	Cà xoắn
96	*S. torvum* Sw.	Cà nông
	42. Sterculiaceae	**Họ Trôm**
97	*Sterculia foetida* L.	Trôm
98	*S. hymenocalyx* K.Schum	Trôm màng
	43. Styracaceae	**Họ Bồ đề**
99	*Styrax tonkinensis* (Pierre) Craib ex Hartw.	Bồ đề
	44. Urticaceae	**Họ Gai**
100	*Boehmeria diffusa* Wedd.	Gai lan
101	*Villebrunea paradoxa* Gagnep.	

Appended Table 9.A1 Continued

No.	SCIENTIFIC NAME	VIETNAMESE NAME
	45. Verbenaceae	**Họ Cỏ roi ngựa**
102	*Callicarpa albida* Blume	Tu hú trắng
103	*Clerodendrum petasites* (Lour.) Moore	Bạch đồng nam
	46. Vitaceae	**Họ Nho**
104	*Ampelopsis heterophylla* (Thunb.) Siebol et Zucc.	Dâu dây khác lá
	B. Monocotyledones	**Lớp Một lá mầm**
	1. Araceae	**Họ Ráy**
1	*Aglaonema siamense* Engl.	Vạn niên thanh
2	*Alocasia macrorrhizos* (L.) G.Don	Ráy to, Ráy dại
3	*Amorphophallus tonkinensis* Engl.	Nưa bắc
4	*Pothos balansae* Engl.	Cơm lện balansae
5	*P. repens* (Lour.) Druce	Dây tràng pháo
6	*Raphidophora chevalieri* Gagnep.	Đuôi phượng lá nhọn
	2. Commelinaceae	**Họ Thài lài**
7	*Commelina begalensis* L.	Cỏ đầu rìu
8	*Floscopa glabratus* Hassk.	Cỏ đầu rìu nhẵn
9	*Pollia macrophylla* (R.Br.) Benth.	
	3. Cyperaceae	**Họ Cói**
10	*Carex tenogyna* Boolt	Cỏ túi nâu
11	*Cyperus cuspidatus* Kunth	Cói mũi cong
12	*Scleria biflora* Roxb.	Đung hai hoa
	4. Dioscoreaceae	**Họ Củ nâu**
13	*Dioscorea bonii* Prain et Burkill	
14	*D. petelotii* Prain et Burkill	
	5. Dracaenaceae	**Họ Huyết giác**
15	*Dracaena angustifolia* Roxb.	Huyết giác lá hẹp
	6. Poaceae	**Họ Hoà thảo**
16	*Apluda mutica* L.	Cỏ trấu
17	*Microstegium vagans* (Nees ex Steud.) A.Camus	Cỏ lá tre
18	*Setaria palmifolia* (Koenig) Stapf	Cá lá dừa
	7. Maranthaceae	**Họ Hoàng tinh**
19	*Phynium dispermum* Gagnep.	Lá dong nếp
20	*P. placentarium* (Lour.) Merr.	Lá dong bánh
	8. Musaceae	**Họ Chuối**
21	*Musa coccinea* Andr.	Chuối rừng
	9. Orchidaceae	**Họ Lan**
22	*Cymbidium lanciofolium* Hook.	Lan kiếm
23	*Dendrobium anosmum* Lindl.	Phi điệp

Appended Table 9.A1 Continued

No.	Scientific name	Vietnamese name
	10. Palmae	**Họ Cau dừa**
24	*Aranga pinnata* (Wurmb) Merr.	Búng báng
25	*Caryota urens* L.	Móc
26	*Livistona saribus* (Lour.) Merr. ex A.Chev.	Cọ
27	*Pinanga baviensis* Becc.	Cau chuột Ba Vì
28	*Rhapis divaricata* Gagnep.	Lụi xoè
	11. Smilacaceae	**Họ Khúc khắc**
29	*Heterosmilax gaudichaudiana* (Kunth) Maxim.	Khúc khắc
	12. Zingiberaceae	**Họ Gừng**
30	*Alpinia bractata* Roxb.	
31	*Amomum villosum* Lour.	Sa nhân
	II. LYCOPODIOPHYTA	**NGÀNH THÔNG ĐẤT**
	1. Lycopodiella cernua (L.) Pie	**Họ Thông đất**
1	*Lycopodium clavatum* L.	Thông đất
	2. Selaginellaceae	**Họ Quyển bá**
2	*Selaginella biformis* A.Braun ex Kuhn	Quyển bá hai dạng
3	*S. monospora* Spring	Quyển bá đơn bào tử
	III. POLYPODIOPHYTA	**NGÀNH DƯƠNG XỈ**
	1. Aspleniaceae	**Họ Tổ điểu**
1	*Asplenium nidus* L.	Tổ điểu to
2	*Asplenium tenuifolium* D.Don	Tổ điểu lá nhỏ
	2. Bechnaceae	**Họ Dương xỉ lá dừa**
3	*Blechnum orientale* L.	Dương xỉ lá dừa
	3. Cyatheaceae	**Họ Dương xỉ thân gỗ**
4	*Cyathea podophylla* (Hook.) Copel.	Dương xỉ thân gỗ
	4. Pteridaceae	**Họ Sẹo gà**
5	*Pteris finotii* H.Christ	Sẹo gà
	5. Schizaeaceae	**Họ Bòng bong**
6	*Lygodium flexuosum* (L.) Sw.	Bòng bong
7	*L. japonicum* (Thunb.) Sw.	Bòng bong Nhật
8	*L. microphyllum* (Cav.) R.Br.	Bòng bong lá nhỏ

Appended Table 9.A2 Checklist of plant species in bamboo forest mixed with woody trees (compiled by Le Tran Chan)

No.	Scientific name	Vietnamese name
	I. ANGIOSPERMAE	**NGÀNH HẠT KÍN**
	A. Dycotyledones	**Lớp Hai lá mầm**
	1. Acanthaceae	**Họ Ô rô**
1	*Thunbergia eberhardtii* Benoist	Dây bông vàng
2	*T. grandiflora* (Rottler) Roxb.	Dây bông xanh
	2. Alangiaceae	**Họ Thôi chanh**
3	*Alangium chinense* (Lour.) Rehder	Thôi canh
	3. Amaranthaceae	**Họ Rau dền**
4	*Achyranthes aspera* L.	Cỏ xước
5	*A. paronychides* A.St. - Hil.	Rau dệu bò
6	*Alternanthera sessilis* (L.) A.DC.	Rau dệu
7	*Celosia argentea* L.	Hoa mào gà
	4. Anacardiaceae	**Họ Đào lộn hột**
8	*Rhus chinensis* Mill.	Muối mềm
	5. Annonaceae	**Họ Na**
9	*Fissitigma thorelii* (Pierre ex Finet et Gagnep.) Merr.	Bô beó trắng
10	*Uvaria microcarpa* Champ. et Benth.	Bù dẻ trườn
	6. Apocynaceae	**Họ Trúc đào**
11	*Kibatalia macrophylla* (Pierre ex Hua) Woodson	Dùi đôi lá to
12	*Strophanthus caudatus* (L.) Kurz	Sùng dê đuôi
13	*Wrightia pubescens* R.Br.	Thừng mức lông mềm
	7. Araliaceae	**Họ Ngũ gia bì**
14	*Aralia armata* (Wall.) Seem.	Đơn châu chấu
15	*Schefflera octophylla* (Lour.) Harms	Chân chim 8 lá
16	*S. petelotii* Merr.	Chân chim núi
17	*Trevesia palmata* (Roxb. et Lindl.) Vis.	Đu đủ rừng
	8. Bignoniaceae	**Họ Núc nác**
18	*Oroxylon indicum* (L.) Vent.	Núc nác
	9. Bombacaceae	**Họ Gạo**
19	*Bombax ceiba* L.	Gạo hoa đỏ
	10. Boraginaceae	**Họ Vòi voi**
20	*Heliotropium indicum* L.	Vòi voi
	11. Caprifoliaceae	**Họ Cơm cháy**
21	*Viburnum lutescens* Blume	Cơm cháy
	12. Cecropiaceae	**Họ Gai dầu**
22	*Poikilospermum suaveolens* (Blume) Merr.	Dái khỉ

Appended Table 9.A2 Continued

No.	Scientific name	Vietnamese name
	13. Combretaceae	**Họ Bàng**
23	*Quisqualis indica* L.	Dây giun Ấń Độ
	14. Convolvulaceae	**Họ Bìm bìm**
24	*Argyreia capitiformis* (Poir.) Ooststr.	Bạc thau đầu
25	*Ipomoea biflora* (L.) Pers.	Bìm hai lá
	15. Cucurbitaceae	**Họ Bầu bí**
26	*Gymnopetalum cochinchinensis* (Lour.) Kantz	Cứt quạ
27	*G. pentaphyllum* (Thunb. ex Murr.) Makino	Cỏ yếm 5 lá
28	*Trichosanthes rubiflos* Cayla	Qua lâu đỏ
	16. Dilleniaceae	**Họ Sổ**
29	*Tetracera scandens* (L.) Merr.	Dây chạc chìu leo
	17. Euphorbiaceae	**Họ Thầu dầu**
30	*Antidesma henryi* Pax et K. Hoffm.	Chòi mòi Henry
31	*Breynia fruticosa* (L.)	Bồ cu vẽ
32	*Croton bonianus* Gagnep.	Cù đèn
33	*Macaranga denticulata* (Blume) Muell	Lá nến
34	*Mallotus apelta* (Lour.) Muell. Arg.	Ba bét trắng
35	*M. barbatus* Muell. Arg.	Ba bét lông
36	*Phyllanthus reticulatus* Poir.	Phèn đen
37	*Aleurites monata* (Lour.) Wils	Trẩu
	18. Labiatae	**Họ Bạc hà**
38	*Gomphostemma chinensis* Oliv.	Sâu cối tàu
39	*Pogostemen stellatus* (Lour.) Kuntze	
	19. Lauraceae	**Họ Long não**
40	*Litsea cubeba* (Lour.) Pers.	Mãng tang
41	*L. glutinosa* (Lour.) C.B.Rob.	Bời lời nhớt
42	*L. monopetala* (Roxb.) Pers.	Bời lời lá tròn
	20. Leeaceae	**Họ Gối hạc**
43	*Leea indica* (Burm.f.) Merr.	Gối hạc ấn
	21. Malvaceae	**Họ Bông**
44	*Abelmoschus moschatus* Medik.	Vông vang
45	*Abutilon indicum* (L.) Sweet	Cối xay
46	*Kydia calycina* Roxb.	Bò Ké
47	*Sida rhombifolia* L.	Ké hoa vàng
48	*Urena lobata* L.	Ké hoa đào
	22. Melastomataceae	**Họ Mua**
49	*Melastoma malabathricum* L.	Mua bà

Appended Table 9.A2 Continued

No.	SCIENTIFIC NAME	VIETNAMESE NAME
50	*Osbeckia chinensis* L.	Mua tép tàu
	23. Moraceae	**Họ Dâu tằm**
51	*Ficus hirta* Vahl	Ngái lông
52	*F. hispida* L.f.	Ngái
	24. Myrsinaceae	**Họ Đơn nem**
53	*Maesa acuminatissima* Merr.	Đơn nem
	25. Myrtaceae	**Họ Sim**
54	*Eucalyptus camaldulensis* Dehnh.	Bạch đàn trắng
	26. Polygonaceae	**Họ Rau răm**
55	*Polygonum chinense* L.	Thồm lồm
	27. Rosaceae	**Họ Hoa hồng**
56	*Rubus alceaefolius* Poir.	Mâm xôi
57	*R. cochinchinensis* Tratt.	Ngấy
58	*R. moluccanus* L.	Mâm xôi góc tù
	28. Rubiaceae	**Họ Cà phê**
59	*Canthium dicoccum* Gaertn.	Căng đôi
60	*Ixora balansae* Pit.	Đơn ba lăng sa
61	*Lasianthus baviensis* Drake	Xú hương Ba Vì
62	*Mussaenda dehiscens* Craib	Bướm bạc
	29. Rutaceae	**Họ Cam**
63	*Clausena excavata* Burm.f.	Hồng bì dại
64	*Euodia lepta* (Spreng.) Merr.	Ba chạc
65	*Micromelum aff. hirsutum* Oliv.	Mắt trâu
	30. Solanaceae	**Họ Cà**
66	*Solanum torvum* Sw.	Cà dại
	31. Sterculiaceae	**Họ Trôm**
67	*Helicteres hirsuta* Lour.	Dó Lông
68	*Sterculia hymenocalyx* K.Schum.	Trôm vàng
	32. Styracaceae	**Họ Bồ đề**
69	*Styrax tonkinensis* (Pierre) Craib ex Hartw.	Bồ đề
	33. Tiliaceae	**Họ Đay**
70	*Grewia paniculata* Roxb. ex DC.	Cũ ke
	33. Urticaceae	**Họ Gai**
71	*Boehmeria diffusa* Wedd.	Gai lan
72	*Elastostema baviensis* Gagnep.	Cao hùng Ba Vì
	35. Verbenaceae	**Họ Cỏ roi ngựa**
73	*Callicarpa albida* Blume	Tu hú

Appended Table 9.A2 Continued

No.	Scientific name	Vietnamese name
74	*Stachytarpheta jamaicensis* (L.) Vahl	Đuôi chuột
	36. Vitaceae	**Họ Nho**
75	*Tetrastigma planicaule* (Hook.f.) Gagnep.	Dây quai bị thân dẹt
	B. Monocotyledones	**Lớp Một lá mầm**
	1. Costaceae	**Họ Mía dò**
1	*Costus speciosus* (Koenig) Sm.	Mía dò
	2. Cyperaceae	**Họ Cói**
2	*Cyperus diffusus* Vahl	Cói hoa xoè
3	*Fimbristylis dichotoma* (L.) Vahl	Cói quăn chẻ đôi
	3. Dioscoreaceae	**Họ Củ nâu**
4	*Dioscorea bulbifera* L.	
5	*D. scortechinii* Prain et Burkill	
	4. Musaceae	**Họ Chuối**
6	*Musa coccinea* Andr.	Chuối rừng
	5. Poaceae	**Họ Hoà thảo**
7	*Dendrocalamus patellaris* Gamble	Giang
8	*bambusa nana* Roxb.	Hóp
9	*Cyrtococcum patens* (L.) A.Camus	
10	*Microstegium vagans* (Nees ex Steud.) A.Camus	
11	*Saccharum spontaneum* L.	Lau
12	*Themeda gigantea* (Cav.) Hack.	
13	*Thysanolaena latifolia* (Roxb. ex Hornem.) Honda	Đót
	6. Zingiberaceae	**Họ Gừng**
14	*Alpinia bractata* Roxb.	
	II. LYCOPODIOPHYTA	**NGÀNH THÔNG ĐẤT**
	1. Lycopodiaceae	**Họ Thông đất**
1	*Lycopodiella cernua* (L.) Pie	Thông đất
	2. Selaginellaceae	**Họ Quyển bá**
2	*Selaginella monospora* Spring	Quyển bá đơn bào tử
	III. POLYPODIOPHYTA	**NGÀNH DƯƠNG XỈ**
	1. Schizacaceae	**Họ Bòng bong**
1	*Lygodium conforme* C.Chr.	Bòng bong to
2	*L. flexuosum* (L.) Sw.	Bòng bong
	2. Gleicheniaceae	**Họ Guột**
3	*Dicranopteris linearis* (Burm.f.) Underw.	Guột

Blank

Appended Table 15.A1 Nutrient balance of the swidden fields (kg/ha/year)

9	PLOT 1	PLOT 2	PLOT 3	PLOT 4	PLOT 5	PLOT 6	PLOT 7	PLOT 8	PLOT
N TOTAL IN 2000	UPLAND RICE						FOREST		
Input									
Seeding	0.46	0.46	0.46	0.46	0.46	0.46	-	-	-
Rainfall	29.30	29.30	29.30	29.30	29.30	29.30	29.30	29.30	29.30
Total input	29.76	29.76	29.76	29.76	29.76	29.76	29.30	29.30	29.30
Output									
Nutrients loss by erosion	9.46	10.61	17.09	30.25	23.19	37.47	2.52	2.21	2.46
Nutrients loss by runoff	10.82	13.54	12.58	19.07	9.64	11.61	6.96	6.15	6.08
Nutrients loss by leaching	7.83	7.02	6.77	6.77	7.43	7.28	3.60	3.49	3.71
Burning	24.87	-	11.09	5.55	9.56	6.12	-	-	-
Harvesting	10.50	5.98	4.68	2.34	4.04	2.58	-	-	-
Total output	63.48	37.15	52.21	63.98	53.86	65.07	13.08	11.84	12.25
Nutrient balance = input – output	–33.72	–7.39	–22.46	–34.23	–24.10	–35.31	16.22	17.46	17.05

N TOTAL IN 2001	PLOT 1	PLOT 2	PLOT 3	PLOT 4	PLOT 5	PLOT 6	PLOT 7	PLOT 8	PLOT 9
	UPLAND RICE	FALLOW	UPLAND RICE		CASSAVA		FOREST		
Input									
Seeding	1.00	-	1.00	1.00	0.01	0.01	-	-	-
Rainfall	29.30	29.30	29.30	29.30	29.30	29.30	29.30	29.30	29.30
Total input	30.30	29.30	30.30	30.30	29.30	29.30	29.30	29.30	29.30
Output									
Nutrients loss by erosion	36.21	0.35	46.05	39.67	22.41	39.78	0.26	0.21	0.34
Nutrients loss by runoff	38.07	28.18	33.81	31.16	23.01	30.30	14.97	11.64	15.07
Nutrients loss by leaching	10.97	11.49	11.68	9.45	12.14	19.06	9.11	18.32	8.48
Burning	-	-	-	-	-	-	-	-	-
Harvesting	7.07	-	3.17	1.45	17.30	16.47	-	-	-
Total output	92.32	40.02	94.70	81.73	74.86	105.60	24.35	30.17	23.90
Nutrient balance = input – output	–62.03	–10.72	–64.41	–51.43	–45.56	–76.30	4.95	–0.88	5.40

Appended Table 15.A1 Continued

N TOTAL IN 2002	PLOT 1	PLOT 2	PLOT 3	PLOT 4	PLOT 5	PLOT 6	PLOT 7	PLOT 8	PLOT 9
	CASSAVA	FALLOW		CASSAVA		FALLOW		FOREST	
Input									
Seeding	0.01	-	0.01	0.01	0.01	-	-	-	-
Rainfall	29.30	29.30	29.30	29.30	29.30	29.30	29.30	29.30	29.30
Total input	29.30	29.30	29.30	29.30	29.30	29.30	29.30	29.30	29.30
Output									
Nutrients loss by erosion	25.30	2.83	23.25	12.08	33.41	1.72	-	-	-
Nutrients loss by runoff	12.79	9.74	10.15	10.15	10.15	9.48	7.08	6.41	10.19
Nutrients loss by leaching	15.71	17.33	17.00	15.13	16.90	17.31	25.40	23.23	22.23
Burning	-	-	-	0.00	-	-	-	-	-
Harvesting	2.11	-	2.11	2.04	2.00	-	-	-	-
Total output	55.90	29.89	52.51	39.40	62.46	28.52	32.48	29.64	32.42
Nutrient balance = input – output	–26.60	–0.60	–23.20	–10.10	–33.16	0.78	–3.18	–0.34	–3.12

N total in 2003	Plot 1	Plot 2	Plot 3	Plot 4	Plot 5	Plot 6	Plot 7	Plot 8	Plot 9
	Cassava	Fallow	Cassava +Melia +Styrax			Fallow	Forest		
Input									
Seeding	0.01	-	0.01	0.01	-	-	-	-	-
Rainfall	29.56	29.56	29.56	29.56	29.56	29.56	29.56	29.56	29.56
Total input	29.57	29.56	29.57	29.57	29.56	29.56	29.56	29.56	29.56
Output									
Nutrients loss by erosion	11.34	-	9.90	-	10.65	-	-	-	-
Nutrients loss by runoff	20.96	11.32	19.76	15.19	25.87	17.59	7.87	18.24	16.82
Nutrients loss by leaching	10.65	13.85	10.97	12.39	9.23	11.73	13.02	13.87	12.64
Burning	-	-	-	-	-	-	-	-	-
Harvesting	1.81	-	1.75	1.68	1.62	-	-	-	-
Total output	44.76	25.16	42.39	29.26	47.37	29.33	20.89	32.11	29.47
Nutrient balance = input – output	−15.19	4.40	−12.82	0.31	−17.81	0.24	8.68	−2.55	0.10

Appended Table 15.A1 Continued

N TOTAL IN 2004	PLOT 1	PLOT 2	PLOT 3	PLOT 4	PLOT 5	PLOT 6	PLOT 7	PLOT 8	PLOT 9
	FALLOW						FOREST		
Input									
Seeding	-	-	-	-	-	-	-	-	-
Rainfall	29.03	29.03	29.03	29.03	29.03	29.03	29.03	29.03	29.03
Total input	29.03	29.03	29.03	29.03	29.03	29.03	29.03	29.03	29.03
Output									
Nutrients loss by erosion	-	-	-	-	-	-	-	-	-
Nutrients loss by runoff	7.20	9.77	13.93	6.97	8.45	9.55	3.24	3.19	3.22
Nutrients loss by leaching	5.43	5.03	6.04	4.85	4.47	6.12	6.57	6.37	6.14
Burning	-	-	-	-	-	-	-	-	-
Harvesting	-	-	-	-	-	-	-	-	-
Total output	12.63	14.80	19.97	11.82	12.92	15.66	9.80	9.56	9.36
Nutrient balance = input – output	16.40	14.23	9.06	17.21	16.11	13.37	19.23	19.48	19.67

N AVAILABLE IN 2000	PLOT 1	PLOT 2	PLOT 3	PLOT 4	PLOT 5	PLOT 6	PLOT 7	PLOT 8	PLOT 9
	UPLAND RICE						FOREST		
Input									
Seeding	-	-	-	-	-	-	-	-	-
Rainfall	20.72	20.72	20.72	20.72	20.72	20.72	20.72	20.72	20.72
Total input	20.72	20.72	20.72	20.72	20.72	20.72	20.72	20.72	20.72
Output									
Nutrient loss by erosion	0.13	0.13	0.86	6.86	1.09	1.92	0.12	0.11	0.12
Nutrient loss by runoff	5.63	7.21	6.87	0.89	5.19	6.32	4.44	3.85	3.79
Nutrient loss by leaching	4.22	3.78	3.65	3.65	4.00	3.92	2.56	2.48	2.64
Burning	-	-	-	-	-	-	-	-	-
Harvesting	-	-	-	-	-	-	-	-	-
Total output	9.98	11.11	11.37	11.40	10.29	12.17	7.13	6.44	6.55
Nutrient balance = input – output	10.74	9.61	9.35	9.32	10.43	8.55	13.59	14.28	14.17

Appended Table 15.A1 Continued

N available in 2001	Plot 1	Plot 2	Plot 3	Plot 4	Plot 5	Plot 6	Plot 7	Plot 8	Plot 9
	Upland rice	Fallow	Upland rice		Cassava		Forest		
Input									
Seeding	-	-	-	-	-	-	-	-	-
Rainfall	20.72	20.72	20.72	20.72	20.72	20.72	20.72	20.72	20.72
Total input	20.72	20.72	20.72	20.72	20.72	20.72	20.72	20.72	20.72
Output									
Nutrient loss by erosion	2.13	0.03	2.70	2.37	1.34	2.35	0.02	0.01	0.02
Nutrient loss by runoff	8.36	7.93	7.24	7.93	6.60	9.13	5.88	4.25	5.23
Nutrient loss by leaching	8.31	8.65	8.83	7.10	9.14	14.37	6.52	13.04	6.05
Burning	-	-	-	-	-	-	-	-	-
Harvesting	-	-	-	-	-	-	-	-	-
Total output	18.80	16.62	18.77	17.39	17.08	25.85	12.42	17.31	11.31
Nutrient balance = input – output	1.92	4.10	1.95	3.33	3.64	–5.13	8.30	3.41	9.41

N available in 2002	Plot 1	Plot 2	Plot 3	Plot 4	Plot 5	Plot 6	Plot 7	Plot 8	Plot 9
	Cassava	Fallow		Cassava		Fallow		Forest	
Input									
Seeding	-	-	-	-	-	-	-	-	-
Rainfall	20.72	20.72	20.72	20.72	20.72	20.72	20.72	20.72	20.72
Total input	20.72	20.72	20.72	20.72	20.72	20.72	20.72	20.72	20.72
Output									
Nutrient loss by erosion	0.72	0.01	0.98	0.28	0.99	1.21	-	-	-
Nutrient loss by runoff	7.83	5.91	6.54	6.17	6.17	5.78	2.72	3.07	4.22
Nutrient loss by leaching	11.48	12.66	12.88	11.05	12.63	12.67	22.00	20.09	19.06
Burning	-	-	-	0.00	-	-	-	-	-
Harvesting	-	-	-	-	-	-	-	-	-
Total output	20.03	18.58	20.40	17.50	19.79	19.66	24.72	23.16	23.27
Nutrient balance = input – output	0.69	2.14	0.32	3.22	0.93	1.06	– 4.00	–2.44	–2.55

Appended Table 15.A1 Continued

N available in 2003	Plot 1	Plot 2	Plot 3	Plot 4	Plot 5	Plot 6	Plot 7	Plot 8	Plot 9
	Cassava	Fallow	Cassava +melia +Styrax			Fallow	Forest		
Input									
Seeding	-	-	-	-	-	-	-	-	-
Rainfall	21.82	21.82	21.82	21.82	21.82	21.82	21.82	21.82	21.82
Total input	21.82	21.82	21.82	21.82	21.82	21.82	21.82	21.82	21.82
Output									
Nutrient loss by erosion	0.62	-	0.52	-	0.58	-	-	-	-
Nutrient loss by runoff	11.76	5.54	11.28	7.89	13.56	9.48	20.09	5.97	6.75
Nutrient loss by leaching	5.74	7.76	5.95	6.95	5.11	6.52	5.94	6.30	6.04
Burning	-	-	-	-	-	-	-	-	-
Harvesting	-	-	-	-	-	-	-	-	-
Total output	18.12	13.31	17.76	14.84	19.26	16.00	26.04	12.27	12.79
Nutrient balance = input – output	3.70	8.51	4.07	6.98	2.57	5.83	–4.22	9.55	9.04

	PLOT 1	PLOT 2	PLOT 3	PLOT 4	PLOT 5	PLOT 6	PLOT 7	PLOT 8	PLOT 9
N AVAILABLE IN 2004			FALLOW					FOREST	
Input									
Seeding	-	-	-	-	-	-	-	-	-
Rainfall	19.62	19.62	19.62	19.62	19.62	19.62	19.62	19.62	19.62
Total input	19.62	19.62	19.62	19.62	19.62	19.62	19.62	19.62	19.62
Output									
Nutrient loss by erosion	-	-	-	-	-	-	-	-	-
Nutrient loss by runoff	1.67	2.58	3.56	1.36	1.80	1.90	0.79	0.71	0.79
Nutrient loss by leaching	5.19	4.81	5.77	4.64	4.28	5.85	5.80	5.62	5.42
Burning	-	-	-	-	-	-	-	-	-
Harvesting	-	-	-	-	-	-	-	-	-
Total output	6.86	7.39	9.33	6.01	6.07	7.75	6.60	6.34	6.21
Nutrient balance = input – output	12.76	12.23	10.29	13.61	13.54	11.87	13.02	13.28	13.41

Appended Table 15.A1 Continued

P TOTAL IN 2000	PLOT 1	PLOT 2	PLOT 3	PLOT 4	PLOT 5	PLOT 6	PLOT 7	PLOT 8	PLOT 9
			UPLAND RICE					FOREST	
Input									
Seeding	0.10	0.10	0.10	0.10	0.10	0.10	-	-	-
Rainfall	6.73	6.73	6.73	6.73	6.73	6.73	6.73	6.73	6.73
Total input	6.83	6.83	6.83	6.83	6.83	6.83	6.73	6.73	6.73
Output									
Nutrients loss by erosion	3.14	1.52	2.15	4.06	3.27	6.17	0.33	0.24	0.32
Nutrients loss by runoff	1.29	1.67	1.54	2.70	1.13	1.47	0.82	0.72	0.71
Nutrients loss by leaching	0.61	0.55	0.53	0.53	0.58	0.57	1.38	1.34	1.43
Burning	6.24	-	2.78	1.39	2.40	1.54	-	-	-
Harvesting	4.07	2.31	1.81	0.91	1.56	1.00	-	-	-
Total output	15.35	6.06	8.83	9.58	8.95	10.75	2.53	2.31	2.46
Nutrient balance = input – output	–8.52	0.77	–1.99	–2.75	–2.11	–3.92	4.20	4.43	4.27

P TOTAL IN 2001	PLOT 1	PLOT 2	PLOT 3	PLOT 4	PLOT 5	PLOT 6	PLOT 7	PLOT 8	PLOT 9
	UPLAND RICE	FALLOW	UPLAND RICE			CASSAVA	FOREST		
Input									
Seeding	0.10	-	0.10	0.10	0.00	0.00	-	-	-
Rainfall	6.73	6.73	6.73	6.73	6.73	6.73	6.73	6.73	6.73
Total input	6.83	6.73	6.83	6.83	6.74	6.74	6.73	6.73	6.73
Output									
Nutrients loss by erosion	3.76	0.15	4.74	4.57	2.65	6.22	0.05	0.04	0.06
Nutrient loss by runoff	4.21	3.68	3.69	3.97	2.95	4.45	1.94	1.63	1.84
Nutrient loss by leaching	2.18	1.00	1.01	0.83	1.05	1.69	0.82	1.59	0.75
Burning	-	-	-	-	-	-	-	-	-
Harvesting	1.40	-	0.63	0.29	2.20	2.09	-	-	-
Total output	11.55	4.83	10.07	9.66	8.85	14.44	2.81	3.25	2.65
Nutrient balance = input – output	–4.72	1.90	–3.24	–2.82	–2.11	–7.71	3.92	3.48	4.08

Appended Table 15.A1 Continued

P total in 2002	Plot 1	Plot 2	Plot 3	Plot 4	Plot 5	Plot 6	Plot 7	Plot 8	Plot 9
	Cassava	Fallow		Cassava		Fallow		Forest	
Input									
Seeding	0.00	-	0.00	0.00	0.00	-	-	-	-
Rainfall	6.73	6.73	6.73	6.73	6.73	6.73	6.73	6.73	6.73
Total input	6.74	6.73	6.74	6.74	6.74	6.73	6.73	6.73	6.73
Output									
Nutrient loss by erosion	1.42	0.03	1.92	0.63	2.41	0.29	-	-	-
Nutrient loss by runoff	1.32	1.09	1.15	1.02	1.02	1.07	0.72	0.61	0.88
Nutrient loss by leaching	1.83	2.04	2.18	1.81	2.04	2.05	2.54	2.30	2.12
Burning	-	-	-	0.00	-	-	-	-	-
Harvesting	0.46	-	0.46	0.45	0.36	-	-	-	-
Total output	5.04	3.16	5.71	3.91	5.83	3.40	3.26	2.91	3.00
Nutrient balance = input – output	1.70	3.58	1.02	2.83	0.91	3.33	3.47	3.82	3.73

P TOTAL IN 2003	PLOT 1	PLOT 2	PLOT 3	PLOT 4	PLOT 5	PLOT 6	PLOT 7	PLOT 8	PLOT 9
	CASSAVA	FALLOW	CASSAVA +MELIA +STYRAX			FALLOW	FOREST		
Input									
Seeding	0.00	-	0.00	0.00	-	-	-	-	-
Rainfall	11.86	11.86	11.86	11.86	11.86	11.86	11.86	11.86	11.86
Total input	11.86	11.86	11.86	11.86	11.86	11.86	11.86	11.86	11.86
Output									
Nutrient loss by erosion	5.83	-	4.35	-	4.49	-	-	-	-
Nutrient loss by runoff	3.45	1.62	2.90	2.12	3.59	2.69	1.56	1.83	2.04
Nutrient loss by leaching	3.37	4.46	3.48	3.98	3.02	3.84	2.02	2.18	2.12
Burning	-	-	-	-	-	-	-	-	-
Harvesting	0.36	-	0.36	0.31	0.29	-	-	-	-
Total output	13.02	6.08	11.09	6.40	11.39	6.53	3.58	4.01	4.16
Nutrient balance = input – output	–1.16	5.78	0.77	5.45	0.46	5.32	8.28	7.84	7.69

Appended Table 15.A1 Continued

P total in 2004	Plot 1	Plot 2	Plot 3	Plot 4	Plot 5	Plot 6	Plot 7	Plot 8	Plot 9
			Fallow					Forest	
Input									
Seeding	-	-	-	-	-	-	-	-	-
Rainfall	1.61	1.61	1.61	1.61	1.61	1.61	1.61	1.61	1.61
Total input	1.61	1.61	1.61	1.61	1.61	1.61	1.61	1.61	1.61
Output									
Nutrient loss by erosion	-	-	-	-	-	-	-	-	-
Nutrient loss by runoff	0.51	0.78	1.06	0.56	0.72	0.74	0.34	0.35	0.29
Nutrient loss by leaching	0.44	0.41	0.49	0.40	0.36	0.50	0.79	0.76	0.74
Burning	-	-	-	-	-	-	-	-	-
Harvesting	-	-	-	-	-	-	-	-	-
Total output	0.95	1.19	1.55	0.96	1.09	1.24	1.13	1.11	1.03
Nutrient balance = input – output	0.66	0.42	0.06	0.66	0.53	0.37	0.49	0.50	0.58

P AVAILABLE IN 2000	PLOT 1	PLOT 2	PLOT 3	PLOT 4	PLOT 5	PLOT 6	PLOT 7	PLOT 8	PLOT 9
				UPLAND RICE				FOREST	
Input									
Seeding	-	-	-	-	-	-	-	-	-
Rainfall	2.38	2.38	2.38	2.38	2.38	2.38	2.38	2.38	2.38
Total input	2.38	2.38	2.38	2.38	2.38	2.38	2.38	2.38	2.38
Output									
Nutrient loss by erosion	0.13	0.06	0.23	0.66	0.27	0.65	0.03	0.02	0.03
Nutrient loss by runoff	0.42	0.53	0.49	0.23	0.36	0.43	0.29	0.26	0.23
Nutrient loss by leaching	0.61	0.55	0.53	0.53	0.58	0.57	1.38	1.34	1.42
Burning	-	-	-	-	-	-	-	-	-
Harvesting	-	-	-	-	-	-	-	-	-
Total output	1.16	1.14	1.25	1.42	1.22	1.66	1.70	1.62	1.68
Nutrient balance = input – output	1.21	1.24	1.13	0.96	1.16	0.72	0.68	0.76	0.70

Appended Table 15.A1 Continued

	Plot 1	Plot 2	Plot 3	Plot 4	Plot 5	Plot 6	Plot 7	Plot 8	Plot 9
P available in 2001	Upland rice	Fallow		Upland rice		Cassava		Forest	
Input									
Seeding	-	-	-	-	-	-	-	-	-
Rainfall	2.38	2.38	2.38	2.38	2.38	2.38	2.38	2.38	2.38
Total input	2.38	2.38	2.38	2.38	2.38	2.38	2.38	2.38	2.38
Output									
Nutrient loss by erosion	0.20	0.01	0.24	0.21	0.12	0.24	0.01	0.01	0.01
Nutrient loss by runoff	0.80	0.69	0.70	0.73	0.56	0.85	0.50	0.43	0.50
Nutrient loss by leaching	1.97	0.89	0.91	0.73	0.94	1.52	0.99	2.02	0.93
Burning	-	-	-	-	-	-	-	-	-
Harvesting	-	-	-	-	-	-	-	-	-
Total output	2.96	1.59	1.84	1.67	1.62	2.60	1.50	2.45	1.44
Nutrient balance = input – output	-0.58	0.79	0.54	0.70	0.75	–0.22	0.88	–0.08	0.94

P AVAILABLE IN 2002	PLOT 1	PLOT 2	PLOT 3	PLOT 4	PLOT 5	PLOT 6	PLOT 7	PLOT 8	PLOT 9
	CASSAVA	FALLOW		CASSAVA		FALLOW		FOREST	
Input									
Seeding	-	-	-	-	-	-	-	-	-
Rainfall	2.38	2.38	2.38	2.38	2.38	2.38	2.38	2.38	2.38
Total input	2.38	2.38	2.38	2.38	2.38	2.38	2.38	2.38	2.38
Output									
Nutrient loss by erosion	0.13	0.01	0.15	0.05	0.19	0.00	-	-	-
Nutrient loss by runoff	0.67	0.53	0.58	0.49	0.49	0.52	0.32	0.29	0.37
Nutrient loss by leaching	0.86	0.96	0.96	0.82	0.94	0.96	1.28	1.17	1.08
Burning	-	-	-	0.00	-	-	-	-	-
Harvesting	-	-	-	-	-	-	-	-	-
Total output	1.66	1.50	1.69	1.36	1.63	1.48	1.60	1.45	1.46
Nutrient balance = input – output	0.72	0.88	0.68	1.01	0.75	0.89	0.77	0.93	0.92

Appended Table 15.A1 Continued

	Plot 1	Plot 2	Plot 3	Plot 4	Plot 5	Plot 6	Plot 7	Plot 8	Plot 9
P available in 2003	Cassava	Fallow	Cassava +melia +Styrax			Fallow	Forest		
Input									
Seeding	-	-	-	-	-	-	-	-	-
Rainfall	3.26	3.26	3.26	3.26	3.26	3.26	3.26	3.26	3.26
Total input	3.26	3.26	3.26	3.26	3.26	3.26	3.26	3.26	3.26
Output									
Nutrient loss by erosion	0.10	-	0.09	-	0.09	-	-	-	-
Nutrient loss by runoff	1.66	0.70	1.49	0.99	1.88	1.15	2.21	1.10	1.26
Nutrient loss by leaching	3.22	4.29	3.34	3.83	2.89	3.68	1.40	1.58	1.69
Burning	-	-	-	-	-	-	-	-	-
Harvesting	-	-	-	-	-	-	-	-	-
Total output	4.98	4.99	4.91	4.82	4.86	4.84	3.61	2.69	2.95
Nutrient balance = input – output	–1.73	–1.73	–1.65	–1.56	–1.60	–1.58	–0.36	0.57	0.30

P available in 2004	Plot 1	Plot 2	Plot 3	Plot 4	Plot 5	Plot 6	Plot 7	Plot 8	Plot 9
	Fallow						Forest		
Input									
Seeding	-	-	-	-	-	-	-	-	-
Rainfall	1.50	1.50	1.50	1.50	1.50	1.50	1.50	1.50	1.50
Total input	1.50	1.50	1.50	1.50	1.50	1.50	1.50	1.50	1.50
Output									
Nutrient loss by erosion	-	-	-	-	-	-	-	-	-
Nutrient loss by runoff	0.13	0.19	0.28	0.14	0.19	0.21	0.13	0.09	0.09
Nutrient loss by leaching	0.42	0.39	0.46	0.37	0.34	0.47	0.68	0.66	0.64
Burning	-	-	-	-	-	-	-	-	-
Harvesting	-	-	-	-	-	-	-	-	-
Total output	0.55	0.58	0.75	0.51	0.53	0.68	0.82	0.75	0.73
Nutrient balance = input – output	0.95	0.92	0.75	0.98	0.96	0.82	0.68	0.75	0.77

Appended Table 15.A1 Continued

K TOTAL IN 2000	PLOT 1	PLOT 2	PLOT 3	PLOT 4	PLOT 5	PLOT 6	PLOT 7	PLOT 8	PLOT 9
			UPLAND RICE					FOREST	
Input									
Seeding	0.10	0.10	0.10	0.10	0.10	0.10	-	-	-
Rainfall	36.85	36.85	36.85	36.85	36.85	36.85	36.85	36.85	36.85
Total input	36.95	36.95	36.95	36.95	36.95	36.95	36.85	36.85	36.85
Output									
Nutrient loss by erosion	145.61	133.36	230.10	348.99	296.06	473.94	44.25	31.01	43.74
Nutrient loss by runoff	108.38	129.43	120.73	240.85	91.22	114.45	84.79	74.53	71.26
Nutrient loss by leaching	18.92	16.97	16.36	16.35	17.94	17.60	10.54	10.22	10.88
Burning	2.46	-	1.10	0.55	0.95	0.61	-	-	-
Harvesting	4.67	2.66	2.08	1.04	1.79	1.15	-	-	-
Total output	280.03	282.42	370.36	607.79	407.97	607.75	139.58	115.76	125.89
Nutrient balance = input – output	–243.08	–245.47	–333.41	–570.84	–371.02	–570.80	–102.73	–78.91	–89.04

K TOTAL IN 2001	PLOT 1	PLOT 2	PLOT 3	PLOT 4	PLOT 5	PLOT 6	PLOT 7	PLOT 8	PLOT 9
	UPLAND RICE	FALLOW	UPLAND RICE			CASSAVA	FOREST		
Input									
Seeding	0.30	-	0.30	0.30	0.00	0.00	-	-	-
Rainfall	36.85	36.85	36.85	36.85	36.85	36.85	36.85	36.85	36.85
Total input	37.15	36.85	37.15	37.15	36.85	36.85	36.85	36.85	36.85
Output									
Nutrient loss by erosion	520.10	5.82	663.73	596.99	321.89	568.15	4.62	3.73	6.04
Nutrient loss by runoff	116.33	96.36	101.88	106.84	79.15	119.14	44.67	38.84	46.65
Nutrient loss by leaching	19.89	17.11	17.49	14.03	18.06	28.58	32.03	69.43	30.94
Burning	-	-	-	-	-	-	-	-	-
Harvesting	2.13	-	0.96	0.44	25.50	24.27	-	-	-
Total output	658.44	119.29	784.05	718.30	444.60	740.14	81.32	112.00	83.63
Nutrient balance = input – output	−621.29	−82.44	−746.90	−681.15	−407.75	−703.29	−44.47	−75.16	−46.78

Appended Table 15.A1 Continued

K TOTAL IN 2002	PLOT 1	PLOT 2	PLOT 3	PLOT 4	PLOT 5	PLOT 6	PLOT 7	PLOT 8	PLOT 9
	CASSAVA	FALLOW		CASSAVA		FALLOW		FOREST	
Input									
Seeding	0.00	-	0.00	0.00	0.00	-	-	-	-
Rainfall	36.85	36.85	36.85	36.85	36.85	36.85	36.85	36.85	36.85
Total input	36.85	36.85	36.85	36.85	36.85	36.85	36.85	36.85	36.85
Output									
Nutrient loss by erosion	199.84	4.77	264.82	87.55	329.60	0.01	-	-	-
Nutrient loss by runoff	47.23	38.73	45.22	37.99	37.99	37.81	15.88	15.02	18.76
Nutrient loss by leaching	12.38	13.79	12.06	11.23	13.25	13.81	16.00	14.64	13.85
Burning	-	-	-	0.00	-	-	-	-	-
Harvesting	2.17	-	2.17	2.10	2.05	-	-	-	-
Total output	261.61	57.29	324.27	138.87	382.89	51.62	31.88	29.65	32.62
Nutrient balance = input – output	–224.76	–20.44	–287.42	–102.01	–346.04	–14.77	4.97	7.20	4.23

K TOTAL IN 2003	PLOT 1	PLOT 2	PLOT 3	PLOT 4	PLOT 5	PLOT 6	PLOT 7	PLOT 8	PLOT 9
	CASSAVA	FALLOW	CASSAVA +MELIA +STYRAX			FALLOW	FOREST		
Input									
Seeding	0.00	-	0.00	0.00	-	-	-	-	-
Rainfall	15.96	15.96	15.96	15.96	15.96	15.96	15.96	15.96	15.96
Total input	15.96	15.96	15.96	15.96	15.96	15.96	15.96	15.96	15.96
Output									
Nutrient loss by erosion	171.42	-	139.91	-	157.09	-	-	-	-
Nutrient loss by runoff	86.30	39.43	72.67	50.69	101.92	72.38	21.32	26.10	37.72
Nutrient loss by leaching	9.03	11.58	9.28	10.36	7.78	9.88	9.71	10.29	9.32
Burning	-	-	-	-	-	-	-	-	-
Harvesting	1.87	-	1.87	1.57	1.51	-	-	-	-
Total output	268.61	51.01	223.73	62.62	268.30	82.26	31.03	36.39	47.05
Nutrient balance = input – output	–252.66	–35.05	–207.77	–46.66	–252.34	–66.31	–15.07	–20.43	–31.09

Appended Table 15.A1 Continued

	Plot 1	Plot 2	Plot 3	Plot 4	Plot 5	Plot 6	Plot 7	Plot 8	Plot 9
K total in 2004				Fallow				Forest	
Input									
Seeding	-	-	-	-	-	-	-	-	-
Rainfall	57.74	57.74	57.74	57.74	57.74	57.74	57.74	57.74	57.74
Total input	57.74	57.74	57.74	57.74	57.74	57.74	57.74	57.74	57.74
Output									
Nutrient loss by erosion	-	-	-	-	-	-	-	-	-
Nutrient loss by runoff	12.05	17.46	22.87	12.65	15.79	13.99	3.63	4.87	4.50
Nutrient loss by leaching	14.76	17.88	22.05	17.77	17.03	22.37	9.04	8.76	8.45
Burning	-	-	-	-	-	-	-	-	-
Harvesting	-	-	-	-	-	-	-	-	-
Total output	26f.81	35.34	44.92	30.43	32.81	36.36	12.67	13.63	12.94
Nutrient balance = input – output	30.93	22.40	12.82	27.31	24.93	21.38	45.07	44.11	44.80

K AVAILABLE IN 2000	PLOT 1	PLOT 2	PLOT 3	PLOT 4	PLOT 5	PLOT 6	PLOT 7	PLOT 8	PLOT 9
				UPLAND RICE				FOREST	
Input									
Seeding	-	-	-	-	-	-	-	-	-
Rainfall	35.11	35.11	35.11	35.11	35.11	35.11	35.11	35.11	35.11
Total input	35.11	35.11	35.11	35.11	35.11	35.11	35.11	35.11	35.11
Output									
Nutrient loss by erosion	1.54	1.28	5.63	94.79	6.74	13.45	1.14	0.87	1.10
Nutrient loss by runoff	89.53	109.97	99.25	5.66	74.59	95.35	72.87	64.54	62.05
Nutrient loss by leaching	10.09	9.05	8.73	8.71	9.57	9.38	5.83	5.65	6.02
Burning	-	-	-	-	-	-	-	-	-
Harvesting	-	-	-	-	-	-	-	-	-
Total output	101.15	120.30	113.62	109.17	90.90	118.18	79.84	71.06	69.18
Nutrient balance = input – output	–66.04	–85.19	–78.51	–74.06	–55.79	–83.07	–44.73	–35.95	–34.07

Appended Table 15.A1 Continued

	PLOT 1	PLOT 2	PLOT 3	PLOT 4	PLOT 5	PLOT 6	PLOT 7	PLOT 8	PLOT 9
K AVAILABLE IN 2001	UPLAND RICE	FALLOW	UPLAND RICE		CASSAVA		FOREST		
Input									
Seeding	-	-	-	-	-	-	-	-	-
Rainfall	35.11	35.11	35.11	35.11	35.11	35.11	35.11	35.11	35.11
Total input	35.11	35.11	35.11	35.11	35.11	35.11	35.11	35.11	35.11
Output									
Nutrient loss by erosion	5.02	0.17	6.35	5.53	3.31	6.40	0.12	0.09	0.15
Nutrient loss by runoff	45.51	39.69	38.88	39.47	31.37	45.76	26.75	26.34	28.89
Nutrient loss by leaching	18.91	16.14	16.57	13.20	17.01	27.13	29.08	64.23	28.32
Burning	-	-	-	-	-	-	-	-	-
Harvesting	-	-	-	-	-	-	-	-	-
Total output	69.44	56.00	61.79	58.20	51.69	79.29	55.95	90.66	57.36
Nutrient balance = input – output	–34.33	–20.89	–26.68	–23.09	–16.58	–44.18	–20.84	–55.55	–22.25

K available in 2002	Plot 1	Plot 2	Plot 3	Plot 4	Plot 5	Plot 6	Plot 7	Plot 8	Plot 9
	Cassava	Fallow		Cassava		Fallow		Forest	
Input									
Seeding	-	-	-	-	-	-	-	-	-
Rainfall	35.11	35.11	35.11	35.11	35.11	35.11	35.11	35.11	35.11
Total input	35.11	35.11	35.11	35.11	35.11	35.11	35.11	35.11	35.11
Output									
Nutrient loss by erosion	1.50	0.08	1.79	0.64	2.18	0.00	-	-	-
Nutrient loss by runoff	38.02	30.53	37.29	29.54	29.54	29.44	12.15	12.89	15.09
Nutrient loss by leaching	11.79	13.15	11.66	10.67	12.66	13.18	14.94	13.66	12.89
Burning	-	-	-	0.00	-	-	-	-	-
Harvesting	-	-	-	-	-	-	-	-	-
Total output	51.31	43.75	50.75	40.85	44.38	42.62	27.09	26.54	27.97
Nutrient balance = input – output	–16.20	–8.64	–15.64	–5.74	–9.27	–7.51	8.02	8.57	7.14

Appended Table 15.A1 Continued

K available in 2003	Plot 1	Plot 2	Plot 3	Plot 4	Plot 5	Plot 6	Plot 7	Plot 8	Plot 9
	Cassava	Fallow	Cassava +melia +Styrax			Fallow	Forest		
Input									
Seeding	-	-	-	-	-	-	-	-	-
Rainfall	14.93	14.93	14.93	14.93	14.93	14.93	14.93	14.93	14.93
Total input	14.93	14.93	14.93	14.93	14.93	14.93	14.93	14.93	14.93
Output									
Nutrient loss by erosion	1.17	-	1.06	-	1.06	-	-	-	-
Nutrient loss by runoff	44.20	22.34	39.35	30.30	48.49	36.01	31.13	17.14	25.30
Nutrient loss by leaching	8.71	11.14	8.95	9.97	7.55	9.58	13.05	13.96	11.91
Burning	-	-	-	-	-	-	-	-	-
Harvesting	-	-	-	-	-	-	-	-	-
Total output	54.08	33.49	49.37	40.27	57.09	45.59	44.18	31.10	37.21
Nutrient balance = input – output	–39.14	–18.55	–34.43	–25.33	–42.16	–30.66	–29.24	–16.17	–22.28

	PLOT 1	PLOT 2	PLOT 3	PLOT 4	PLOT 5	PLOT 6	PLOT 7	PLOT 8	PLOT 9
K AVAILABLE IN 2004	FALLOW						FOREST		
Input									
Seeding	-	-	-	-	-	-	-	-	-
Rainfall	55.29	55.29	55.29	55.29	55.29	55.29	55.29	55.29	55.29
Total input	55.29	55.29	55.29	55.29	55.29	55.29	55.29	55.29	55.29
Output									
Nutrient loss by erosion	-	-	-	-	-	-	-	-	-
Nutrient loss by runoff	9.57	13.91	20.57	8.75	13.01	12.22	2.82	3.47	3.42
Nutrient loss by leaching	14.59	17.73	21.87	17.63	16.89	22.19	8.95	8.67	8.36
Burning	-	-	-	-	-	-	-	-	-
Harvesting	-	-	-	-	-	-	-	-	-
Total output	24.16	31.64	42.44	26.38	29.91	34.40	11.77	12.14	11.79
Nutrient balance = input – output	31.13	23.65	12.84	28.91	25.38	20.88	43.52	43.14	43.50

Appended Table 15.A2 Nutrient balances (kg/0.102 ha/year) in paddy fields adjacent to upland stream outlet

	Total		
	N	P	K
2000			
Input			
Seeding	0.20	0.06	0.19
Rainfall	2.99	0.69	3.76
Organic fertilizer	1.01	0.76	7.32
Chemical fertilizer	3.45	6.69	0.00
Water inflow	97.33	9.68	337.98
Total	104.98	17.88	349.25
Output			
Harvesting	9.76	3.12	11.19
Herbivores	0.48	0.21	0.70
Water outflow	70.19	8.81	337.35
Total	80.42	12.14	349.24
Nutrient balance	24.56	5.74	0.01
2001			
Input			
Seeding	0.11	0.02	0.03
Rainfall	2.99	0.69	3.76
Organic fertilizer	-	-	-
Chemical fertilizer	11.73	10.00	0.00
Water inflow	227.23	25.23	379.90
Total	242.06	35.93	383.69
Output			
Harvesting	7.18	1.81	4.18
Herbivores	0.48	0.21	0.70
Water outflow	267.27	22.34	391.56
Total	274.93	24.36	396.45
Nutrient balance	–32.87	11.58	–12.76

Appended Table 15.A2 Continued

	TOTAL		
	N	P	K
2002			
Input			
Seeding	0.11	0.02	0.03
Rainfall	2.99	0.69	3.76
Organic fertilizer	-	-	-
Chemical fertilizer	11.73	10.00	-
Water inflow	77.87	7.95	188.98
Total	92.70	18.65	192.77
Output			
Harvesting	11.65	3.19	13.45
Herbivores	0.12	0.03	0.13
Water outflow	79.01	9.74	174.01
Total	90.78	12.96	187.59
Nutrient balance	1.92	5.69	5.18
2003			
Input			
Seeding	0.11	0.02	0.03
Rainfall	3.02	1.21	1.63
Organic fertilizer	-	-	-
Chemical fertilizer	11.73	10.00	-
Water inflow	64.48	7.57	62.52
Total	79.34	18.79	64.18
Output			
Harvesting	9.54	2.15	12.56
Herbivores	0.07	0.02	0.09
Water outflow	62.36	15.17	63.42
Total	71.97	17.33	76.07
Nutrient balance	7.37	1.46	–11.89

Appended Table 15.A2 Continued

	Total		
	N	P	K
2004			
Input			
Seeding	0.11	0.02	0.03
Rainfall	3.02	1.21	1.63
Organic fertilizer	-	-	-
Chemical fertilizer	11.73	10.00	-
Water inflow	64.48	7.57	62.52
Total	79.34	18.79	64.18
Output			
Harvesting	9.54	2.15	12.56
Herbivores	0.07	0.02	0.09
Water outflow	62.36	15.17	63.42
Total	71.97	17.33	76.07
Nutrient balance	7.37	1.46	–11.89

Appended Table 17.A1 Rules used for classifying general farming system type by commune

General village level farming system type	% of landscape under ag.	Ratio of upland ag. to paddy	NLSI upland	NLSI paddy	Landscape pattern guidelines for visual interpretation[a]
Pure systems					
Rotational swidden system (RSA)	Less than 25%	Greater than 10:1	> 0.27	> 0.40	Little paddy (very scattered); upland ag. (0.3 ha to 10 ha) patches; regrowth and trees intermixed with patches of upland ag. and paddy.
Permanent paddy system (PRP)	Greater than or equal to 40%	Less than 1:7	Not important	< 0.20	Little upland ag.; paddy (large areas); other land cover (large continuous areas)
Permanent upland agriculture system (PUA)	Greater than or equal to 40%	Greater than (+/–) 8:1	< 0.20	Not important	Little paddy; permanent upland ag. (large areas); other land cover (large continuous areas)
Mixed Systems					
Systems with PRP and PUA	Greater than 25%	Roughly equal; can be more 7:1 or 1:7	< 0.20 (can be > 0.20, determined by visual analysis)	< 0.20	Paddy and upland ag. (large and aggregated); other types of land cover (aggregated). Few to no scattered upland ag. fields.
Systems with PUA and RSA	Between 25% and 50%	Greater than 10:1	Between (+/–) 0.20 and 0.28	N/A	Large continuous upland ag. in parts of commune; patches of upland ag. interspersed with other land cover in other parts of commune. Little if any paddy.
Systems with PUA, PRP and RSA (can be composite swidden agriculture)	Between 25% and (+/–) 40%	Usually roughly equal; can be more 3:1 or 1:3	Between 0.20 and 0.28 (can be > 0.28; determined by visual analysis)	> 0.28	Large continuous upland ag. found in parts of commune; large continuous paddy found in parts of commune; patches of upland ag. (swidden) interspersed with other land cover. (If in close proximity (500m) of small and/or medium size paddy then can be composite swidden agriculture).
Composite swidden agriculture or mixed composite swidden agriculture with RSA in the same commune	Less than 25%	Greater than 1.5:1 (can be 1:1 or have more paddy than swidden; refer to visual analysis)	> 0.27	> 0.27	Patches of upland ag. 0.3 to 10 ha (swidden) interspersed with other land cover (regrowth and trees); in close proximity (500m) of small to medium size paddy. If rotational swidden fields not in close proximity to paddy, can be areas of 'only' RSA.

Note. a: When pattern and metrics do not agree, more weight is given to the visual interpretation of pattern (Adapted from Leisz et al. (2005))

Appended Table 17.A2 Land cover, landscape, topographic and population density information for each of the general farming system types identified in the NMR (Adapted from Leisz et al. 2005)

General village level farming system type	% of landscape under ag.	Ratio of upland ag. to paddy	NLSI upland	NLSI paddy	Topography	Landscape pattern guidelines for visual interpretation[a]	Pop density (sq km)
Pure systems							
Rotational swidden system (RSA)	Less than 25%	Greater than 10:1	> 0.27	> 0.40	No flat valley bottoms suitable for paddy; slope between 10 and 45 degrees	Little paddy (very scattered); upland ag. (0.3 ha to 10 ha) patches; regrowth and trees intermixed with patches of upland ag. and paddy.	53
Permanent paddy system (PRP)	Greater than or equal to 40%	Less than 1:7	Not important	< 0.20	Large flat valleys	Little upland ag.; paddy (large areas); other land cover (large continuous areas)	1,411
Permanent upland agriculture system (PUA)	Greater than or equal to 40%	Greater than (+/–) 8 : 1	< 0.20	Not important	No flat valley bottoms suitable for paddy; slope between 10 and 45 degrees	Little paddy; permanent upland ag. (large areas); other land cover (large continuous areas)	50
Mixed Systems							
Systems with PRP and PUA	Greater than 25%	Roughly equal; can be more 7:1 or 1:7	< 0.20 (can be > 0.20, determined by visual analysis)	< 0.20	Large flat valley bottoms suitable for paddy; slope between 10 and 45	Paddy and upland ag. (large and aggregated); other types of land cover (aggregated). Few to no scattered upland ag. fields.	220
Systems with PUA and RSA	Between 25% and 50%	Greater than 10:1	Between (+/–) 0.20 and 0.28	N/A	No flat valley bottoms suitable for paddy; slope between 10 and 45 degrees	Large continuous upland ag. in parts of commune; patches of upland ag. interspersed with other land cover in other parts of commune. Little if any paddy.	26

Systems with PUA, PRP and RSA (can be composite swidden agriculture)	Between 25% and (+/–) 40%	Usually roughly equal; can be more 3:1 or 1:3	Between 0.20 and 0.28 (can be > 0.28; determined by visual analysis)	> 0.28	Small to medium areas of flat valley bottoms suitable for paddy; surrounding mountain/hill slopes between 10 and 45 degrees	Large continuous upland ag. found in parts of commune; large continuous paddy found in parts of commune; patches of upland ag. (swidden) interspersed with other land cover. (If in close proximity (500m) of small and/or medium size paddy then can be composite swidden agriculture).	110
Composite swidden agriculture or mixed composite swidden agriculture with RSA in the same commune	Less than 25%	Greater than 1.5:1 (can be 1:1 or have more paddy than swidden; refer to visual analysis)	> 0.27	> 0.27	Small areas of flat valley bottoms suitable for paddy; surrounding mountain/hill slopes between 10 and 45 degrees	Patches of upland ag. 0.3 to 10 ha (swidden) interspersed with other land cover (regrowth and trees); in close proximity (500m) of small to medium size paddy. If rotational swidden fields not in close proximity to paddy, can be areas of "only" RSA.	48

Note. a: When pattern and metrics do not agree, more weight is given to the visual interpretation of pattern

Bibliography

Abadie, M. (1924), *Les Races du Haut-Tonkin de Phong-Tho à Lang Son*, Paris: Société d'Editions Géographiques, Maritimes, et Coloniales.

Adger, W.N. and K. Brown (1994), *Land Use and the Causes of Global Warming*, Chichester: John Wiley & Sons.

Agarwal, A. and A. Chak (1991), *Floods, Flood Plains and Environmental Myths: State of India's Environment: A Citizen's Report*, #3, New Delhi: Centre for Science and Environment.

.... and C.C. Gibson (1999), 'Enchantment and disenchantment: The role of community in natural resource conservation', *World Development*, 27 (4), pp. 629-649.

.... and E. Ostrom (2001), 'Collective action, property rights, and decentralization in resource use in India and Nepal', *Politics & Society* 29, pp. 485-514.

Alcorn, J. (1996), 'Forest use and ownership: patterns, issues, and recommendations', in J. Schelhas and R. Greenberg (eds), *Forest Patches in Tropical Landscapes*, Washington, D.C.: Island Press, pp. 233-257.

Andriesse, J.P. and R.M. Schelhaas (1987), 'A monitoring study of nutrient cycles in soils used for shifting cultivation under various climatic conditions in tropical Asia', *Agriculture, Ecosystems, and the Environment*, 19, pp. 285-332.

Anthes, R.A. (1984), 'Enhancement of convective precipitation by mesoscale variations in vegetative covering in semiarid regions', *Journal of Climate and Applied Meteorology*, 23, pp. 541-554.

ASB (1994), *Alternatives to Slash-and-Burn: A Global Initiative,* Nairobi: International Centre for Research on Agroforestry.

Bandy, D.E., Garrity, D.P., and P.A. Sanchez (1993), 'The worldwide problem of slash-and-burn agriculture', *Agroforestry Today,* 5 (3), pp. 2-6.

Banfield, E.G. (1958), *The Moral Basis of a Backward Society*, The Free Press: New York.

Bế Viết Đặng, Nguyễn Văn Huy, and Chu Thái Sơn (eds) (1992), *Các dân tộc Tày, Nùng ở Việt Nam* (The Tay and Nung ethnic groups in Vietnam), Hà Nội: Viện dân tộc học.

Botkin, D. (1990), *Discordant Harmonies: A New Ecology for the Twenty-First Century*, New York: Oxford University Press.

Bouahom, B., Bùi Hữu Phát, Cù Viết Hà, Lê Văn Thư, Lê Văn Lanh, and Phạm Xuân Thao (1993), 'The role of livestock in the agroecosystem', in Lê Trọng Cúc and A.T. Rambo (eds), *Too Many People, Too Little Land: The*

Human Ecology of a Wet Rice-Growing Village in the Red River Delta of Vietnam. Report of the SUAN-EWC-CRES Field Research Workshop in Thai Binh Province, July 1991. Program on Environment Occasional Paper No. 15, Honolulu: East-West Center, pp. 117-129.

Braun, B. (2002), *The Intemperate Rainforest: Nature, Culture, and Power on Canada's West Coast*, Minneapolis: University of Minnesota Press.

Brons, J.E. and R. Ruben (2005), *Farm Supplementary Activities, Poverty and Income Distribution: An Income Composition Analysis in Burkina Faso*, The Hague: Wageningen University Dissertation No 3764, Development Economics Group.

Brosius, J.P. and A.L. Tsing (1998), 'Representing communities: histories and politics of community based natural resource management', *Society & Natural Resources*, 11 (2), pp. 157-169.

Browder, J. (1996), 'Reading colonist landscapes: social interpretations of tropical forest patches in an Amazonian agricultural frontier', in J. Schelhas and R. Greenberg (eds), *Forest Patches in Tropical Landscapes*, Washington, D.C.: Island Press, pp. 285-299.

Brown, S. and A.E. Lugo (1990), 'Tropical secondary forests', *Journal of Tropical Ecology*, 6, pp. 1-32.

Bryant, R. (1997), *The Political Ecology of Forestry in Burma, 1824-1994*, Honolulu: University of Hawaii Press.

Bùi Thị Gia, Phạm Tiến Dũng, Đặng Việt Quang and P. Calkins (2005), 'Production and marketing of agro-forest products in Northwest Uplands of Vietnam; A case study in Tat village, Tan Minh, Da Bac, Hoa Binh province', in Trần Đức Viên, Nguyễn Viết Tùng, Calkins, P., Allbritton, A., Hiesinger, M.A., Vũ Thị Thao, Phạm Thị Mai Hương, Trần Nam Anh, Lê Thị Thanh Phương, and Đặng Việt Quang (eds), *Marketing and Agroforestry Development in Vietnam's Uplands*, Hà Nội: Nhà xuất bản Nông nghiệp Hà Nội, pp. 178-211.

Calder, I. (1999), *The Blue Revolution: Land Use and Integrated Water Resources*, London: Earthscan.

Cannon, C.H., Peart, D., and M. Leighton (1998), 'Tree species diversity in commercially logged Bornean rainforest', *Science*, 281, pp. 1366-1368.

Chu Thái Sơn and Nguyễn Chí Huyền (1974), 'Thành phần dân tộc người Thổ ở huyện Đà Bắc, Hòa Bình' (Ethnic composition of the Tho people of Da Bac district, Hoa Binh), *Tạp chí Dân tộc học*, 2, pp. 42-52.

Collins, B.S. and S.T. Pickett (1987), 'Influence of canopy opening on the environment and herb layer in a northern hardwood forest', *Vegetation*, 70, pp. 3-10.

Collins, M., Sayer, J., and T. Whitmore (1991), *The Conservation Atlas of Tropical Forests: Asia and the Pacific,* Toronto: Simon and Schuster.

Conklin, H.C. (1957), *Hanunoo Agriculture: A Report on an Integral System of Shifting Cultivation in the Philippines*, Rome: FAO.

Conway, G. (1984), 'What is an agroecosystem and why is it worthy of study?', in A.T. Rambo and P.E. Sajise (eds), *An Introduction to Human Ecology Research on Agricultural Systems in Southeast Asia,* Los Baños: University of the Philippines at Los Baños, pp. 25-38.

.... (1986), *Agroecosystem Analysis for Research and Development,* Bangkok: Winrock International.

Cuisinier, J. (1948), *Les Mường. Géographie Humaine et Sociologie* (The Muong : Human geography and sociology), Paris: Institut d'Ethnologie.

Cupet, P. (2000), *Travels in Laos and among the Tribes of Southeast Indochina,* Translated by W. E. J. Tips. Bangkok: White Lotus Press.

Đặng Nghiêm Văn and Chu Thái Sơn (2000), Các dân tộc thiểu số ở Việt Nam (*Ethnic Minorities in Vietnam),* Hà Nội: Nhà xuất bản Thế Giới.

Đào Thế Tuấn (1998), Những thách thức trong sản xuất nông nghiệp ở vùng núi phía bắc của Việt Nam (*The Crisis in Agriculture in the Northern Uplands of Vietnam),* Hà Nội: Viện khoa học Nông nghiệp Việt Nam (VASI).

Devall, B. and G. Sessions (1985), *Deep Ecology,* Salt Lake City: G.M. Smith.

Dillon, J.L. and D.J. McConnell (1997), *Farm Management for Asia: A Systems Approach,* (FAO Farm Systems Management Series-13), *http://www.fao.org/docrep/W7365E/w7365e00.HTM*

Donovan, D., Rambo, A.T., Fox, J., Lê Trọng Cúc, and Trần Đức Viên (eds) (1997), *Development Trends in Vietnam's Northern Mountain Region,* Hanoi: National Political Publishing House, 2 volumes.

Dove, M.R. (1983), 'Review of "Ethnographic Atlas of Ifugao": Implications for theories of agricultural evolution in Southeast Asia', *Current Anthropology,* 24 (4), pp. 516-519.

.... (1985), *Swidden Agriculture in Indonesia: The Subsistence Strategies of the Kalimantan Kantu'*. Berlin/New York: Mouton.

Durrenberger, E.P. (1981), 'The economy of a Shan village', *Ethnos*, 46 (1-2), pp. 64-79.

Ehrlich, P. (1968), *The Population Bomb*, New York: Ballantine Books.

Ewel, J. (1980), 'Tropical succession: Manifold routes to maturity', *Biotropica*, 12 (2), pp. 2-7.

Fairhead, J. and M. Leach (1996), *Misreading the African Landscape: Society and Ecology in a Forest-Savanna Mosaic,* Cambridge: Cambridge University Press.

FAO (1993), *Forest Resources Assessment 1990: Tropical Countries*, Rome: FAO.

FAO (2000), *FRA 2000 on Definitions of Forest and Forest Change*, Forest Resources Assessment Programme, Rome: FAO.

Ford Foundation (1997), *Forestry for Sustainable Rural Development: A Review of Ford Foundation-supported Community Forestry Programs in Asia*, New York: Ford Foundation.

Forsyth, T. (1996), 'Science, myth, and knowledge: Testing Himalayan environmental degradation in Thailand', *Geoforum*, 27 (3), pp. 375-392.

.... (2003), *Critical Political Ecology: The Politics of Environmental Science*, London: Routledge.

Fox, J., Krummel, J., Yarnasarn, S., Ekasingh, M., and N. Podger (1995), 'Land use and landscape dynamics in northern Thailand: Assessing change in three upland watersheds', *Ambio*, 24, pp. 328-334.

....................., Đào Minh Trường, Rambo, A.T., Nghiêm Phương Tuyến, Lê Trọng Cúc, and S.J. Leisz, (2000), 'Shifting cultivation: A new old paradigm for managing tropical forests', *BioScience,* 50 (6), pp. 521-528.

....................., Rambo, T., Donovan, D., Le Trong Cuc, Giambelluca, T., Ziegler, A., Plondke, D., Tran Duc Vien, Leisz, S., and Dao Minh Truong (2003), 'Linking household and remotely sensed data for understanding forest fragmentation in northern Vietnam', In J. Fox, Rindfuss, R., Walsh, S., and Mishra, V., (eds), *People and the Environment: Approaches for Linking Household and Community Surveys to Remote Sensing and GIS*, Boston: Kluwer Academic Publishers, pp. 201-221.

Fukui, H. (1993), *Food and Population in a Northeast Thai Village,* Monographs of the Center for Southeast Asian Studies, Kyoto University, English Language Series, 19, Honolulu: University of Hawaii Press.

Fukuyama, F. (1999), *The Great Disruption: Human Nature and the Reconstitution of Social Order,* New York: Simon & Schuster.

Ganjanapan, A. (1994), 'The Northern Thai land tenure system: Local customs versus national laws', *Law & Society Review*, 28, pp. 609.

Geertz, C. (1963), *Agricultural Involution: The Process of Ecological Change in Indonesia*, Berkeley: University of California Press.

Geist, H.J. and E.F. Lambin (2002), 'Proximate causes and underlying driving forces of tropical deforestation', *BioScience*, 52 (2), pp. 143-150.

Giambelluca, T.W. (2002), 'Hydrology of altered tropical forest', *Hydrological Processes*, 16, pp. 1665-1669.

....................., Ziegler, A.D., Nullet, M.A., Đào, T.M., and L. T. Trần (2003),

'Transpiration in a small tropical forest patch', *Agricultural and Forest Meteorology*, 117, pp.1-22.

Giorgi, F. (1989), 'Two-dimensional simulations of possible mesoscale effects of nuclear war fires, I. Model description', *Journal of Geophysical Research*, 94, pp. 1127-1144.

Gomez-Pompa, A. and C. Vasquez-Yanes (1974), 'Studies on secondary succession of tropical lowlands: The life cycle of secondary species,' Paper read at First International Congress of Ecology, at The Hague.

Grandstaff, T.G. (1988), 'Environment and economic diversity in Northeast Thailand', in T. Charoenwatana and A.T. Rambo (eds), *Sustainable Rural Development in Asia: Selected Papers from the Fourth SUAN Regional Symposium on Agroecosystem Research*, Khon Kaen, Thailand: KKU-USAID Farming Systems Research Project and SUAN, pp. 11-22.

Hamilton, L. (1987), 'What are the impacts of deforestation in the Himalayas on the Ganges-Brahmaputra lowlands and delta? Relations between assumptions and facts', *Mountain Research and Development,* 7, pp. 256-263.

.... (1988), 'Forestry and watershed management', in J. Ives and D. Pitt (eds), *Deforestation: Social Dynamics in Watershed and Mountain Ecosystems*, London: Routledge, pp. 99-124.

Harvey, D. (1996), *Justice, Nature, and the Geography of Difference*, Oxford: Blackwell.

Harwood, R.R. (1979), *Small Farm Development*, Boulder: Westview Press.

Hashim, G.M., Coughlan, K.J., and J.K. Syers (1998), 'On-site nutrient depletion: an effect and a cause of soil erosion', in: F.W.T. Penning de Vries, F. Agus and J. Kerr (eds), *Soil Erosion at Multiple Scales*, New York: CAB International, pp. 207-221.

Henderson-Sellers, A., Dickinson, R.E., Durbidge, T.B., Kennedy, P.J., McGuffie, K. and A.J. Pitman (1993), 'Tropical deforestation: Modeling local- to regional-scale climate change', *Journal of Geophysical Research*, 98, pp. 7289-7315.

Hirsch, P. (1993), *Political Economy of Environment in Thailand,* Manila: Journal of Contemporary Asian Publishers.

Hoàng Thị Sen and Lê Văn An (2006), 'Creating opportunities for change: Strengthening the social capital of women and the poor in upland communities in Hue, Vietnam', In R. Vernooy (ed), *Social and Gender Analysis in Natural Resources Management; Learning Studies and Lessons from Asia*, New Delhi: SAGE publications, pp. 155-180.

Hollings, C. (1973), 'Resilience and stability of ecological systems', *Annual Review of Ecology and Systematics,* 4, pp. 1-23.

.... (1979), 'Myths of ecological stability: resilience and the problem of failure', In *Studies on Crisis Management*, G. Smart and W. Standbury (eds), Montreal: Butterworth, pp. 102-133.

Hölscher D., Möller R.F., Denich M. and H. Fölster (1996), 'Nutrient input-output budget of shifting agriculture in Eastern Amazonia', *Nutrient Cycling in Agroecosystems,* 47, pp. 49-57.

International Food Policy Research Institute (IFPRI) (2003), *Income Diversification and Poverty in the NMR of Vietnam,* Hanoi: Agricultural Publishing House.

Ireson, C.J. and W.R. Ireson (1996), *Cultivating the Forest: Gender and the Decline of Wild Resources among the Tay of Northern Vietnam,* Honolulu: East-West Center Working Papers Indochina Series No. 6.

Ives, J. and B. Messereli (1989), *The Himalayan Dilemma: Reconciling Conservation and Development,* London: Routledge/UNU.

Jamieson, N.L., Lê Trọng Cúc, and A.T Rambo (1998), *The Development Crisis in Vietnam's Mountains*, East-West Center Special Report No 6, Honolulu: East-West Center.

Jewitt, S. (1995), 'Europe's "others"? Forestry policy and practices in colonial and postcolonial India', *Environment and Planning D: Society and Space,* 13, pp. 67-90.

Johnson, C. and T. Forsyth (2002), 'In the eyes of the state: Negotiating a 'rights-based approach' to forest conservation in Thailand', *World Development*, 30, pp. 1591-1605.

Kapos, V. (1989), 'Effects of isolation on the water status of forest patches in the Brazilian Amazon', *Journal of Tropical Ecology*, 5, pp. 173-185.

Kass, D.C.L., and E. Somarriba (1999), 'Traditional fallow in Latin America', *Agroforestry Systems,* 47, pp. 13-36.

Keister, L.A. (2000), *Wealth in America: Trends in Wealth Inequality,* Cambridge: Cambridge University Press.

Khổng Diễn (ed) (1996), *Những đặc điểm kinh tế- xã hội các dân tộc miền núi phía bắc (Economic and social characteristics of ethnic groups in the Northern Mountain Region)*, Hà Nội: Nhà xuất bản Khoa học Xã hội.

KKU-Ford Cropping Systems Project (1982), *An Agroecosystem Analysis of Northeast Thailand,* Khon Kaen: Faculty of Agriculture, Khon Kaen University.

Kono, Y. and A.T. Rambo (2004), 'Some key issues relating to sustainable agro-resources management in the mountainous region of mainland Southeast Asia', *Southeast Asian Studies (Kyoto),* 41 (4), pp. 550-565.

Kunstadter, P. (1978), 'Subsistence agriculture economics of Lua and Karen hill farmers, Mae Sariang District, Northwestern Thailand', in P. Kunstadter,

E.C. Chapman, and S. Sabhasri (eds), *Farmers in the Forest: Economic Development and Marginal Agriculture in Northern Thailand*, Honolulu: The University Press of Hawaii, pp. 74-133.

.... and E.C. Chapman (1978), 'Problems of shifting cultivation and economic development in Northern Thailand,' in P. Kunstadter, Chapman, E.C., and S. Sabhasri (eds), *Farmers in the Forest: Economic Development and Marginal Agriculture in Northern Thailand*, Honolulu: The University Press of Hawaii: 3-23.

...................., E.C. Chapman, and S. Sabhasri (eds) (1978), *Farmers in the Forest: Economic Development and Marginal Agriculture in Northern Thailand,* Honolulu: The University Press of Hawaii.

Laurance, W. F., and R.O. Bierregaard (eds) (1997), *Tropical Forest Remnants: Ecology, Management, and Conservation of Fragmented Communities*, Chicago: University of Chicago Press.

Lawrence, D., Peart, D., and M. Leighton (1998), 'The impact of shifting cultivation on a rainforest landscape in West Kalimantan: Spatial and temporal dynamics', *Landscape Ecology,* 13, pp.135-148.

Lê Trần Chấn and Ngô Thế Ân (2000), *Vegetation map with scale 1:500*, Hanoi: unpublished CARES report.

Lê Trọng Cúc, Gillogly, K., and A.T. Rambo (eds) (1990), *Agroecosystems of the Midlands of Northern Vietnam: A Report on a Preliminary Human Ecology Field Study of Three Districts in Vinh Phu Province,* EAPI Occasional Paper No. 12. Honolulu: East-West Center.

.... and A.T. Rambo (eds) (1993), *Too Many People, Too Little Land: The Human Ecology of a Wet Rice-Growing Village in the Red River Delta of Vietnam.* Report of the SUAN-EWC-CRES Field Research Workshop in Thái Bình Province, July 1991. Program on Environment Occasional Paper No. 15, Honolulu: East-West Center.

.... and A.T. Rambo (eds), (1995), *Các vấn đề sinh thái nhân văn ở Việt Nam* (Issues in human ecology in Vietnam), Hanoi: Agriculture Publishing House.

.... and A.T. Rambo (eds) (1999), *Composite swidden farmers of Ban Tat: A case study of the environmental and social conditions in a Tay ethnic minority community in Hoa Binh Province, Vietnam*, Review draft, Research Report No. 1, Center for Natural Resources and Environmental Studies (CRES), Hanoi: Vietnam National University.

.... and A.T. Rambo (eds) (2001), *Bright Peaks, Dark Valleys; A Comparative Analysis of Environmental and Social Conditions and Development Trends in Five Communities in Vietnam's Northern Mountain Region,* Hanoi: National Political Publishing House.

.... , Rambo, A.T., Fahrney, K., Trần Đức Viên, Romm, J., and Đặng Thị Sy, (eds) (1996), *Red Books, Green Hills: The Impact of Economic Reform on Restoration Ecology in the Midlands of Northern Vietnam,* Honolulu: East-West Center.

Lebbie, A.R. and M. Schoonmaker Frudenberger (1996), 'Sacred groves in Africa: Forest patches in transition', in J. Schelhas and R. Greenberg (eds), *Forest Patches in Tropical Landscapes*, Washington, D.C.: Island Press, pp. 300-326.

Leisz, S.J., Nguyễn Thị Thu Hà, Nguyễn Thị Bích Yên, Nguyễn Thanh Lâm, and Trần Đức Viên (2005), 'Developing a methodology for identifying, mapping and potentially monitoring the distribution of general farming system types in Vietnam's Northern Mountain Region', *Agricultural Systems,* 85, pp. 340-363.

.... , Đào Minh Trường, Lê Trần Chấn, and Lê Trọng Hải (2001), 'Land-cover and land-use', in Lê Trọng Cúc and A.T. Rambo (eds), *Bright Peaks, Dark Valleys: A Comparative Analysis of Environmental and Social Conditions and Development Trends in Five Communities in Vietnam's Northern Mountain Region,* Hanoi: National Political Publishing House, pp. 85-122.

Li, T.M. (2002), 'Engaging simplifications: community-based resource management, market processes and state agendas in upland Southeast Asia', *World Development,* 30, pp. 265-283.

Lindert, P.H. (2000), *Shifting Ground: The Changing Agricultural Soils of China and Indonesia,* Cambridge, MA: The MIT Press.

Marten, G.C. (2001), *Human Ecology: Basic Concepts for Sustainable Development,* London: Earthscan Publications.

.... and A.T. Rambo (1988), 'Guidelines for writing comparative case studies of Southeast Asian rural ecosystems', in K. Rerkasem and A.T. Rambo, (eds), *Agroecosystem Research for Rural Development,* Chiang Mai: Chiang Mai University Multiple Cropping Centre, pp. 263-285.

Matlack, G.R. (1993), 'Microenvironment variation within and among deciduous forest edge sites in the eastern United States', *Biological Conservation,* 66, pp. 185-194.

McCaskill, D. and K. Kampe (eds) (1997), *Development or Domestication? Indigenous Peoples of Southeast Asia,* Chiang Mai: Silkworm Books.

Miller, S. (2001), *An Analysis of the Use of Plants for the Treatment of Postpartum Ailments Among Tay Females in the Highland of Northern Vietnam*, M.S. Thesis, Department of Anthropology, Kent University.

Mission 'Economie et Humanisme' (1959), *Etude sur les Conditions de Vie et les Besoins de la Population du Viet-Nam. (Study on the living conditions and the needs of the population of Vietnam)*, Saigon: Banque Nationale du Viet-Nam.

Murcia, C. (1995), 'Edge effects in fragmented forests: Implications for conservation', *Trends in Ecology & Evolution,* 10 (2), pp. 58-62.

Myers, N. (1984), *The Primary Source: Tropical Forests and our Future*, New York: Norton.

.... (1992), 'Tropical forests: The policy change', *Environmentalist,* 12 (1), pp. 15-27.

National Centre for Social Sciences and the Humanities (2001), *National Human Development Report 2001: Doi Moi and Human Development,* Hanoi: National Political Publishing House.

Ngô Đình Quế, Vũ Tấn Phương, Lê Quốc Huy, Đinh Thanh Giang, Nguyễn Thanh Tùng, Nguyễn Văn Thắng (2003), *Nghiên cứu xây dựng các tiêu chí và chỉ tiêu trồng rừng theo cơ chế phát triển sạch ở Việt nam* (Development of indicators and criteria for forest plantations in Vietnam following the clean development mechanism), Hà nội: Viện Khoa học Lâm nghiệp Việt nam.

Ngô Thị Thuận, Đặng Văn Tiến, Phạm Thị Hương Dịu, and Mai Lan Phương (2004), *Relationship among household characteristics, agricultural products, and local infrastructure to variation in price of several major products in Tat hamlet, Tan Minh commune, Da Bac district, Hoa Binh province*, Report to the Ford Foundation under the grant to the HUA on capacity building program, Hanoi: Hanoi University of Agriculture.

Nguyễn Đình Hòa (1966), *Vietnamese-English Dictionary*, Rutland, Vermont: Charles E. Tuttle.

Nguyễn Huệ and Thái Phiên (2005), 'Quản lý đất dốc, kinh nghiệm từ nghiên cứu lâu năm' (Sloping land management, view from long-term research), Báo cáo thường niên, Hà Nội: Viện thổ nhưỡng và nông hóa, pp. 33-43.

Nguyễn Thanh Lâm (2004), *Agricultural Sustainability of a Composite Swiddening System: A Case Study in the Uplands of Northern Vietnam*, Khon Kaen: Unpublished Ph.D. dissertation, Khon Kaen University.

...................., Patanothai, A., and A.T. Rambo (2004), 'Recent changes in the composite swidden farming system of a Da Bac Tay ethnic minority community in Vietnam's Northern Mountain Region', *Southeast Asian Studies (Kyoto),* 42 (3), pp. 273-293.

...................., Patanothai, A., Limpinuntana, V., and P. Vityakon (2005), 'Land-use sustainability of composite swiddening in the uplands of Northern Vietnam: Nutrient balances of swiddening fields during the cropping period and changes of major soil nutrient over the swidden cycle', *International Journal of Agricultural Sustainability*, 3 (1), pp. 57-68.

Nguyễn Văn Dung, Cadisch, G., Trần Đức Viên, Nguyễn Thanh Lâm, Trần Mạnh Tường, Patanothai, A., and A.T. Rambo (2006), *A nutrient balance analysis of the sustainability of a composite swiddening agro-ecosystem in Tat hamlet*, Hanoi: Unpublished CARES report.

.................., Trần Đức Viên, Nguyễn Thanh Lâm, Trần Mạnh Tường, and G. Cadisch (2008), 'Analysis of the sustainability within the composite swidden agroecosystem in northern Vietnam. 1. Partial nutrient balances and recovery times of upland fields', *Agriculture, Ecosystems and Environment*, 128 (1-2), pp. 37-51.

Nibbering, J.W. (1991), 'Crisis and resilience in upland land use in Java', in J. M. Hardjono (ed), *Indonesia: Resources, Ecology, and Environment*, Singapore: Oxford University Press, pp. 104-132.

Padoch, C., Harwell, E., and A. Susanto (1998), 'Swidden, sawah, and in-between: Agricultural transformation in Borneo', *Human Ecology,* 26 (1), pp. 3-20.

Patanothai, A. (1996), *Soil under Stress: Nutrient Recycling and Agricultural Sustainability in the Red River Delta of Northern Vietnam*, Honolulu: East-West Center.

.... (ed) (1998), *Land Degradation and Agricultural Sustainability: Case Studies from Southeast and East Asia,* Khon Kaen, Thailand: Regional Secretariat, the Southeast Asian Universities Agroecosystem Network (SUAN), Khon Kaen University.

Peluso, N.L. (1992), *Rich Forests, Poor People: Resource Control and Resistance in Java*, Berkeley: University of California Press.

Phạm Công Hoan (1995), *The Role of Traditional Social Institutions in Community Management of Resources among the Hmong of Vietnam*, East-West Center Working Papers Indochina Series, No. 5, Honolulu: East-West Center.

Phạm Hoàng Hộ (1993), *Cây cỏ Việt nam* (Flora of Vietnam), Hà Nội: Nhà xuất bản Nông nghiệp.

Phạm Thị Mai Hương, Nguyễn Thanh Lâm, and Trần Đức Viên (2005), *Development of Market Channels Associated with Natural Resources Management Issues in Vietnam's Northern Mountain Region: A Case Study in Tat Hamlet,* Hanoi: Unpublished CARES report.

Pinedo-Vasquez, M., and C. Padoch (1996), 'Managing forest remnants and forest gardens in Peru and Indonesia', in J. Schelhas and R. Greenberg (eds), *Forest Patches in Tropical Landscapes*, Washington, D.C.: Island Press, pp. 327-342.

Poll, J.P., and F. Orazem (1984), *Production Economics: Theory with Application,* New York: John Wiley & Sons.

Potter, L., Brookfield, H., and Y. Byron (1994), 'The Sundaland region of Southeast Asia', in J.X. Kasperson, R.E. Kasperson and B.L. Turner II (eds), *Regions at Risk: Comparisons on Threatened Environments*, Tokyo: United Nations University Press, pp. 460-518.

Ramakrishnan, P.S. (1992), *Shifting Agriculture and Sustainable Development: An Interdisciplinary Study from North-East India*, Man and the Biosphere Series Vol. 10, Paris and Carnforth, U.K.: UNESCO and Parthenon.

Rambo, A.T. (1983), *Conceptual Approaches to Human Ecology,* Research Report No. 14, Honolulu: East-West Environment and Policy Institute.

.... (1992), 'Report of a reconnaissance visit to Hoa Binh Province, Northern Vietnam, 22-27 January 1992', Honolulu: East-West Center Environment and Policy Institute. Unpublished typescript.

.... (1997a), 'Development trends in Vietnam's Northern Mountain Region', in D. Donovan, A.T. Rambo, J. Fox, Lê Trọng Cúc and Trần Đức Viên (eds), *Development Trends in Vietnam's Northern Mountain Region*, Vol. 1, Chapter 2, Hanoi: National Political Publishing House, pp. 5-52.

.... (1997b), *the Fallacy of Global Sustainable Development,* Honolulu: East-West Center Issues Paper No. 30.

.... (1997c), 'Culture, environment and human settlement in the uplands of Northern Vietnam', *Vietnam Social Sciences*, 5 (61), pp. 68-72.

.... (1998), 'The composite swiddening agroecosystem of the Tay ethnic minority of the Northwestern Mountains of Vietnam', in A. Patanothai (ed), *Land Degradation and Agricultural Sustainability: Case Studies from Southeast and East Asia*, Khon Kaen, Thailand: Regional Secretariat, the Southeast Asian Universities Agroecosystem Network (SUAN), Khon Kaen University, pp. 43-64.

.... (2001), 'Book review of Bùi Minh Đạo. *Trồng trọt Truyền thống của các Dân tộc ở Tây Nguyên* (Traditional agriculture of indigenous ethnic groups in the Central Highlands of Vietnam', *Southeast Asian Studies (Kyoto)*, 39 (3), pp. 437-439.

.... (2002), 'An agroecosystem analysis of five communities in Vietnam's Northern Mountain Region', Paper presented at the IIAS Workshop on Vietnamese Peasants' Activity, An Interaction between Culture and Nature, Leiden, 28-30 August 2002.

.... (2004), 'Development policies and development trends in Vietnam's Northern Mountain Region', in Furukawa, H., Nishibuchi, M., Kono, Y., and Kaida, Y., (eds), *Ecological Destruction, Health, and Development: Advancing Asian Paradigms,* Kyoto Area Studies on Asia

Vol. 8, Center for Southeast Asian Studies, Kyoto University, Kyoto: Kyoto University Press and Trans Pacific Press, pp. 415-438.

.... (2007), 'Observations on the role of improved fallow management in swidden agricultural systems', in M. Cairns (ed), *Voices From the Forest: Integrating Indigenous Knowledge into Sustainable Upland Farming*, Washington, D.C.: Resources for the Future Press, pp. 780-801.

.... and Neil L. Jamieson (2003), 'Upland areas, ethnic minorities, and development', in Hy Văn Lượng, (ed), *Postwar Vietnam: Dynamics of a Transforming Society,* Lanham: Rowman & Littlefield, pp. 139-170.

.... and Lê Trọng Cúc (1997), 'Some observations on the role of livestock in composite swidden systems in northern Vietnam', In *Upland Farming Systems in the Lao PDR—Problems and Opportunities for Livestock*, ACIAR Proceedings No. 87, Canberra: Australian Centre for International Agricultural Research, pp. 71-78.

.... and P.E. Sajise (1984), 'Introduction', in A.T. Rambo and P.E. Sajise, (eds), *An Introduction to Human Ecology Research on Agricultural Systems in Southeast Asia,* Los Bãnos: University of the Philippines at Los Bãnos, pp. 1-25.

.... and Trần Đức Viên (2001a), 'Social organization and the management of natural resources: A case study of Tat hamlet, a Da Bac Tay ethnic minority settlement in Vietnam's Northwestern Mountains', *Southeast Asian Studies (Kyoto),* 39 (3), pp. 299-324.

.... and Trần Đức Viên (2001b). Chapter 6: Social Organization. In Le Trong Cuc and A.T. Rambo (eds), *Bright Peaks, Dark Valleys; A Comparative Analysis of Environmental and Social Conditions and Development Trends in Five Communities in Vietnams' Northern Mountain Region.* National Political Publishing House, pp. 169-199.

Rao, Y. (1989), 'Forest resources of tropical Asia', Paper presented at the Environment and Agriculture Environmental Problems Affecting Agriculture in Asia - Pacific Region, World Food Day Symposium, Bangkok, Thailand.

Rigg, J. and N. Sakunee (2001), 'Embracing the global in Thailand: Activism and pragmatism in an area of deagrarianisation', *World Development,* 29 (6), pp. 945-960.

Rindfuss, W., R.S., Mishra, V., Fox, J. and G. Dolcemacolo (2003), 'Linking household and remotely sensed data: Methodological and practical problems', in J. Fox, R. Rindfuss, S. Walsh, and V. Mishra (eds), *People and the Environment: Approaches for Linking Household and Community*

Surveys to Remote Sensing and GIS, Boston: Kluwer Academic Press, pp. 1-30.

Robbins, P. (1998), 'Paper forests: imagining and deploying exogenous ecologies in arid India', *Geoforum,* 29 (1), pp. 69-86.

Rocheleau, D. and L. Ross (1995), 'Trees as tools, trees as text: struggles over resources in Zambrana-Chacuey, Dominican Republic', *Antipode,* 27 (4), pp. 407-428.

Ruben, R. and V.D.M. Berg (2001), 'Nonfarm employment and poverty alleviation of rural farm households in Honduras', *World Development,* 29 (3), pp. 549-560.

Saberwal, V.K. (1998), *Pastoral Politics: Shepherds, Bureaucrats, and Conservation in the Western Himalaya*, New Delhi: Oxford University Press.

Sakurai, Y. (2003), 'Comprehensive Vietnamese village studies: From ten years of research in a small village', *IIAS Newsletter* No. 32, November 2003, p. 11.

Sanchez, P.A. (1976), *Properties and Management of Soil in the Tropics,* Toronto: Wiley-Interscience.

.... (1999), 'Improved fallow comes of age in the tropics', *Agroforestry Systems,* 47, pp. 3-12.

Schelhas, J. (1996), 'Land-use choice and forest patches in Costa Rica', in J. Schelhas and R. Greenberg (eds), *Forest Patches in Tropical Landscapes*, Washington, D.C.: Island Press, pp. 258-284.

.... and R. Greenberg, (eds) (1996), *Forest Patches in Tropical Landscapes,* Washington, D.C.: Island Press.

Schmidt-Vogt, D. (1998), 'Defining degradation: the impacts of swidden on forests in northern Thailand', *Mountain Research and Development,* 18 (2), pp. 135-149.

Schmitter, P., Hertel, M., Dercon, G., Hilger, T., Lê Hà, T., Thanh, N.H.,Viên, T.D., Văn Dung, N., and G. Cadisch (2008), 'Sedimentation in paddy fields: A neglected issue associated with land use intensification', Abstract of paper presented at EUROSOIL 2008, European Confederation of Soil Science, Vienna, Austria.

Segal, M., Avissar, R., McCumber, M., and R. Pielke (1988), 'Evaluation of vegetation effects on the generation and modification of mesoscale circulation', *Journal of Atmospheric Sciences,* 45, pp. 2268-2292.

Sikor, T. and Đào Minh Trường (2000), *Sticky Rice and Collective Fields: Community-based Development among the Black Thai,* Hanoi: Agricultural Publishing House.

.... and Đào Minh Trường (2002), 'Agricultural policy and land use changes in a Black Thai commune of Northern Vietnam, 1952-1997', *Mountain Research and Development,* 22 (3), pp. 248-255.

Skole, D.L. and C. Tucker (1993), 'Tropical deforestation and habitat fragmentation in the Amazon: satellite data from 1978 to 1988', *Science,* 260, pp.1905-1910.

Smaling, E.M.A., Oenema, O., and L.O. Fresco (eds) (1999), *Nutrient Disequilibria in Agroecosystems: Concepts and Case Studies,* Wallington: CABI Publishing.

Southwich, C.H. (1976), *Ecology and the Quality of Our Environment*, Second edition, NewYork: Van Nostrand.

Soule, M. and G. Lease (eds) (1995), *Reinventing Nature? Responses to Postmodern Deconstructionism*, Washington D.C.: Island Press.

Spencer, J. (1966), *Shifting Cultivation in Southeast Asia*, University of California Publications in Geography No. 19, Berkeley and Los Angeles: University of California Press.

Trần Công Hùng (1999), 'The roles of Ba Na women in socio-economic activities in Kon Rung hamlet, Ayun village, Mang Yang district, Gia Lai province', in Lê Trọng Cúc, Nghiêm Phương Tuyến, and A.T. Rambo (eds), *Working Papers on Culture and Environment in Vietnam's Uplands,* 1, Hanoi: CRES, Vietnam National University, pp. 10-20.

Trần Đức Viên (1998), 'Soil erosion and nutrient balance in swidden fields of the composite swiddening agroecosystem in the Northwestern Mountains of Vietnam', in A. Patanothai (ed), *Land Degradation and Agricultural Sustainability: Case Studies from Southeast and East Asia,* Khon Kaen, Thailand: Regional Secretariat, the Southeast Asian Universities Agroecosystem Network (SUAN), Khon Kaen University, pp. 65-86.

.... (2001), *Kinh nghiệm quản lý đất bỏ hóa sau nương rẫy ở Việt Nam* (Experience in managing swidden fallows in Vietnam), Hà Nội: Nhà Xuất bản Nông Nghiệp.

.... (2003), 'Culture, environment and farming systems in Vietnam's Northern Mountain Region', *Southeast Asian Studies (Kyoto),* 41 (2), pp. 180-205.

.... (2004), 'Changes in the composite swiddening system in Tat hamlet in Vietnam's Northern Mountains in response to integration into the market system', in H. Furukawa, Nishibuchi, M., Kono, Y., and Y. Kaida (eds), *Ecological Destruction, Health, and Development: Advancing Asian Paradigms,* Kyoto: Kyoto University Press, pp. 453-465.

.... (2007), 'Indigenous fallow management with *Melia azedarach* Linn. in northern Vietnam', In M. Cairns (ed), *Voices from the Forest: Integrating Indigenous Knowledge into Sustainable Upland Farming,* Washington, D.C.: Resources for the Future Press, pp. 435-443.

......................, Leisz, S.J., Nguyễn Thanh Lâm, and A.T. Rambo (2006), 'Using traditional swidden agriculture to enhance rural livelihoods in Vietnam's uplands', *Mountain Research and Development* 26 (3): 192-196.

......................, Nguyễn Văn Dung, Phạm Tiến Dũng, and Nguyễn Thanh Lâm (2004), 'A nutrient balance analysis of the sustainability of a composite swiddening agroecosystem in Vietnam's Northern Mountain Region', *Southeast Asian Studies (Kyoto),* 41 (4), pp. 491-502.

......................, Nguyễn Viết Tùng, Calkins, P., Allbritton, A., Hiesinger, M.A., Vũ Thị Thao, Phạm Thị Mai Hương, Trần Nam Anh, Lê Thị Thanh Phương, and Đặng Việt Quang (eds) (2005), *Marketing and Agroforestry Development in Vietnam's Uplands,* Hanoi: Agricultural Publishing House.

......................, Rambo, A.T., and Nguyễn Thanh Lâm (eds) (2006), *Composite Swiddening Farming Systems in Vietnam's Northern Mountain Region*, Unpublished final technical report of the HUA-Ford Foundation research project, Hanoi: CARES, Hanoi University of Agriculture.

Trần Thị Nhung, Vũ Quỳnh Anh, Hoàng Thu Hà, and Nguyễn Bích Thảo (1999), 'The roles of rural Vietnamese women in household economic development and natural resources management in upland areas in Phú Thọ province', in Lê Trọng Cúc, Nghiêm Phương Tuyến, and A.T. Rambo (eds), *Working Papers on Culture and Environment in Vietnam's Uplands,* 1, Hanoi: CRES, Vietnam National University, pp. 21-38.

Trịnh Thị Tiện (2003), *The Role of Non-Timber Forest Products in Household Livelihood of the Tay People in Hoabinh Province, Vietnam,* M.Sc. Thesis in Agricultural Systems, Chiang Mai University.

Van Noordwijk, M., Widianto, Heinen M. and K. Hairiah (1991), 'Old tree root channels in acid soils in the humid tropics: Important for crop root penetration, water infiltration and nitrogen management', in R.J. Wright, V.C. Baligar and R.P. Murrmann (eds), *Plant-soil Interactions at Low pH*, The Netherlands: Kluwer Academic Publishers, pp. 423-430.

Vandergeest, P. and N.L. Peluso (1995), 'Territorialization and state power in Thailand', *Theory and Society,* 24, pp. 385-426.

Viện dân tộc học (1978), *Các dân tộc ít người ở Việt Nam: Các tỉnh phía bắc* (Minority ethnic groups in Vietnam's northern provinces), Hà Nội: Nhà xuất bản khoa học xã hội.

Vityakon, P. (1994), *Report on the paddy fields as a sub-component of the composite swidden agroecosystem,* Unpublished research report, Department of Soil Science, Khon Kaen University.

Võ Đại Hải, Nguyễn Đình Quế, and Phạm Ngọc Thương (2003), *Canh tác nương rẫy và phục hồi rừng sau nương rẫy ở Việt* Nam (Shifting cultivation and forest regeneration during the fallow period in Vietnam), Nghệ An: Nhà xuất bản Nghệ An.

Vũ Quyết Thắng (1991), *The impacts of dam/reservoir construction on mountainous rural development in Vietnam*, paper presented at the CRES-EWC International Seminar on Human Ecology in Vietnam, Hanoi.

Vũ Tiến Hinh and Phạm Ngọc Giao, 1998. *Điều tra qui hoạch rừng* (Forest inventory and planning), Hà nội: Nhà xuất bản Nông nghiệp.

Whitmore, T.C. (1997), 'Tropical forest disturbance, disappearance, and species loss', in W.F. Laurance and R.O. Bierregaard (eds), *Tropical Forest Remnants: Ecology, Management, and Conservation of Fragmented Communities*, Chicago: University of Chicago Press, pp. 3-12.

Wiens, J. (1977), 'On competition and variable environments', *American Scientist,* 65, pp. 590-597.

Williams-Linera, G. (1990), 'Vegetation structure and environmental conditions of forest edges in Panama', *Journal of Ecology,* 78, pp. 356-373.

Yan, H., and R. Anthes (1987), 'The effect of latitude on sea breeze', *Monthly Weather Review,* 115, pp. 936-956.

Young, A., and N. Mitchell (1994), 'Microclimate and vegetation edge effects in a fragmented podocarp broadleaf forest in New Zealand', *Biological Conservation,* 67, pp. 63-72.

Ziegler, A.D., Giambelluca, T.W., Tran, L.T., Vana, T.T., Nullet, M.A., Fox, J., Viên, T.D., Pinthong, J., Maxwell, J.F., and S. Evett (2004), 'Hydrological consequences of landscape fragmentation in mountainous Northern Vietnam: evidence of accelerated overland flow generation', *Journal of Hydrology*, 287, pp. 124-146.

......................, Trần, L.T., Giambelluca, T.W., Sidle, R.C., Sutherland, R.A., Nullet, M.A., and T.D. Vien (2006), 'Effective slope lengths for buffering hillslope surface runoff in fragmented landscapes in northern Vietnam.' *Forest Ecology and Management*, 224: 104-118.

......................, Giambelluca, T.W., Plondke, D., Leisz, S., Trần, L.T., Fox, J., Nullet, M.A., Vogler, J.B., Truong, D.M., and T.D. Viên (2007a), 'Hydrological consequences of landscape fragmentation in mountainous northern Vietnam: Buffering of Hortonian overland flow', *Journal of Hydrology*, 337, pp. 52-67.

......................, Giambelluca, T.W., Sutherland, R.A., Nullet, M.A., and T.D. Viên (2007b), 'Soil translocation by weeding on steep-sloped swidden fields in northern Vietnam', *Soil and Tillage Research*, 96: 219-233.

Zimmerer, K. (2000), 'The reworking of conservation geographies: Non-equilibrium landscape and nature-society hybrids', *Annals of the Association of American Geographers,* 90 (2), pp. 356-369.

Index

buffalo 49, 56, 90-95, 141, 143, 149

cash cropping 41, 141, 148-149, 151, 157, 242
cattle 49, 56, 90-95, 141, 143, 149
chickens (*see* 'poultry')
composite swidden agriculture
  carrying capacity of 152
  components of 47, 48-62
  definition of xvii, 5-8, 47
  distribution in NMR of 309-311, 316
  future of 242
  interaction of system components of 62-65, 100, 239, 243, 275-276, 281
  occurrence in Southeast Asia of 7
  recent changes in 46, 134-153, 158
  sustainability of 7, 8, 239, 241-242, 243, 282-283, 303, 307-308, 332-333, 338
  types of 46-48, 144-150, 185, 242, 309-316
cooperative system 135, 166-169
crop species diversity 50-52, 57, 77, 117

Đà Bắc Tày ethnic group
  ethnic identity of 27-28
  language of 28
  dress style of 28-30
  house types of 30
  kinship system of 33
  religion of 31

deforestation 241, 284-285, 302-303, 308, 342
diversification of farming systems 144, 185
diversification of income sources 40-41, 48, 150, 157, 184-185, 192-194, 198-201
division of labor 33-35, 92, 141, 142, 226-236
ducks (*see* 'poultry')

environmental degradation 41, 152, 241
economic efficiency of agriculture 194-198
emergent properties of system 323, 327-328
equitability of social system 35-37, 333-335
exchange labor 33, 180-181

fallow fields 45, 58-59
  floral diversity of 79
  nutrient status of 269-270
  plant biomass of 81
  soil erosion from 263-267
  successional dynamics of 80-82
  useful plants in 79-80
fallow cycle 40, 58-59, 140, 151-152, 240, 271-274
fallow management 59, 85-88
firewood 129, 130, 220, 246
fishponds 46, 56-57, 104-111, 139, 141, 197-198

forest 46, 59-62, 119-133, 208, 240-241, 348-350
forest products 59, 82-85, 124, 126-132, 140, 208, 218-222
forest protection 172, 173

gender roles 33-35, 141, 142, 226-236
green manure 49, 73, 129, 249

habitat fragmentation 22, 60, 62, 88, 308, 348-350
homegardens 57-58, 112-115, 117, 139, 141-142
household income 131-132, 190-192
household labor force 188-189
human ecology xvii-xviii
hunting 60, 128

integration of community into larger system 335-337
interactions among agroecosystem components 45, 62-65, 100, 239, 243, 274-277, 281

land allocation to households 135, 140, 166, 173-174, 347-348
land management 139, 140, 171-172, 347-348
livelihood strategies 41, 198-200
livestock 45, 56, 90-103, 140, 141-142, 189
  conflict with swidden farming of 95, 102
  diseases of 94-95
  marketing of 222-224
  role in household economy of 91, 98-100
  species diversity of 98
logging 41, 46, 60, 219

manure 49-50, 73, 99, 223, 249
market economy 38, 142, 151, 202
markets 38-39, 165
marketing channels 202-225

non-timber forest products (*see* 'forest products')
Northern Mountain Region (NMR) xviii, xxxi-xxxii, 160-161, 317
nutrient balance analysis 239-240, 243-283
nutrient loss from biomass burning 259, 268-269

off-farm labor 157, 194

paddy fields 48-50, 66-75, 141, 248-250
  area of 48-49, 139, 187-188, 290
  expansion of area of 139, 142
  farmer classification of quality of 71-73
  management of soil fertility of 73-74
  nutrient balance of 240, 279-282
  pest management in 74-75
  soils of 67-70, 72-73, 251
  sustainability of 282-283
  water supply to 49, 70-71
pigs 56, 95-96, 141, 143, 197-198
policies impacting swidden farming 40, 143, 151-152
population 26, 135, 140, 142, 150, 299-300, 314
poultry 56, 96-98, 141, 143, 197-198
prices of agricultural and forest products 142, 169, 203, 205, 214-217, 220 ff
productive assets of households 187-189